Industrial Solvents

Selection, Formulation and Application

Paulo Garbelotto
Coordinator

Industrial Solvents

Selection, Formulation and Application

English version by
Robin Dianne Zang Santos

EDITORA BLUCHER 50 anos

Rhodia

Original title: Solventes industriais:
seleção, formulação e aplicação

The edition in Portuguese (Brazil) has been published
by Editora Edgard Blücher Ltda.

EDITORA EDGARD BLÜCHER LTDA.
Rua Pedroso Alvarenga, 1245 – 4° andar
04531-012 – São Paulo, SP – Brasil
Tel.: (55_11) 3078-5366
e-mail: editora@blucher.com.br
site: www.blucher.com.br

ISBN 978-85-212-0479-4

Presentation

Inform, shape, educate: three timeless words which are invariably attributed to leaders.

For more than three decades, Rhodia has dedicated investments in technological advances in the field of solvents with the objective of supporting industrial development in Brazil. The story of this book is the translation of that spirit.

It all started with the ideal of sharing our accumulated knowledge, not only in the plant but also in the market, applications, physical properties, analytical techniques, care and handling, and safety, environmental and responsibility concepts.

Fourteen members from different areas in the company accepted the challenge. Hours, days, weeks, and months of voluntary dedication were given to the project, with the objective of bringing together the most current existing information on the subject.

The result of this work is practical and assertive concepts written using a light style, but without losing the inherent depth of the subject as a true state-of the-art classic.

I invite you, reader, to travel through this world of industrial solvents.

Rhodia Management, Brazil

Preface

A solvent, many times treated with a simple reaction medium, in reality represents a universe of enormous complexity, in which its secrets are explained in this book. The word "solvent" expresses the ability to dissolve a species or entity, allowing its atomic/molecular constituents to penetrate in the liquid phase and be transported, fulfilling its destiny or purpose. Along the way, all the forces work together in the species to be dissolved: the solute, as well as the solute-solvent interface, and between the solvent molecules. This is why the role of the dipolar, hydrogen bond, and the van der Waals interactions need to be understood well and calculated correctly. In addition to these forces, the solvent's entropic nature also takes part. The scientific approach of the solvent's role is therefore essential to planning its technological process performance. This is one of the important aspects this book offers the reader, allowing a deeper understanding of the world of solvents.

The implications of this topic are immense, starting with water, the universal solvent, and without it, life would not exist. In fact, from the largest oceans to the smallest cell, more important chemical transformations are processed in the liquid phase. In the industry, the solvent is a part of almost all formulations used in the most different production stages. In the majority of them, the solvent plays a transitory role and must be removed at the end of the process. However, this must be done carefully as to not compromise the result. This is the case with coatings where the use of a solvent mixture must be well-planned based on physical parameters so that the evaporation of the most volatile components does not cause solute precipitation in the remaining mixture. Scientific planning is another important point the reader will learn about in this book.

A large part of solvent applications occurs in an open environment like in painting residences and buildings, printing processes, and industrial and domestic cleaning activities. Thus, volatile components will go into the atmosphere; others, water soluble, will contribute to water and soil contamination. Resulting problems are immense, justifying increasing preoccupation for the preservation of the quality of life and the environment, now being reflected in national and international rules that have begun regulating the use of solvents. In this book, in addition to these regulations, the reader will find a fundamental introduction to procedures called Green Chemistry, addressing the correct choice from options favoring quality of life and sustainability.

Finally, different analytical and monitory technical aspects, as well as economic and commercial aspects are being presented, giving a wide current scope of industrial solvents. In this magnificent book, the authors, recognized competent professionals working at Rhodia, have shared their expertise and knowledge, as well as addressing relevant topics regarding healthy, safety, and environmental management.

Henrique E. Toma

Professor of Chemistry Institute
at Universidade de São Paulo

About the authors

Alessandro Rizzato

Alessandro Rizzato has a Bachelor's and Master's degree in chemistry from the Universidade Estadual Paulista (UNESP). He has a Ph.D. in chemistry from the UNESP and the Université de Bourgogne, Dijon – France. He has also done graduate work in the chemistry of nanostructured and mesoporous ceramic material. He has worked for two years in the Rhodia Research and Technology Center in Paulinia as a researcher in the Physic-Chemical Laboratory; since May 2006 he is a Six Sigma Black Belt and currently he is Marketing Innovation Specialist at Rhodia.

Ana Cristina Leite

Ana Cristina Leite has a degree in chemical engineering from Universidade Estadual de Campinas (UNICAMP). She had done graduate work in marketing at the Fundação Getúlio Vargas. She has worked for 10 years at the Rhodia Latin America chemistry divisions.

Cristina Maria Schuch

Cristina Maria Schuch has a Bachelor's and Master's degree in chemistry from the Universidade Federal do Rio Grande do Sul (UFRGS). She has a Ph.D. in chemistry from the Universidade Estadual de Campinas (UNICAMP). She has worked in the chemical industry since 2001 and currently is a senior researcher at the Rhodia Research Center in Paulinia, in the Department of Analytical Chemistry.

Danilo Zim

Danilo Zim has a Ph.D. in chemistry from the Universidade Federal do Rio Grande do Sul (UFRGS) and has post doc at the Massachusetts Institute of Technology (MIT). He has worked in conception and improvement of chemical processes since 2003 and currently is a Research Scientist at the Rhodia Research Center in Paulinia.

Denílson José Vicentim

Denílson José Vincentim has a Bachelor's degree in chemistry from the Pontifícia Universidade Católica de Campinas (PUC Campinas). He has worked at the Rhodia Research Center in Paulinia since 1987. Currently he is a Development and Applications Specialist for the coating and varnish, printing ink, and thinner markets.

Edson Leme Rodrigues

Edson Leme Rodrigues has a Ph.D. in chemical engineering from the Universidade Federal de São Carlos (UFSCar). Since 2000, he has worked in the development of chemical and petrochemical processes. Currently, he is a researcher at the Rhodia Research Center in Paulinia in development and applications for the esther, polyester, and polyurethane markets, and also works in the pilot phase of the conception of chemical processes.

Fabiana Marra

Fabiana Marra has a Bachelor's degree in chemistry from the Faculdade Oswaldo Cruz. She has done graduate work in industrial administration at the Fundação Vanzolini da Universidade de São Paulo as well as completed graduate coursework in Marketing at the Fundação Getúlio Vargas. She has 17 years of experience working in coating industries and with their suppliers, and currently she is responsible for the marketing of the Rhodia solvent business unit.

Fernando Zanatta

Fernando Zanatta has a degree in pharmacy from the Universidade Metodista de Piracicaba. He is a Specialist in Product Stewardship and in Quality Management Systems. At Rhodia, he works as a company specialist in Product Stewardship for Latin America.

Hidejal Santos

Hidejal Santos is a Biologist and Civil/Sanitation Engineer, with a degree from Pontífica Universidade Católica de Campinas. He has worked at the Rhodia Research Center in Paulinia-CPP since 1978. In 1992 he began working at the CPP in the area of technology for the environment. Currently, he is a Research and Development Consultant for the environment and Guarantor for the Industrial Hygiene and Process Risk Analysis areas.

Léo Santos

Léo Santos has a Bachelor's and Master's degree in chemistry from the Universidade Estadual Paulista (UNESP), a Ph.D. from UNESP and the Université de Montpellier II – Montpellier – France. He has done graduate work in materials chemistry: polymers, surfaces, and amorphous state. He has worked for three years in the Rhodia Research and Technology Center in Paulinia as a researcher in the Physical-Chemistry Laboratory; and since 2006, he is one of people responsible for the Industrial Chemical Analysis area in the Chemical Plant at Paulinia.

Maria Luiza Teixeira Couto

Maria Luiza Teixeira Couto has a Master's degree in geosciences from UNICAMP and a chemical engineering degree from the Universidade Federal de Minas Gerais. She currently works in the environmental area at the CPP.

Rosmary De Nadai

Rosmary De Nadai has a Bachelor's degree in chemistry from the Universidade Estadual de Campinas (UNICAMP). She is a specialized in environmental management, studying at UNICAMP. Currently, she works in the environmental area at the Rhodia Research Center in Paulinia (CPP).

Sérgio Martins

Sérgio Martins has a Bachelor's degree in chemistry studies with technological attributions from UFSCar. He has a Master's degree in chemistry from USP. He has worked at the Rhodia Research Center in Paulinia since 1994, and during this time was responsible for the Molecular Characterization and Process Chemistry laboratories. Currently he is a Development and Applications Specialist for the adhesive, leather, mining, and coating and varnish markets.

Content

8 Solvents and Their Applications

10 A Segment Committed to Sustainability

1 Overview

The Industrial Solvent Market

This chapter has the objective of presenting an overview of solvents including: their history, markets, the main products and how they are inserted into some productive chains

Ana Cristina Leite
Fabiana Marra

The world of solvents is very broad and the applications are present in the daily lives of everyone. Industrial solvents represent the greatest manipulated volume and are key components in several processes and products: coatings, packaging, adhesives, cleaning products, purification of drugs, and chemicals in general, to name a few examples.

This chapter has the objective of presenting an overview of industrial solvents including: their history, markets, main products and how they are inserted in some productive chains.

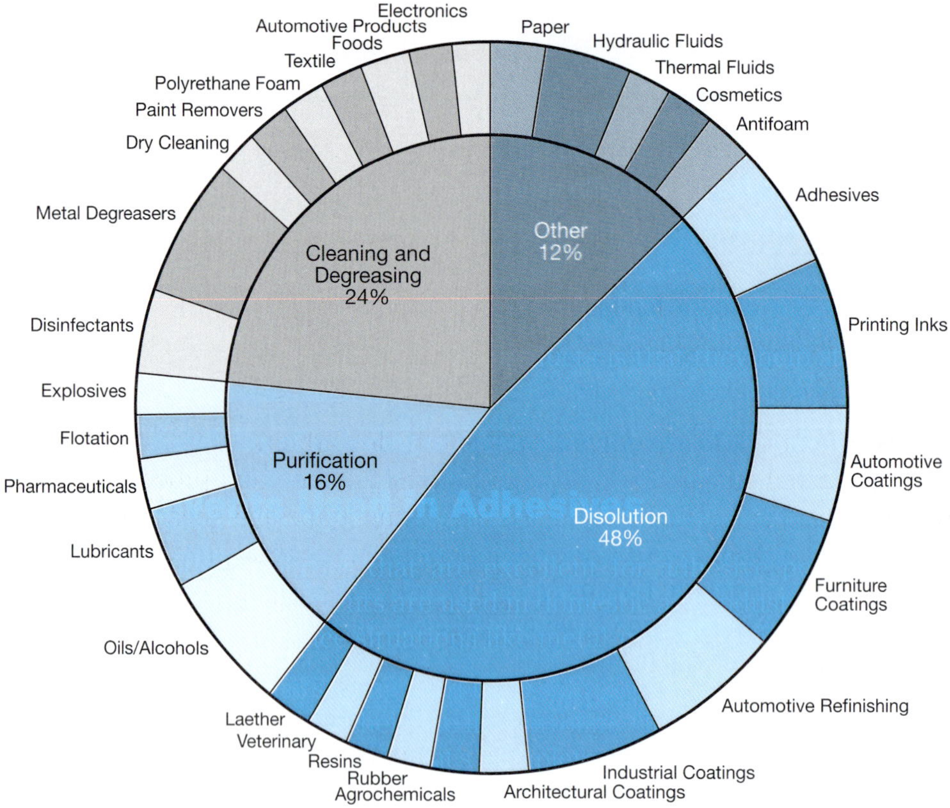

Figure 1.1. Solvents: a family of products which are present in a broad range of end markets

1.1. History

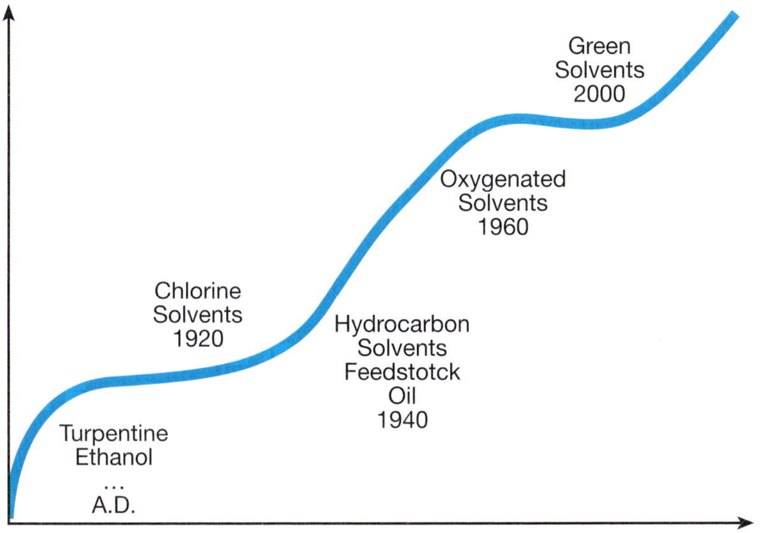

Figure 1.2. The history of solvents.

1.1.1. The Beginning

It is believed the first solvents to be used were hydrocarbons, such as turpentine, which was extracted from wood and ethanol from fermentation processes. Greek and Roman civilizations had already been fermenting grape and sugar cane for obtaining ethanol and discovered that, despite resembling water in appearance and behavior, it could dissolve oils and resins.

As early as BCE, there were records of the use of solvents for medical purposes among the Assyrians. There is also data that prove the Egyptians used them to synthesize substances for cosmetic purposes. Scientific investigations by L´Oreal chemists and Louvre scientists in Paris were conclusive in determining that the black color present in the makeup of second millennium relics was synthetic: made up of components that are nonexistent in nature.

Antoine de Chiris and Roure Bertrand Fils left their mark in the history of solvents in 1900 at the Paris World Fair, where they were awarded the grand prize for their presentation of essences extracted using volatile solvents.

The first solvent from a petrochemical source was produced in 1920.

By the end of World War I in 1918, there was a need for production on a greater scale. Coating solvents started being used to meet the demands of a market in a hurry to rebuild itself.

In the automotive industry, the introduction of machinery that enabled more agility in the painting process of cars required a faster drying process. Vegetable oil-based resins and wood resins were substituted by nitrocellulose-based resins. Interestingly enough, at the end of the war, gunpowder factories became idle and their production equipment found use in the production of these new resins.

Phenol resins came next as the first synthetic resins (1920). In 1930, alkyd resins started being commercialized.

The use of organic industrial solvents was propelled and modeled by the evolution of the type of resins and production technology of coatings. The force behind this revolution was undoubtedly a necessity of more modern fast-drying resins [1].

1.1.2. The First Wave: the Substitution Process of Organochloride Solvents

The first solvents to be produced were called organochlorides. The ICI (International Coatings Industry) was a pioneer in the production and use of these solvents.

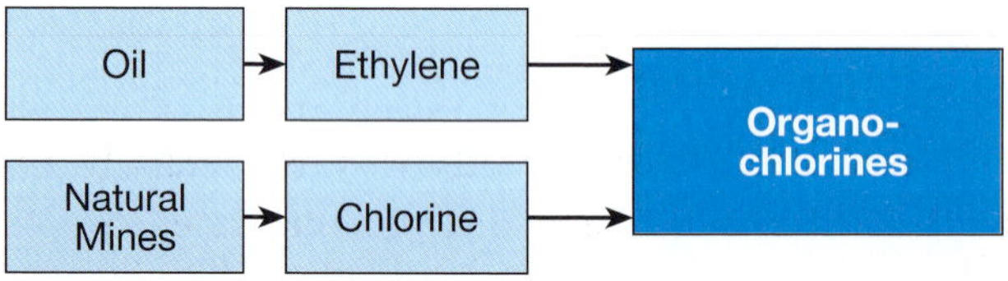

Figure 1.3. Raw materials of the main organochlorides.

The solvent trichloroethylene (TCE) was first produced in 1910, but only became relevant in applications in 1930. Carbon tetrachloride substituted gasoline in the application of cleaning materials and was considered the main organic solvent of this application until 1960, when the 1,1,1 - trichloroethane (TCA) was introduced.

Organochloride solvents were used for several years to remove wax, oils, and other substances from several surfaces. They were high performing and considered safe for workers because they were nonflammable. The most used organochlorides were: 1,1,2-trichloro-1,2,2-trifluoroethane (CFC 113), 1,1,1-trichloroethane (TCA), trichloroethylene (TCE), tetrachloroethylene also called perchloroethylene or PERC, and also dichloromethane.

Organochloride solvents were denominated as dense non-aqueous phase liquids (DNAPL) for being odorless or tasteless and for having a higher density than water.

Despite the existing measurement techniques, organochloride solvents were not targets of detection and measurement until the 1980s. Only after analysis of residue disposal methods was it understood how these products contaminated the subsoil.

In the 1970s, scientists identified some chlorofluorocarbons (CFCs) that underwent chemical changes in the upper layers of the atmosphere leading to the depletion of the ozone layer. As a result, in 1987, 45 nations signed an agreement to restrict the production and use of these substances. In 1992, it was voted that the use of ozone-depleting substances, found in the ODSL (Ozone Depleting Substance List) Class 1, would be banned as of January 1, 1996.

Despite this, the use of substances such as CF113 and TCA, which were extremely important to the industry in metal degreasing applications, continued in the following years until alternative techniques were developed.

1.1.3. The Second Wave: the Restriction Process for the Use of Hydrocarbons

The first hydrocarbon commercialized as a solvent was benzene. In 1849, it was produced from coal and only in 1941, was it produced from oil. In the beginning, its main application was in gasoline, but by the middle of World War II, its application was extended to the chemical industry. Currently it is used mainly as a raw material in the production of ethylbenzene, cumene, and cyclohexane.

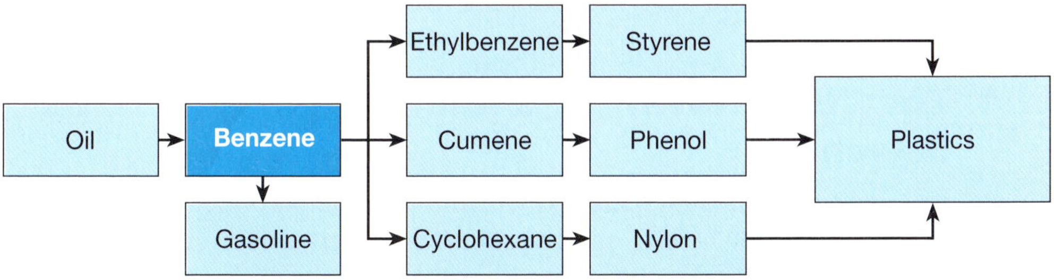

Figure 1.4. Benzene and its main applications.

In 1989 the EPA (Environmental Protection Agency) restricted the use of benzene to industrial applications. Studies were carried out on the negative health impacts that the benzene by-products, toluene and xylene, could have on workers; and the results were published in the 1990s.

Also in the 1990s, the Montreal Protocol and the Clean Air Act, one of the EPA regulations, were internationally accepted, and a gradual reduction process of ozone depleting solvents was put in place. In the United States, the emission of volatile organic compounds (VOCs) was limited and the use of hydrocarbons, organochloride compounds, and other products were restricted in compliance with the Clean Air Act.

The industry has willingly made efforts to reduce the use of some products such as benzene, toluene, methyl ethyl ketone (MEK), methyl isobutyl ketone (MIBK), chloroform, and methylene chloride, for example. Environmental protection agencies are aware of these initiatives, which are mainly observed in developed countries. The greatest reduction indices have occurred in the segments of coatings and varnishes and industrial cleaning products.

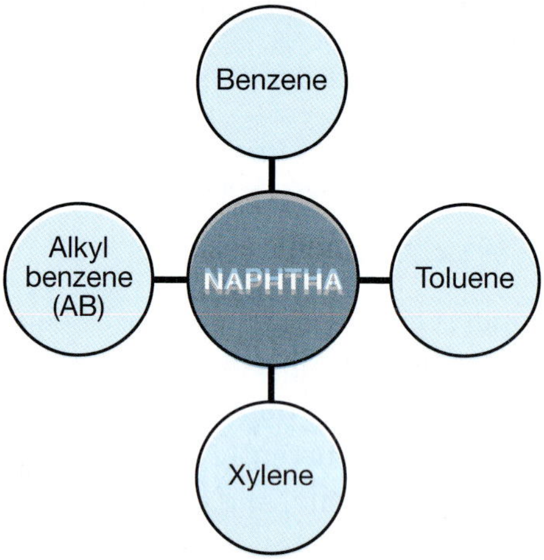

Figure 1.5. The most used hydrocarbons.

1.1.4. The Third Wave: the Evolution of Oxygenated Solvents

The evolution of the use of chemical products and their impact on human health has fueled the introduction of health, safety and environmental legislation and forced manufacturers and users to develop other kinds of products. This effort gave rise to the third wave of solvents: oxygenated solvents. In this category, there are substances with chemical functions such as alcohols, ketones, esters, glycols, glycol ethers, and others.

Before those legislations existed, more than half of the total amount of solvent-borne coatings were made from aliphatic and aromatic hydrocarbons. After the Montreal Protocol, several companies, mainly in developed countries, started substituting them with oxygenated solvents whenever it was technically possible. Currently the world consumption of industrial solvents is about 20 million tons, and 70% of those are oxygenated.

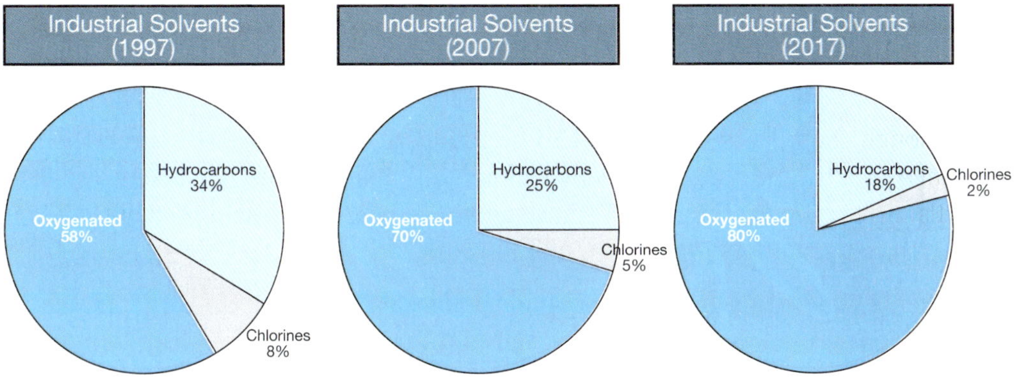

Figure 1.6. Profile of industrial solvents: organochlorides, hydrocarbons, and oxygenated (1997-2007).

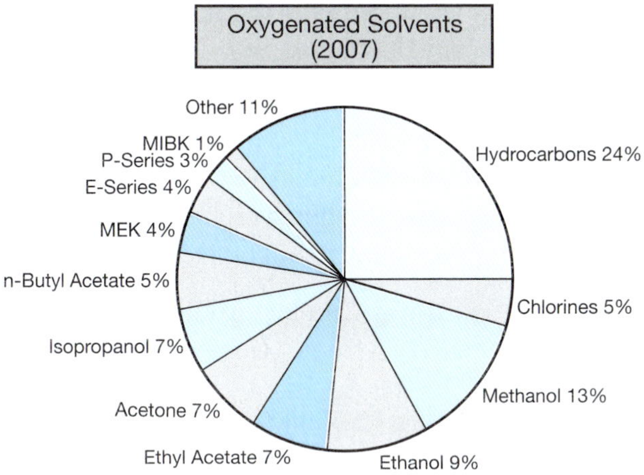

Figure 1.7. World's most consumed solvents.

1.1.5. The 21st Century: the Green Revolution

The use of petrochemical solvents is fundamental in several chemical processes and daily applications but have caused serious environmental impacts. Due to the Montreal Protocol, it has been noted the need for reevaluating chemical processes, identifying and quantifying the use of volatile organic compounds (VOCs) and their impact on the environment.

In the last decades, an increasing demand for less environmentally hazardous solvents has been noticed which has led to the creation of a new group of products: the green solvents.

The evolution of the solvent market presents, on average, the same indices as the world GDP. Although during a specific period, organochloride and hydrocarbon solvents have had negative growth indices, oxygenated solvents have grown to twice as high as the GDP and the so-called "green solvents" have grown to four times as high.

Although the ideal solvent is yet to be found, it is necessary to search the ideal balance of efficiency, cost, and environmental impact. Some biosolvents such as methyl soyate, ethyl lactate, and D-limonene have been fairly accepted and used in several market segments for at least fifteen years.

Solvents originating from petrochemical sources, however, are predominant in the world, but growth level is directly related to hazard level for human beings and the environment.

1.2. Solvents: a Multimarket Product

As seen in Figure 1.8, due to their characteristics, solvents are used in several markets and applications.

In these markets, it is worth mentioning the coating market, not only for being the first to use solvents but for being the one which still uses them the most, be it as coatings or varnishes, printing ink, paint remover, thinners, etc.

However, in Latin America, the consumer profile presents a different distribution due to the fact that some markets are not as developed as in other regions, such as the pharmaceutical market. In this segment, solvents are basically used in the processing of active ingredients, which are mostly imported.

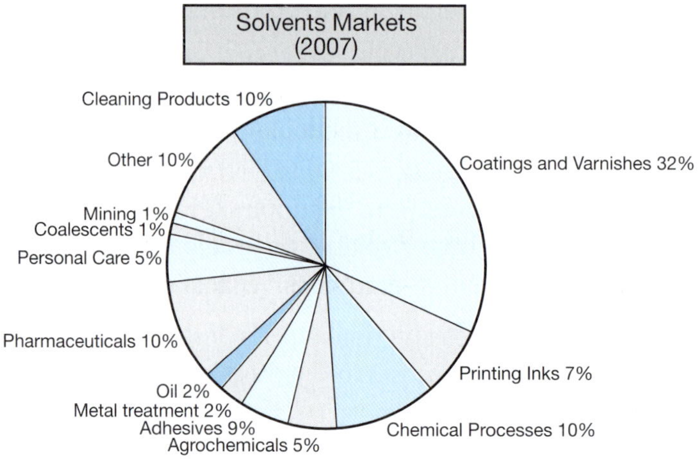

Figure 1.8. Main end markets for industrial solvents (worldwide).

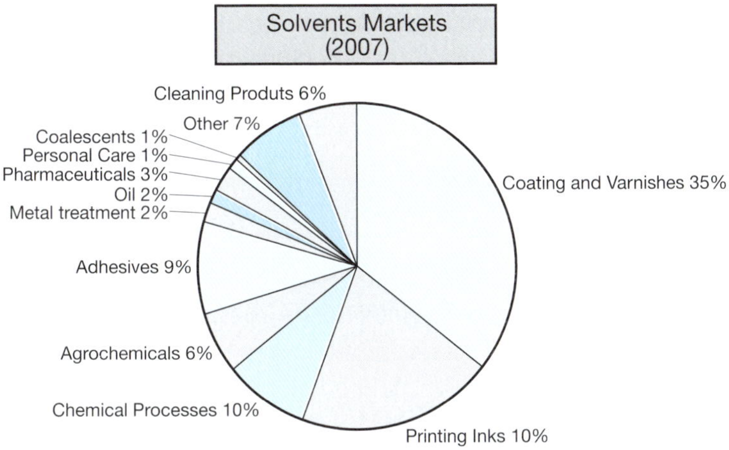

Figure 1.9. Main end markets for industrial solvents (Latin America).

1.2.1. The Coating and Varnish Market

In Europe, the United States, and Japan the coating and varnish market grows at a rate of 2% a year and is linked to the economic development of the region. In the less developed regions, market growth is around 6% per year.

It is estimated that in 2007 world consumption of coatings and varnishes was around 7 million tons.

The coating and varnish market is divided as follows:

- Decorative and Architectural Coatings;
- Automotive Coatings for OEMs (Original Equipment Manufacturers);
- Furniture Coatings;
- Industrial Coatings: (appliances, electronics, etc.).

Since the beginning of the 1970s, the majority of coatings which were produced for the architectural segment followed technology based on low solid solvents and waterborne coatings. Toward the end of the decade, pressure from laws aiming at diminishing VOCs in industrial operations, the need for saving energy and the increase of solvent cost favored the introduction of new technologies such as:

- *Waterborne* (thermocured emulsions, colloidal, and water-soluble dispersions);
- *High solids*;
- Two-component Systems;
- Powder Coatings;
- UV Cured.

These technologies experienced a strong growth in the 1990s when the *Montreal Protocol* and the *Clean Air Act* were internationally accepted.

Currently the predominant technologies are: *solventborne and waterborne.*

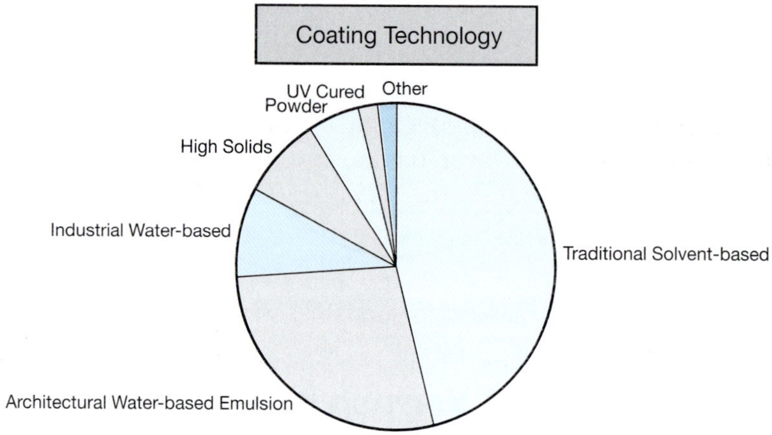

Figure 1.10. Profile indicating current technologies in the coating market.

The most highly used technology is the solventborne coating because it presents the best cost x performance relationship. It is estimated that in the next five years, legislation for air pollution will favor the growth of other technologies, such as high solids, UV cured, and waterborne.

In the United States and Europe, a consolidation of coating-producing companies is taking place; in other regions, small and mid-size producers are representing a significant portion of the market. In Asia, for example, 60% of the market is in the hands of small and mid-sized producers, while in Europe and in the United States, this figure is 20% and 35% respectively.

1.2.2. The Printing Ink Market

The printing ink market is divided into two large groups:

- printing/advertising;
- packaging industry.

The most common printing processes are: lithography, flexography, rotary printing, letterpress and screen.

The basis of excellence for the printing ink industries rests on two pillars:

- product technical quality;
- customer technical service.

This market differs from others because a large portion of the material is already sold before it is even produced. For 2007, the estimated demand was approximately 3.5 million tons, with a growth index of 4%, including the industrially developed countries.

Since it is an activity that produces a high amount of VOC emissions, it has been pressed by many legislations. However, technology changes slowly since substituting the current technologies for waterborne technology or UV-cured has an extremely high cost.

1.2.3. The Adhesive Market

There are several types of adhesives:

- waterborne;
- solventborne;
- 100% solid (*hot melt* technology and pressure sensitive adhesives);
- powder;

Adhesives have a great variety of applications in the textile industry, in flexible packaging, and in several types of structures in which it is necessary that the adhesive has high adhesion power.

This market is growing faster than the world GDP because adhesives are versatile products with the widest applications. The market has a need for products with high adhesion power because new substrates are constantly developed; be it for product design issues or for environmental issues.

Despite the fact that the adhesive industry is not among those producing the most VOCs, it still needs to adapt to new regulations.

The adhesive market may be divided according to the type of adhesive resin and the final user. Below is the market divided according to the type of resin:

- Natural Polymers;
- Water Soluble Polymers;
- Solventborne Polymers;
- *Hot melt*;
- Reactives;
- Dispersion/Emulsion.

It is estimated that in 2007 the demand for adhesives was 15.5 million tons. The segment of solventborne adhesives was third in importance and has a demand of 1.8 million tons.

In the last few years, the increasing number of regulations aiming at protecting human health and the environment have led to the decrease of consumption of solvents; for some applications, solventborne adhesives have been substituted by others produced through waterborne or hot melt technology.

New reductions in the consumption of solvents will be more difficult in the future because currently this technology is used in specific applications in which a high technical performance is required. In these cases, the substitute technologies have still not met the needed requirements.

Although the reduction of solvents is not an expected fact in the near future, in this market there is a clear trend of evaluation and substitution of current solvents by friendlier ones.

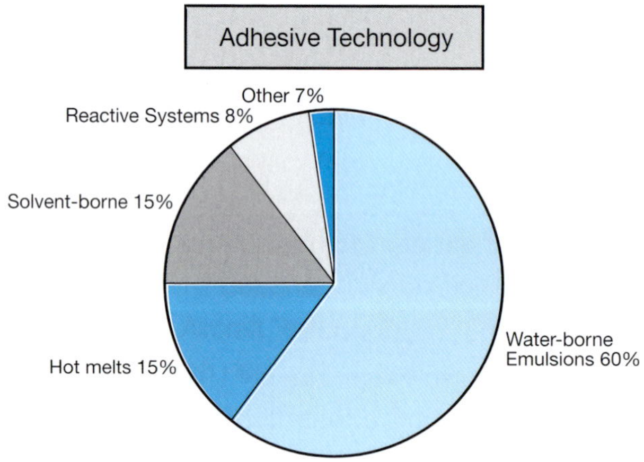

Figure 1.11. Profile of current technologies in the adhesive market.

1.3. Solvents: an Important Piece in the Petrochemical Chain

Solvents are a part of the primary product or by-product of the most important production chains of our time. Whether they are from petrochemical sources – oil or natural gas - or renewable sources, these products, are always present.

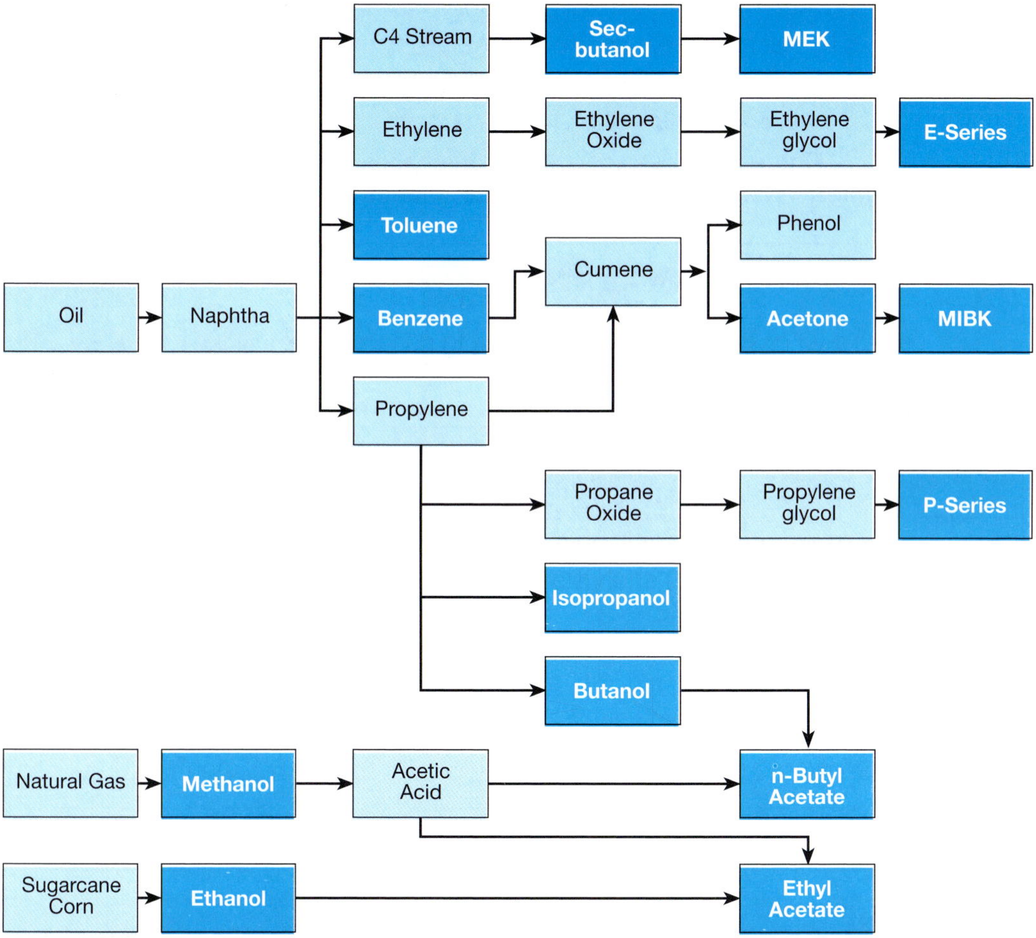

Figure 1.12. Simplified flowchart on how principal solvents are obtained.

It is important to note that some of the main molecules recognized as solvents do not have this application as the only one. Examples include acetone, methanol, and ethanol.

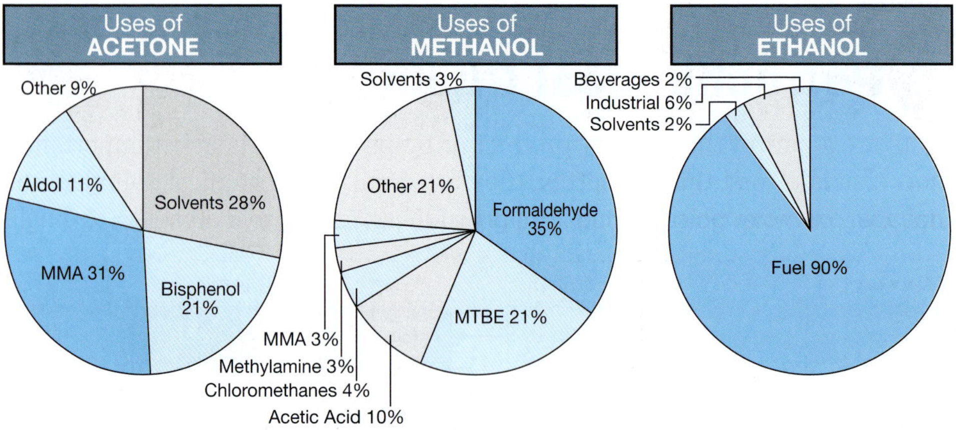

Figure 1.13. Uses of acetone, methanol, and ethanol (MMA = Methyl Methacrylate; MTBE = Methyl ter-Butyl Ether).

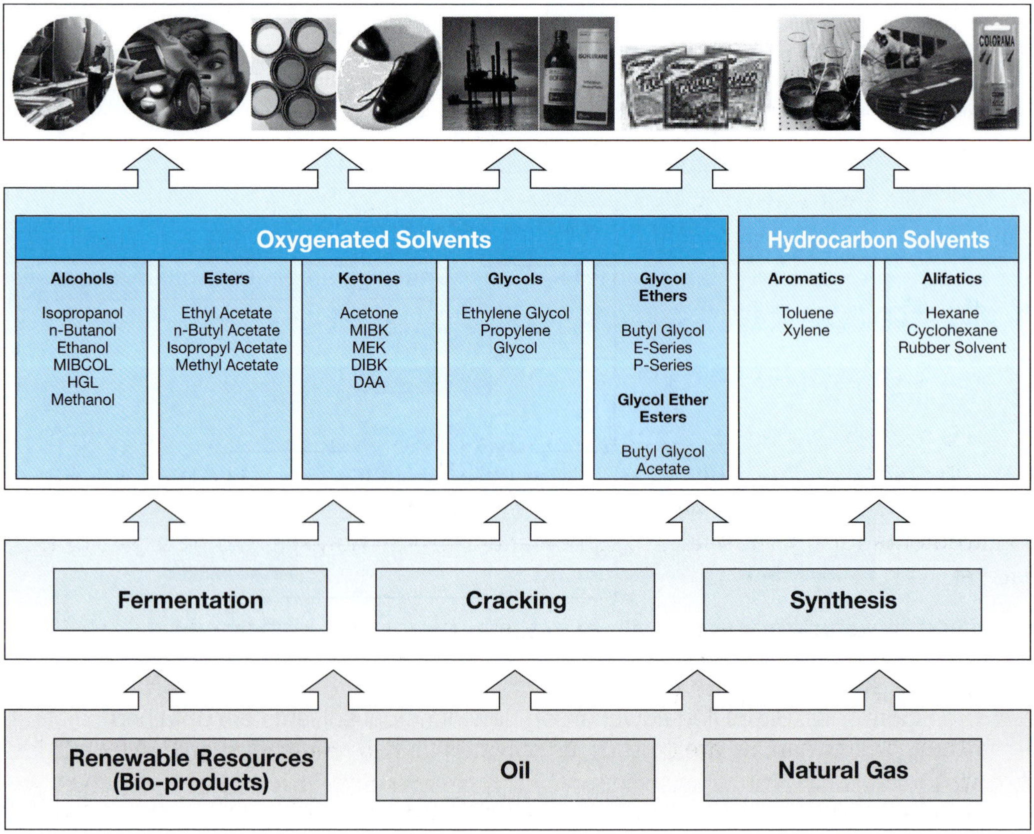

Figure 1.14. Raw materials to end markets (MIBC = Methyl Isobutyl Carbinol; HGL = Hexylene Glycol; MIBK = Methyl Isobutyl Ketone; MEK = Methyl Ethyl Ketone; DIBK = Diisobutyl Ketone; DAA = Diacetone Alcohol; IPA = Isopropanol).

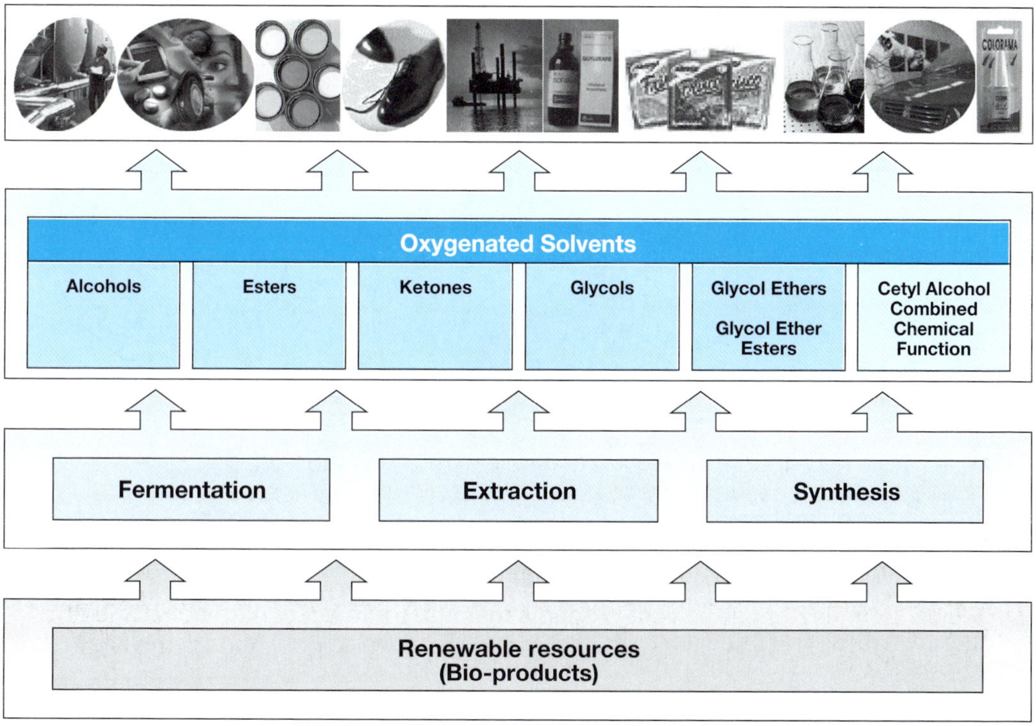

Figure 1.15. From beginning to end – a green perspective.

1.4. Solvents: *Commodities* and Specialties

If one is aware that the size of the world market of solvents is around twenty million tons, one may conclude that they are undoubtedly commodities, i.e., products which are commercialized in large quantities and in which producers must meet a series of requirements for operational excellence and a production scale to be a part of this market.

Some solvents are commercialized in vast quantities (more than 500,000 tons/year for each). Examples include toluene, methanol, ethanol, ethyl acetate, n-butyl acetate, acetone, MEK, and IPA (isopropanol).

Others, more than 50, are considered solvents and are commercialized in quantities that are lower than 200,000 tons/year each. Examples include diacetone alcohol, hexylene glycol, products of series E and P individually, methylisobutylcarbinol (MIBC), cyclohexanol, n-butanol, n-propanol, cyclopentanone, mesityl oxide, t-butyl acetate, isopropyl acetate, methyl soyate, ethyl lactate, n-methyl pyrrolidone, etc. The majority of these products may be considered a chemical specialty. There are few producers in the world, their production and commercialization scale are less relevant, and they normally meet the specific needs of some applications, which makes them more of a niche application.

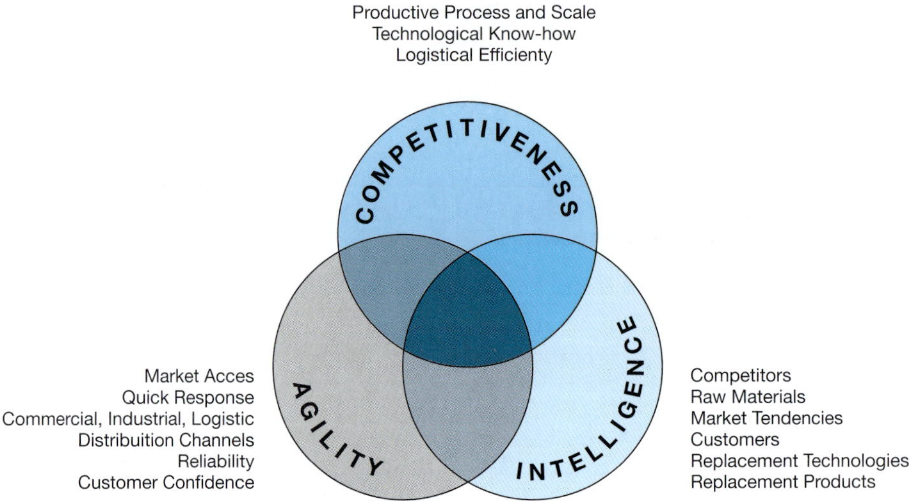

Figure 1.16. Factors all solvent products must take into account to act efficiently.

1.5. Solvents: an Old and Current Technology

Depending on the application, more than seven possible substituting technologies can be indicated.

Despite the fact that some applications have used solvents since BCE and solventborne technologies are among the oldest ones, this technology may be considered modern and current because it has evolved over the years. It came about in 1920 with the organochlorides and has evolved to meet the requirements of green chemistry and created a new class and a new concept of solvent: the green solvents, at the beginning of the 21st century.

Meeting the largest part of selection criteria: performance, cost, low initial investment, little consumption of energy in the drying and application processes, and aiming at complying with the occupational health and environmental regulations, solventborne technologies have been reborn and reinvented in this new century.

There is no ideal technology. Analyzing the history and weighing the advantages and disadvantages of solventborne technology, one may assume they will be present in the market in centuries to come.

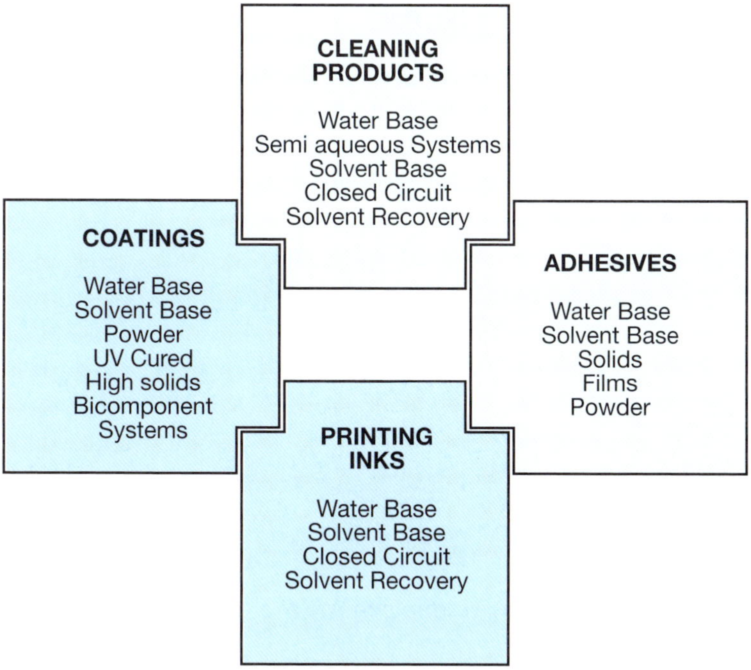

Figure 1.17. Some Replaceable Technologies.

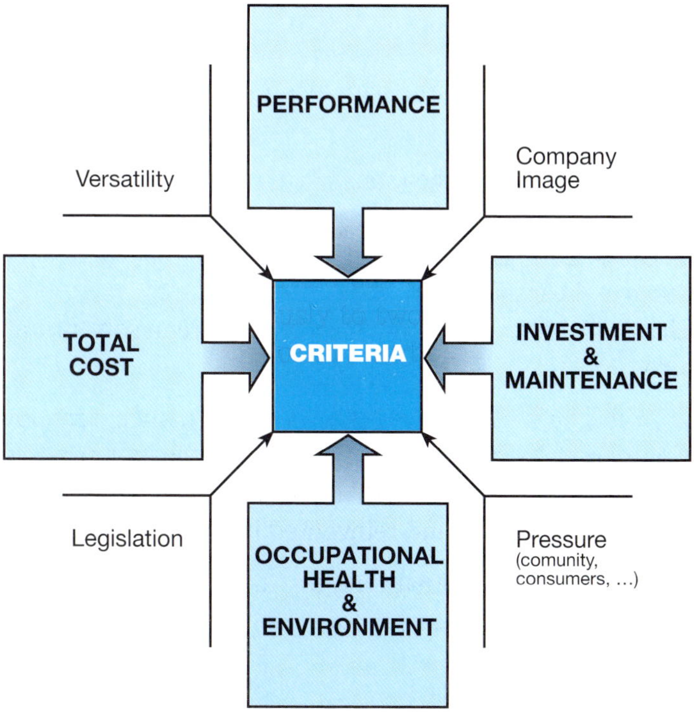

Figure 1.18. Criteria for Choosing Technology.

Bibliographical References

Dieter Stoye, Werner Freitag Paints, Coatings and Solvents Second, Completely Revised Edition, p. 2.

www.dnapl.group.shef.ac.uk

www.chemlink.com.au

www.webelements.com

2 General Aspects

This chapter gives a general approach to the role of solvents in several applications as well as a classification from different aspects: chemical composition, acid-base behavior, solute-solvent interaction, and physical properties.

Edson Leme Rodrigues

Solvents are used by all types of industries for processing, manufacturing, and formulating products. In industrial chemical processes, solvents are used in several process operational stages such as separation (gases, liquids and/or solids), reactions (reaction medium, reagent, carrier), washing, and many other operations. As part of formulated products, they are used in coatings, fabric, rubber products, adhesives, and thousands of other industrial and domestic products. In the cleaning product industry, they are used as cleaning agents or used in cleaning products. This is not the complete list, and solvents are used for other industrial, domestic, and research purposes.

Allied with a wide range of benefits that solvents offer, environmental, health, and safety aspects, including human and eco-toxicity aspects, process safety and residual management should be considered.

Therefore, guidelines orienting solvent development try to reduce the impact on the environment, putting a special emphasis on selection of the most appropriate solvent and the development of new solvents.

2.1. Definition

Solvents are composed of chemicals found normally in a liquid state at room temperature and 1 atmospheric pressure and are capable of dissolving, suspending or extracting other substances without chemically altering them.

A solution is made up of at least two components, a substance called solute and is dissolved in a solvent. Solutions are obtained by mixing solid, liquid and/or gaseous components with a liquid. When two liquid components are mixed together arbitrarily, the component with the molar excess is considered the solvent.

When we talk about solution properties, our main interest is in the solute properties. Thus, choosing a solvent is based particularly on solute properties that we want to develop.

The following characteristics are linked to solvents[1]:

- Clean, colorless appearance;
- Volatilize without leaving a residual;
- Chemically inert;
- Light or pleasant odor;
- Low water content (anhydrous);
- Constant physical properties;
- Low toxicity;
- Biodegradability.

2.2. Classification

2.2.1. Classification of Solvents According to Chemical Structure and Function

Three large classes of solvents are recognized: aqueous, nonaqueous, and organic. Although nonaqueous and organic classifications are both nonaqueous, the term organic solvent is generally applied to a large group of compounds containing carbon, which are found in industrial use and as a means for chemical synthesis. Organic solvents are generally classified by functional group, which are present in a molecule, such as alcohols, ketones, esters, ethers, glycol ethers, amines, hydrocarbons, and halogens. These groups indicate types of physical-chemical interaction that can occur between the solute and solvent.

Nonaqueous solvents include inorganic substances and some compounds containing carbon with low molecular weight, like acetic acid, methanol, and dimethyl sulfoxide. Nonaqueous solvents can be solids (for example, lithium iodide melt-LiI), liquids (H_2SO_4), or gases (NH_3) at room temperature conditions.

2.2.2. Classification of Solvents in Terms of Acid-Base Behavior

Brönsted-Lowry Acid-Base Theory

According to the Brönsted-Lowry acid-base theory, acids and bases are proton donors and recipients, respectively, as shown in the equilibrium equation (2.1):

$$\underbrace{HA^{z+1}}_{\text{Acid}} \rightleftharpoons \underbrace{A^{z}}_{\text{Conjugate base}} + H^{+} \tag{2.1}$$

where $Z = 0, \pm 1\ldots$

The majority of solvents have acid or base properties, and the acid or base strength depends on the medium in which they are dissolved.

The equilibrium shown in equation 2.2 will be established when an HA acid is dissolved in an SH base solvent.

$$\underbrace{HA^{z+1}}_{\text{Acid}} + \underbrace{SH}_{\text{Solvent}} \rightleftharpoons SH_2^{+} + A^{z} \tag{2.2}$$

The acid strength of the HA in the solvent SH is given by the Ka acidity constant as shown in the equation (2.3).

$$K_a = \frac{[SH_2^+]*[A^z]}{[HA^{z+1}]} \tag{2.3}$$

In an SH acid solvent, the acid-base equilibrium represented by the equation (2.4) will be established.

$$\underbrace{A^z}_{\text{Base}} + \underbrace{SH}_{\text{Solvent}} \rightleftharpoons + HA^{z+1} + S^- \tag{2.4}$$

The increase in solvent basicity or acidity moves the equilibrium (2.2) and (2.4) to the right, respectively.

Solvents that are self-ionizing have both acid and base characteristics (ex. water). These solvents are called amphiprotic solvents, contrary to aprotic solvents, which are not "self-ionizable" (ex. aliphatic hydrocarbons, carbon tetrachloride)[2].

This classification of solvents was primarily proposed by Brönsted, who distinguished four types of solvents based on their acid and base properties. Davies[2] extended the Brönsted classification and distinguished solvents with dielectric constants greater and lesser than 20, proposing 8 classes of solvents. A more simplified classification was presented by Kolthoff and is shown in Table 2.1[3].

Table 2.1. Classification of organic solvents according to acid-base behavior

Solvent Classification		Relative Acidity[(a)]	Relative Basicity[(a)]	Examples
Amphiprotics	Neutral	+	+	H_2O, CH_3OH, $(CH_3)_3COH$, $HOCH_2CH_2OH$
	Protégénic	+	–	H_2SO4, HCOOH, CH_3COOH
	Protophilic	–	+	NH_3, $HCONH_2$, $CH_3CONHCH_3$, $H_2N\text{-}CH_2CH_2\text{-}NH_2$
Aprotic	Dipolar Protophilic	–	+	$HCON(CH_3)_2$, CH_3SOCH_3, Pyridine, 1,4-dioxane, $(C_2H_5)_2O$, tetrahydrofuran
	Dipolar Protophobic	–	–	CH_3CN, CH_3COCH_3, CH_3NO_2, $C_6H_5NO_2$, sulfolane
	Inert	–	–	Aliphatic Hydrocarbons, C_6H_6, $Cl\text{-}CH_2\text{-}Cl$, CCl_4

(a) Acid-base index (–) weaker or (+) stronger than water.

Water is a prototype of an amphiprotic solvent and all other solvents with similar acid-base properties are called *neutral solvents*. Solvents that are stronger acids and weaker bases than water are called *protogenic solvents*. Contrarily, solvents with greater basicity and weaker acidity in relation to water are called protophilic solvents.

Looking at the equation (2.2), one can see that ionization of an acid depends on solvent basicity. However, acid ionization does not depend only on solvent basicity, but also on its dielectric constant and its solvation ability. The dependence of the acid and base constant for a compound in relation to basicity and acidity, respectively, of solvents leads to a distinction of solvent ionization capacity. For example, in methanol, the hypochlorous acid is completely ionized but nitric acid is only partially ionizable.

Several acidity and basicity scales are created for water and other solvents. However, there is no simple acid and base scale, equally valid and usual for all solvents and that is applicable for both equilibrium and kinectic situations. A review of the different acidity functions was written by Boyd[4].

Solvent acidity and basicity can be measured in different ways[5]. Along side the experimental methods for acid-base equilibrium constant measurements, possible approximations are what determine solvent basicity and acidity through the variation of a physical property (like infrared or UV/Vis absorption or RMN) of a standard substrate molecule when transferred from a reference solvent to another.

Lewis Acid/Base Theory

According to Lewis, acids are electron pair acceptors (EPA) and bases are electron pair donors (EPD) related by the equilibrium shown in the equation (2.5).

$$\underbrace{\text{A}}_{\substack{\text{Acid}\\ \text{EPA}\\ \text{eletrophilico}}} \quad + \quad \underbrace{:\text{D}}_{\substack{\text{Base}\\ \text{EPD}\\ \text{nucleophilico}}} \quad \rightleftharpoons \quad \underbrace{\text{A}-\text{D}}_{\substack{\text{Acid-base}\\ \text{complex}}} \tag{2.5}$$

Solvents can be classified as EPD and EPA based on their chemical makeup and reaction medium compounds [6]. However, not all solvents can be classified in this class. For example, aliphatic hydrocarbons do not have EPD and EPA properties. An EPD solvent preferentially solvates electron pair acceptor ions or molecules. The reverse is true for EPA solvents.

An extension of the Lewis acid/base classification was proposed by Pearson [7], which subdivides them into two groups, strong and weak, according to electronegativity and polarizability. The concept is called HSAB.

Strong acids (ex. H^+, Li^+, Na^+, BF_3, $AlCl_3$, HX hydrogen bond donors) and strong bases (e.g. F^-, Cl^-, HO^-, RO^-, H_2O, ROH, R_2O, NH_3) are derived from small atoms with high electronegativity and generally low polarizability. Weak acids (ex. Ag^+, Hg^+, I_2, 1,3,5-trinitrobenzene) and weak bases (ex. H^-, I^-, R^-, RS^-, RSH, R_2S, alkenes, C_6H_6) are usually derived from large atoms with low electronegativity and are normally polarizable. The benefit from this division was a simple rule for the Lewis acid/base complex stability: strong and weak acids prefer to mix with strong and weak bases, respectively. This HSAB supports the description of several chemical phenomena on a qualitative path and finds many applications in organic chemistry.

The application of the HSAB concept for solutions supports the rule that strong solutes dissolve in strong solvents and weak solutes dissolve in weak solvents. This supports the rule of Thumb that similar dissolves similar.

2.2.3. Solvent Classification in Terms of Solute-Solvent Interaction

Solvents can be divided into three groups according to specific interaction with the solvent:

- Polar Protics;
- Dipolar Aprotics;
- Nonpolar Aprotics.

The distinction between them is mainly in the solvent's polarity and in its ability to form hydrogen bonds. The word 'protic' in this context refers to a hydrogen atom bonded to an electronegative atom, and the word aprotic means molecules that do not have an –O–H,–N–H, F–H bond, etc.

Polar Protic Solvents

Polar protic solvents are compounds that have a hydrogen atom bonded to an electronegative atom (F–H, –O–H, –N–H, etc.) and are apt to form hydrogen bonds. Due to this ability, these solvents are particularly good anion solvating agents. With the exception of acetic acid and its homologues, the dielectric constant is greater than 15. Water, ammonia, alcohols, carboxylic acids, and primary amides belong to this class.

Dipolar Aprotic Solvents

Dipolar aprotic solvents have high dielectric constants ($\varepsilon > 15$) and dipole moment ($\mu > 8{\cdot}3 \cdot 10^{-30}$ C · m). These solvents do not cause hydrogen bonds to form since their C–H bonds are not polarized enough. However, they are generally good EPD solvents and therefore cation solvating agents. Included among the main dipolar aprotic solvents are acetone, ethyl acetate, acetonitrile, benzonitrile, N,N-dimethyl acetamide, N,N-dimethylformamide, dimethylsulfone, dimethyl sulfoxide, 1-methyl-2-pyrrolidone, nitrobenzene, cyclic carbonates with propyline carbonate and sulfolanes.

Nonpolar Aprotics Solvents

Nonpolar aprotic solvents are characterized by a low dielectric constant ($\varepsilon < 15$), a low dipole moment ($\mu < 8 \cdot 3 \cdot 10^{-30}$ C · m) and an inability to form hydrogen bonds. The solvents slightly interact with the solute, where only induction and dispersion effects can operate. This group of solvents includes aliphatic and aromatic hydrocarbons and their halogen derivatives, tertiary amine, and carbon disulfide.

This is not a rigid classification. There are several solvents that cannot be classified in any of these three groups. Examples include ethers, carboxylic esters, primary and secondary amines, and N-monosubstituted amides with N-methylacetamide. However, this classification is prominently related to ion solvation.

Protic polar solvents are good anion solvating agents and this tendency is more pronounced in the high charge density (charge by volume relation) of the anion to be solvated and consequently, in its hardness according to the HSAB concept (item 2.2.2). It can be noted that the increase of strength of the solvating agent decreases the anion nucleophilic reactivity.

Oppositely, with aprotic dipolar solvents, the anion solvating agent is mainly through dipole-ion and induced dipole-ion forces. This solvation will be important for soft anions with low charge density. However, these solvents are generally bad anion solvating agents. As a result, anion reactivity is exceptionally high in the aprotic dipolar solvents, and the reaction constant of the Sn2 nucleophilic substitution reactions can increase by 10 times when we change a protic solvent for an aprotic solvent.

In summary, protic polar solvents are better anion solvating agents than aprotic dipolar solvents, and the reverse is true for cation solvation. This observation is important in the solvent selection for reactions.

2.2.4. Classification of Solvents Using Physical Constants

The following physical constants can be used for characterizing solvent properties[6]:

- Melting and Boiling Point;
- Vapor Pressure;
- Heat of Vaporization;
- Refractive Index;
- Density;
- Viscosity;
- Surface Tension;
- Dipole Moment;
- Dielectric Constant;
- Polarizability;
- Specific Conductivity.

The solvents are classified according to the low, medium, and high boiling point level (Table 2.2). Knowing the solvent boiling point is important in the extraction and distillation process. However, knowing solvent volatility in the coating and adhesive application is more important.

Table 2.2. Classification of Solvents Based on Boiling Point, Relative Evaporation Temperature, and Viscosity[6]

Property	Low	Medium	High
Boiling Point at 760mmHg	< 100 °C	100 – 150 °C	> 150 °C
R.E.R.*	< 10	10 – 35	> 35
Viscosity at 20 °C	< 2 Cps	2 – 10 cP	> 10 cP

* Relative Evaporation Rate.

There is no general correlation between solvent evaporation rate and boiling point. Volatility generally decreases with the increase in boiling point if the solvents belong to the same chemical class. Solvents which tend to form hydrogen bonds (ex. water, alcohols, amines) are less volatile than other solvents with the same boiling point as long as the energy has been supplied to break the hydrogen bonds before becoming a vapor state.

The evaporation rate depends on the following properties [1]:
1) Vapor Pressure at Processing Temperature;
2) Specific Heat;
3) Enthalpy of Vaporization;
4) Degree of Molecular Association;
5) Speed of Heat Supply;
6) Surface Tension;
7) Solvent Molecular Mass;
8) Atmospheric Turbulence;
9) Atmospheric Humidity.

Since these factors depend on the others, it is difficult to give a theoretical prediction for evaporation rate. In practice, evaporation time of a solvent's given quantity is experimentally determined under identical external conditions and compared to that of diethyl ether or butyl acetate. In Brazil, butyl acetate is used as a reference. The relative evaporation rate (R.E.R.) of solvents using butyl acetate can be determined according to ASTM D3539. Solvents can be divided into 3 groups according to their R.E.R.: low, medium and high volatility (Table 2.2).

Density and the refractive index are used to assess, and with some restrictions, determine the solvents purity. Solvent density is generally measured at 20 °C and referenced at water density at 4°C (relative density d_4^{20}). The density of the majority of organic solvents decreases with the increase of temperature. Relative density of homologous glycol esters and ethers decreases with the increase of molecular weight, however, relative density of ketones and alcohols increase (Table 2.3).

Refractive index (n_D) is measured in a refractometer using a sodium-vapor lamp (589.0 and 589.6 nm). The refractive index value is strongly determined by the substance's hydrocarbon skeleton. Aliphatic esters, ketones and alcohols have a refractive index between 1.32 and 1.42. In homologous series, the refractive index increases with the increase of the carbon chain length, and decreases with the increase of ramifications. Aliphatic-cycle and aromatic structures increase the refractive index.

Viscosity of the homologous series of solvents increases with the increase of molecular weight. Solvents with hydroxyl groups show greater viscosity due to the hydrogen bond formation. Solvent viscosity strongly influences solution viscosity. Solvents can be divided into three groups according to their viscosity: low, medium or high (Table 2.2).

Solvent surface tension for coatings is important for the evaporation rate, surface covering formation, and substrate and pigment wetting characteristics.

Table 2.3. Influence of Molecular Weight in Ester, Ether Glycol, Alcohol and Ketone Relative Density

Éster	d_4^{20}	Alcohol	d_4^{20}
Methyl Acetate	0.934	Methanol	0.791
Ethyl Acetate	0.901	Ethanol	0.789
Propyl Acetate	0.886	Propanol	0.804
Butyl Acetate	0.881	Butanol	0.810
Amyl Acetate	0.876	Amylic Alcohol	0.815
Glycol Ether	d_4^{20}	**Ketones**	d_4^{20}
Methyl Glycol	0.966	Acetone	0.792
Ethyl Glycol	0.931	Methyl Ethyl Ketone	0.805
Propylglycol	0.911	Methyl Propyl Ketone	0.807
Butyl Glycol	0.902	Amyl Methyl Ketone	0.816

Solvents and solutes can be widely classified as polar (hydrophilic) and nonpolar (lipophilic). Polarity can be measured as a dielectric constant or dipole moment of a compound. A solvent's polarity determines which types of compounds it is apt to dissolve and with which other solvents or liquid compounds it is immiscible. Using the rule of Thumb, "similar dissolves similar," polar solvents dissolve polar compounds better and nonpolar solvents dissolve nonpolar compounds better.

2.3. Solute-Solvent Interaction

Solutions are homogeneous mixtures of two or more substances. A homogeneous mixture is a physical combination of two or more pure substances distributed uniformly in the mixture. This means if we take a part of the solution (aliquots), the proportion of each pure mixture in that aliquots will be the same as the solution. These proportions in relation to the total volume are called concentrations.

When a substance (solute) is dissolved in a solvent or mixture of solvents, the strength of attraction between solute molecules decreases because solvent molecules penetrate between solute molecules and surround each molecule, thus isolating each solute molecule. This process is called *solvation* and results in the distribution of the solute in solution at the molecular level.

Solvation strength and the number of solvent molecules in the solvation layer depend on the solubility parameter, dipole moment, hydrogen bonding, polarizability, and molecular size of the solute and solvent.

In aqueous solutions, water's polar molecules will interact with solute molecules, thus solvating it. If the bonds that keep solute molecules together are weak enough, the weakest bond in the solute molecule will be broken by the electric strength of attraction in the water molecules for the solute molecules. If this happens, solute molecules will then be called solvated ions. In some cases, the ionization process can almost be complete while in other cases this process occurs in a limited manner. The solution that has not been ionized has solute molecules solvated in the solution and they are called non-electrolytes. Such solutions do not conduct an electric current because of the absence of ions. Solutions in which molecules were ionized during the solution process are called electrolytes and conduct electric current because of the presence of ions.

2.3.1. Energy Transference in Solution Formation

When solutions are formed, solvent molecules surround solute molecules. This solvation process involves a certain quantity of thermal energy. Some solutions absorb energy when they are formed, thus it can be said that these solutions have heat or enthalpy of an endothermic solution. The majority of aqueous solutions involving solid or liquid solutes have an endothermic solution heat.

$$\text{Solute} + \text{Solvent} + \text{Thermal Energy} \Rightarrow \text{Solution} \tag{2.6}$$

Other solutions involving gaseous solutes in water release thermal energy during the formation process of the solution. These solutions are said to have exothermic solution heat.

$$\text{Solute}(g) + \text{Solvent} \Rightarrow \text{Solution} + \text{Thermal Energy} \tag{2.7}$$

2.3.2. Solubility of Solutions

Solubility of a particular solute in a solvent is the maximum amount of solute that will dissolve in a specific quantity of a solution or solvent. This represents the level of saturation of the solution where no more solute will dissolve in the solution. The saturation condition creates a dynamic physical equilibrium among the solute, solvent, and solution.

$$\text{Solute} + \text{Solvent} \rightleftharpoons \text{Solution} \tag{2.8}$$

Equilibrium involves two processes: forward reaction and reverse reaction. When the rate of the forward reaction equals the rate of the reverse reaction, the system is in dynamic equilibrium.

2.3.3. Factors that Affect Solubility

Some factors can affect solubility, even in the same solution.

Solute/Solvent Interaction: solute and solvent molecule polarity affects solubility. Generally, polar solute molecules will be dissolved by polar solvents, and nonpolar solutes by nonpolar solvents. When a solution is formed by a polar solute and solvent, the resulting force of attraction of this dissolution is known as dipole-dipole forces. This is one type of intermolecular force that differs from those called the London Dispersion Forces where an effective nuclear charge of a solute's molecular atoms interacts with electrons in a solvent's molecular atoms. This type of interaction occurs in solutions obtained through nonpolar solvents.

Temperature: temperature affects solubility. If the solution formation process absorbs energy, solubility will increase with the increase of temperature. If the solution process releases energy, solubility decreases with the increase of temperature.

At a given temperature, the solution can reach its solubility limit, maintaining a dynamic equilibrium. According to Le Chatelier's Principle, when a change is externally applied to the equilibrium, the equilibrium is temporarily upset and will shift to counter-act the change. An external change in a solution can be a temperature. According to the principle, increasing the equilibrium's temperature will always favor the equilibrium of the endothermic process. The majority of liquid and solid solutes dissolved in water have an endothermic solution process which will be favored by a temperature increase resulting in an increase in the solubility limit. There are some exceptions, for example, $Ce_2(SO_4)_3$, a solid, has its solubility in water reduced with the increase of temperature. On the other hand, some solutions have the heat of an exothermic solution, like the majority of solutions with gaseous solutes. Thus, the solubility of all gas solutions in water decreases with the increase of temperature.

Pressure: pressure change on a solution does not affect solubility limits of solids or liquids in water. However, gaseous solutes are affected. If the pressure of a gas is increased on a gas solution, then the solubility will increase in linear proportion. This is expressed in Henry's Law.

$$C = k \cdot P \tag{2.9}$$

where: k = Henry constant for gas
P = Partial pressure of gas on a solution
C = Concentration of a gas in a solution

Molecular Weight: in homologous compounds, with greater molecular weight of a molecule, solubility will lessen. Large molecules are more difficult to be surrounded by the solvent during the solvation process. In the case of organic compounds, the increase of ramifications will increase solubility (even molecular weight) since the ramifications reduce the volume of molecules and favor solvation by the solvent.

2.3.4. Classification of Solutions

Solutions can be unsaturated, saturated, or supersaturated. *Unsaturated solutions* are those that are below the solubility limit of the solute in the solvent. *Saturated solutions* are those that are at the solubility limit. *Supersaturated solutions* are those solutions that are above the solubility limit. Supersaturated solutions are metastable. In these solutions, the excess solute is maintained without crystallization. Any change in the solution (a single solute crystal grain or the introduction of a single body in the solution) can cause crystallization of the solute excess, thus re-establishing the saturated solution.

2.4. The Role of the Solvent in Industry

2.4.1. Solvents in Chemical Reactions

A large number of chemical reactions occur in the solution and solvents have many functions during chemical reactions. They can be used as a reaction medium to keep reagents together, and as a carrier to separate chemical compounds in solution. As a reaction medium, they can solvate reagents thus aiding in dissolution. They facilitate the collision of reagents favoring the transformation of reagents in products, or increase the energy of particle collision so that the reaction happens faster. Solvents can also be used for other purposes. For example, in endothermic reactions, heat can be supplied through a heated inert solvent having a high heat capacity, while in exothermic reactions heat is absorbed. Further, solvents can offer a means of controlling temperature. Similarly, reactions in a gas phase, those which are normally at a high temperature and/or pressure, can be carried out in a liquid phase under significantly low temperature and/or pressure. As carriers, solvents can be used for indirectly influencing a reaction by removing one or more products from the reaction. The selection of an appropriate solvent is guided by the theory and the experiment[8]. Generally, a good solvent will have the following characteristics:

- It should be inert in reaction conditions;
- It should dissolve the reagents;
- It should have an appropriate boiling point;
- It should be easily removed at the end of the reaction.

2.4.2. Solvents in Extraction Processes

Extraction by solvent can be defined as a mass transporting process of a phase to another with the purpose of separating one or more mixture components.

The liquid-liquid extraction, also known as liquid extraction or solvent extraction, is the separation of the constituents of a liquid solution, called feed, by close contact with another appropriate liquid, immiscible or partially miscible, called solvent, which must have the capacity of preferentially extracting one or more desired components (solute). Coming from this contact are two new liquids: the raffinate, which is a residual solution from the feed, low in solvent, with one or more of one of the solutes removed by extraction; and the extract, rich in solvent, containing the extracted solute[9].

With the extraction of components present in solid states, mechanisms involved are: lixiviation, washing, diffusion and dialysis. The success of a solid-liquid extraction, in large part, depends on the previous treatment of the solid state so as to maximize the contact area between the solid and the solvent.

The desired properties of solvents used for extracting liquid-liquid or solid-liquid are[9]:

- Low flammability;
- Thermal and chemical stability;
- Inertia in relation to equipment (minimizing maintenance expenses);
- High level of purity for making operational characteristics uniform;
- High volume/low price;
- Easy recovery;
- High selectivity;
- High solubility of the solute of interest;
- High ability of penetration in the solid state;
- Easy diffusion of soluble solid in solvent (low viscosity);

A modern variant of the liquid-liquid extraction is the supercritical fluid or liquefied gas extraction. Liquefied gases near the critical point or supercritical fluids close to the critical point show good solvent characteristics. They have higher densities, closer to the density of common water than that of normal gas. They show low viscosity, closer to that of a gas than to a liquid. In general, supercritical extraction has the advantage of the possibility of changing the solvent's selectivity by changing the system's pressure or temperature. It also allows an easier elimination of the solvent through simple system depressurization. The main disadvantages are: compression for liquefying the gas near its critical pressure and high cost of the equipment[10].

2.4.3. Solvents Used in Paints and Coatings

In paints, the solvent dissolves or disperses different components used in formulations (like resins and pigments), giving the coating an appropriate viscosity or consistency for a uniform application. Once the coating is applied, the solvents evaporate, allowing the resin and pigments to produce a coating film and drying in good time.

To produce a coating, the main properties of the solvents are[11]:

- Specific Weight;
- Flammability;
- Solvency Capacity;
- Distillation;
- Evaporation Rate;
- Toxicological Aspects.

2.4.4. Solvents Used in Printing Inks

The solvent is the main component in terms of quantity for the majority of printing inks used in lithography, offset, letterpress, flexography, and printing processes using engravings and screens.

Solvency power and solvent volatility are important properties influencing the ink type, which can be used for different printing processes.

Solvents are used to control ink viscosity, allowing an appropriate flow without damaging printing rollers. Further, like ink, solvents help optimize drying time in the printing process.

Solvents or solvent mixtures used in lithography and flexography inks and for engravings are assessed in terms of printing stability, ink printability and drying in the substrate. Additionally, flexography and engraving used for packaging require that solvents are chosen carefully so as to avoid odor and odor retention problems.

2.4.5. Solvents Used in Cleaning Products

Solvents are used in a variety of cleaning products, helping to increase effectiveness. They can act in different ways:

Solubilization of Impurities

Solvents help reduce the need of any type of abrasive, soften fabric, and dissolve dirt.

Surface Tension Reduction

Solvents also help disinfect using surface tension reduction. This characteristic, working with other product ingredients, helps the disinfectant penetrate the empty spaces in the cleaning objective material, making it more effective in attacking bacteria.

Increase Product Stability

Another benefit of solvents in cleaning products is that they help stabilize the product, thus increasing its shelf life.

2.4.6. Solvents Used in Adhesives

Solvents provide properties that are excellent for adhesive performance. From shoe soles to car tires, solvents are used in domestic and industrial applications, as well as in high-performance situations like metal-metal bonding

In addition to solvents being used to dissolve solid adhesives and prepare surfaces for the bonding process, their use in controlling drying time is a critical attribute for the most adhesives. The solvent's complete removal is essential for obtaining optimum bonding efficiency.

2.4.7. Solvents Used in Manufacturing Pharmaceuticals

Solvents are used in thousands of pharmaceutical products. They have two functions in the production process. Solvents can be used as a reaction medium, making molecules available for the manufacturing of medicines, and they can also be used for extraction and purification.

Frequently, solvents have a role at the start of the pharmaceutical manufacturing process. Antibacterial creams and corticosteroids often use solvents at the start of the process to manufacture active ingredients. Some products have their final forms in creams, lotions or liquids, thus solvents can also be used in blending product components. Another use of solvents is helping the consistency of the final product.

Other pharmaceutical items using solvents include pet products. Solvents are frequently found in pet shampoos and oral medication. Pet shampoos use solvents to dissolve medication or for moistening and softening purposes for animal skin and hair.

2.4.8. Solvents Used in the Production of Agricultural and Food Products

Solvents are used in the preparation and isolation of active ingredients for many pesticides and agricultural products. The presence of a solvent in pesticides promotes slow uniform drying allowing adequate penetration and high spraying efficiency, thus reducing the amount of a pesticide needed.

Solvents have a key role in the delivery of pesticides, herbicides, and insecticides with regard to their respective applications. The exact selection of a solvent produces optimal drying time (slow enough to allow adequate penetration) and good spraying efficiency.

Solvents are used in many aspects of food preparation and packaging. They can be used to extract fats, oils, and aromas from nuts, seeds, and other raw material. They are also used in liquid formulation of aromas and essences. Printing inks and adhesives for food packaging use appropriate solvents. In cleaning products, solvents help clean and sterilize food preparation areas.

2.4.9. Solvents Used in Producing Personal Care Products

Solvents are an important part of beauty and cosmetic products. Many products use solvents to dissolve ingredients and allow them to work properly.

Solvents Used in Hair Care Products

Solvents present in hair treatment products allow a means for mixing hair coloring/dyes before applying them to hair. Along with other ingredients, solvents in shampoos and conditioners help the product leave hair soft and manageable. Solvents also help other products in a variety of uses, such as straightening and curling.

Solvents can be found in many products, such as mousse, gels, and hair sprays. Alcohols are commonly found in these products.

Solvents Used in Skin Care Products

Solvents can help in the development of lotion, creams, rouges, shaving creams, and other products used to maintain elasticity, soften skin, to give a healthy appearance. Solvents are used to deliver antibacterial agents or to give an appropriate consistency to skin products. They also have a function of delivering fragrances. Some skin cleaning products, like astringents, contain solvents that help dissolve and remove traces of make-up. From facial lotions to shaving gels, solvents have an important role in product performance. .

Solvents Used in Cosmetics and Make-up

Solvents can have multiple functions in eye make-up. They can carry an active ingredient designed for resistance of bacterial growth, giving an adequate consistency and helping make-up stay on longer. Solvents can also help make-up highlight facial features.

The majority of perfumes have an alcohol base. The alcohol allows the perfume to spray in different directions. When alcohol evaporates, it leaves the fragrance over an applied area. Solvents are also used at the initial stage of some perfume manufacturing processes. Oil fragrants from fruit, flowers, roots, or barks can be extracted and purified using solvents. The general process uses a solvent to dissolve the oil (fragrance) and carry it into the product.

2.4.10. Solvents for Automotive Use

Solvents help in automobile maintenance in several ways. They help in cleaning injection systems and carburetors, keeping the engine cool, and are used in cleaning carpets and upholstery.

Additives used in gasoline contain solvents in the composition. These additives always keep the injection nozzle clean, maintaining the electronic injection system efficient. Further, they increase useful life of the catalyst, protect against corrosion, and guarantee perfect fuel combustion.

2.4.11. Solvents Used in Microchip Production

Solvents have an important role in the microelectronic industry. Integrated circuits or microchips use electronic-grade solvents in their manufacturing. Using electronic-grade solvents helps minimize defects in the circuit. Electronic-grade means there is a low level of metal ions in the solvent. These metal ions can cause a short-circuit resulting in a bad microchip.

Electronic grade solvents are used for dissolving a photo-sensitive polymer which is used for coating a silicon wafer to produce the micro circuitry. Solvents are also used to clean wafer surfaces and circuits. Common solvents used in microchip production are alcohols, esters, and ketones.

Photolithography is a process used in manufacturing integrated circuit chips. In this process, there is an initial deposit of a layer of SiO_2 on a silicon wafer, and then this wafer is coated with a very fine film (0.5 to 1.0 micrometer, 10^{-6} meters thick) of a solubilizated photo-sensitive organic polymer. The wafer is exposed to UV light, using a mask (a kind of model, mold used to shape the circuit to the desired form). In this case, the polymer irradiated sections that are not covered with the mask react and their solubility is changed, becoming more soluble than the original material in the areas covered with the mask. Then, the wafer is washed with an appropriate solvent. This solvent will have the responsibility of removing the polymer from the exposed areas, that is, the areas not covered by the mask. After this process, the silicon surface is exposed showing the circuit format. Recovering areas and using chemical solvents is carried out repeatedly so that the circuit is completely produced.

2.4.12. Solvents Used in Aerosols

Aerosol sprays are present in a large variety of products, as well as reduce waste and can get to small spaces effectively. They are used for coatings, cleaning agents, air fresheners, personal care items, and insecticides. A key part of many aerosol sprays is the solvent, which helps improve product performance and extend shelf life.

The choice of a solvent or solvent system that meets performance needs is not so simple. The primary function of solvents in aerosol products is maintaining the formulation uniform in order to ensure product ingredient proportions are the same throughout all its use.

Aerosol formulations contain three main components: propellant, solvent, and active ingredients. The majority of propellants have poor solvency capability, thus the solvent is used to couple the active ingredients in a solution with propellant. A second solvent that works to keep active ingredients in the solution can also be used.

Another important function of the solvent is to help produce a spray with a particle size so that it has the most efficient application. Different spray characteristics are necessary for paint, creams, and insecticides. The solvent's ability to reduce propellant vapor pressure affects the particle size of the spray.

In some cases, the solvent is essential to the product's performance. An example is aerosols for paints. Many aerosol paint formulations have three types of solvents with different evaporation rates: rapid, medium, and slow. The solvent that evaporates quickly gives a lower initial viscosity in order to make the application easier; the solvent that evaporates at a medium rate stops dripping; and the solvent with a slow evaporation rate is the last to leave the system. This last one finalizes the flow and promotes a uniform thickness, which is a solvent's finalizing function by its polymer.

Bibliographical References

1. B. Elvers, S. Hawkins, W. Russey, G. Schulz. Ullmann's Encyclopedia of Industrial Chemistry: Volume A 24- Silicon Compounds, Inorganic to Stains, Microscopic. Fifth Edition. Wiley VCH, November 1993. Pages 581.
2. M. M. Davies. Brönsted Acid-base behavior in Inert Organic Solvents, in J. J. Lagowski, The Chemistry of Nonaqueous Solvent Systems. New York, London: Academic Press, 1970. Volume. III.
3. I. M. Kolthoff. Acid-Base Equilibria in Dipolar Aprotic Solvents. Anal. Chem. 46, 1974, in C. Reicfortest, Solvent effects in organic chemistry. Weinheim, New York: Verlag Chemie, 1978.
4. J. F. Coetzee and C. D. Ritchie (editors). Soluto-Solvent Interactions. Volume 1. Chapter: Acidity Functions, edited by R. H. Boyd. New York, London : M. Dekker, 1969.
5. A. K. Covington and T. Dickinson (editors). Physical Chemistry of Organic Solvent System. Chapter: Acid-Base Behaviour, edited by E. J. King. London, New York: Plenum Press, 1973.
6. C. Reicfortest. Solvent effects in organic chemistry. Weinheim, New York: Verlag Chemie, 1978.
7. R. G. Pearson. Hard and soft acid and bases. Journal of the American Chemical Society, ACS Publications, Volume 85, Issue 22, Pages 3533-3539, November 1963.
8. R. Gani, C. Jiménez-González, A. ten Kate, P. A. Crafts, J. H. Atherton, J. L. Cordiner. A Modern Approach to Solvent Selection. Chemical Engineering: Technical & Pratical, Pages 30-42, March 2006.

9. Robert E. Treybal. Mass Transfer Operation. 3rd Edition, 1981. Singapure: Mac-Graw Hill Book Company. Pages 800.

10. G. Brunner. Gas Extraction: An Introduction to Fundamentals of Supercritical Fluids and the Application to Separation Processes. Hamburgo Alemanha: Springer, December 1994. Pages 399.

11. C. A. T. V. Fazano. Tintas: Métodos de controle de pinturas em superfícies. 5. ed. São Paulo: Hemus, 1998.

3 Main Classes of Solvents

Solvents can be classified according to the chemical function present in their molecules. Based on this aspect, topics such as reactivity, incompatibility with the medium, and production methods for main solvents will be approached in this chapter.

Danilo Zim

3.1. Aliphatic Solvents

3.1.1. Definition

Hydrocarbons are molecules formed by carbon and hydrogen atoms exclusively. Based on the structure of the carbon atom chain, it is possible to classify hydrocarbons as aromatics or aliphatics. Aromatic hydrocarbons have a special structure: a benzene ring with six carbon atoms bonded together in a cyclic structure which alternates resonating single and double bonds. Aliphatic hydrocarbons are those that have no aromatic ring. These can be further classified as cyclic or acyclic (also called open or closed chain hydrocarbons), linear or branched, and saturated or unsaturated.

Linear hydrocarbons have a single chain of carbon atoms organized so that each carbon atom is bonded to, at most, two other carbon atoms. Branched hydrocarbons have a carbon atom bonded to three or four other carbon atoms at one or more points on the carbon chain.

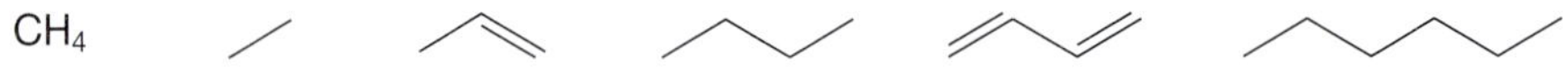

methane ethane propene n-butane butadiene n-hexane

Figure 3.1. Examples of linear hydrocarbons.

methylpropene methylpropane 2,2-dimetyllbutane 2,3-dimethylbutane

Figure 3.2. Examples of branched hydrocarbons.

Cyclic or closed chain hydrocarbons have a chain in which its carbon atoms are bonded together forming a cycle. Acyclic or open chain hydrocarbons have no cyclic structure.

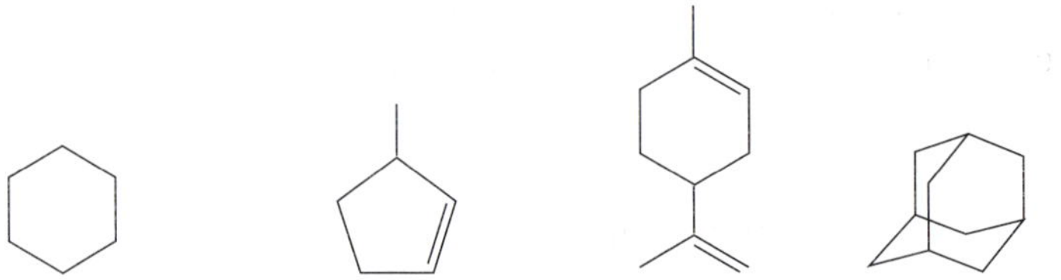

cyclohexane 3-methylcyclopentene limonene adamantane

Figure 3.3. Examples of cyclic hydrocarbons.

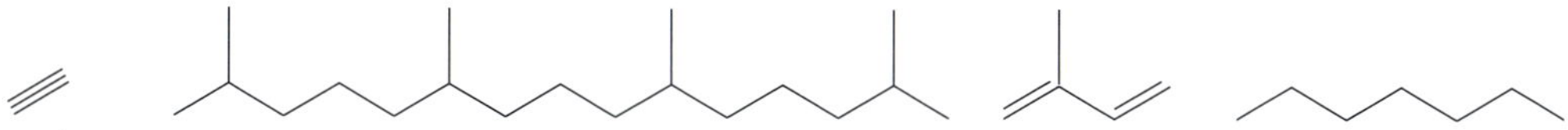

acetylene Pristane isoprene n-heptane

Figure 3.4. Examples of acyclic hydrocarbons.

Saturated hydrocarbons, also called alkanes, only have single bonded atoms in their structure whereas unsaturated hydrocarbons have at least a double or triple bond between their carbon atoms. Hydrocarbons with double bonds are called alkenes or olefins, and those with triple bonds are called alkynes.

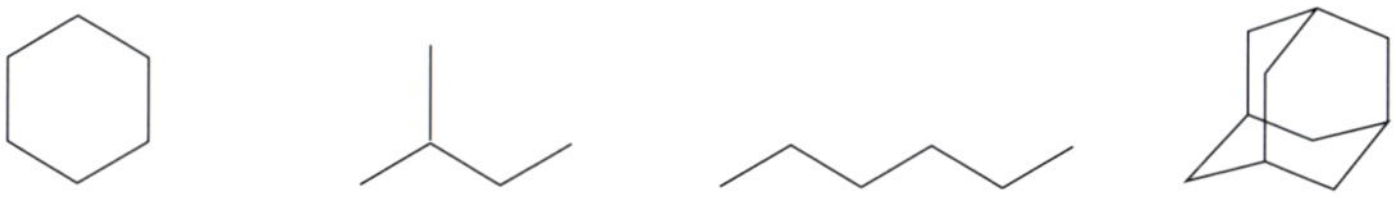

cyclohexane 2,2-dimethylbutane n-hexane adamantane

Figure 3.5. Examples of saturated hydrocarbons.

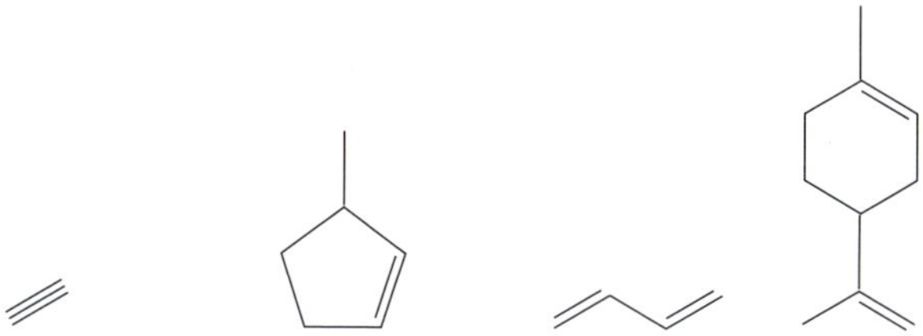

acetylene 3-methylcyclopentene butadiene limonene

Figure 3.6. Examples of unsaturated hydrocarbons.

The carbon chain atom structure not only defines the classification to which the molecule belongs, it also defines the physical properties as well as the reactivity of the molecule.

3.1.2. Characteristics and Reactivity

Due to the small difference in electronegativity between carbon and hydrogen, the only two components in hydrocarbons, there are no strong polar bonds in this class of molecules. Thus, intermolecular forces acting on this kind of compound are induced dipole-induced. All hydrocarbons are very nonpolar compounds and as such they have some typical properties.

Aliphatic hydrocarbons are practically immiscible in water. The mixture between hydrocarbons and water tend to form two distinct phases: an organic upper phase and an aqueous lower phase because hydrocarbons are less dense than water. They are also soluble with each other in any proportion and tend to solubilize other nonpolar compounds (like greases and waxes).

Low-weight molecular hydrocarbons (up to 4 carbon atoms) are gases at room temperature and 1 atmospheric pressure while linear hydrocarbons with more than 5 carbon atoms tend to be liquids or solids. In the following figure, melting and boiling points for the first 14 linear alkanes are shown.

Hydrocarbon reactivity is basically a function of the presence or absence of unsaturation in a molecule.

Saturated hydrocarbons, i.e., alkanes, are practically inert at room temperature and 1 atmospheric pressure. They do not suffer attack even from strong acids or bases. They also do not undergo hydrolysis, polymerization, or decomposition reaction under such conditions. They are resistant to strong oxidants, even in hydrogen peroxide or nitric acid, and do not undergo reduction even in the presence of metal hydrides. Halogenation and cracking reactions of alkanes are only possible at high temperatures. The only alkanes that show some differentiated reactivity are cyclic alkanes with high ring tension, like cyclopropane that tends to behave similarly to alkenes.

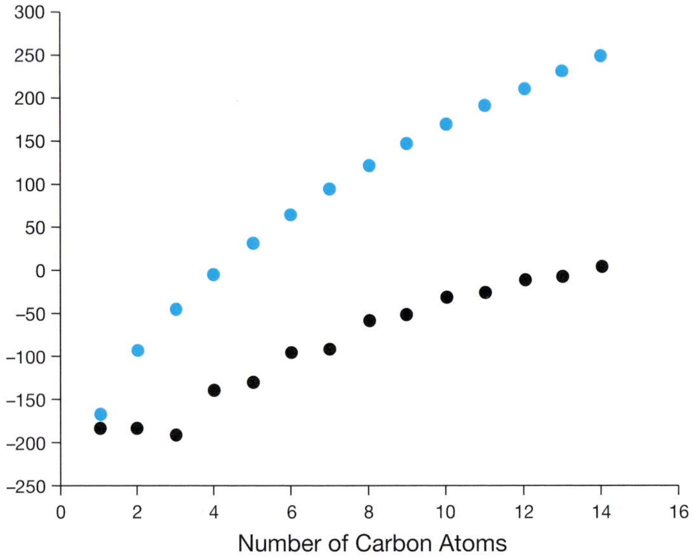

Figure 3.7. Melting point (black) and boiling point (blue) for the first 14 linear alkanes.

Perhaps the only reaction that represents an exception to low reactivity of alkanes is combustion. Alkanes, especially those with low molecular weight, are highly flammable. Fossil fuels, like natural gas, gasoline, kerosene, and diesel, are basically made up of alkanes. Alkane combustion reactions release large amounts of energy and this is the main application of this type of compound.

Figure 3.8. Complete combustion reaction of propane, the main constituent of liquefied petroleum gas (LPG).

On the other hand, unsaturated hydrocarbons, i.e. alkenes and alkynes, do not present the same stability as alkanes. The presence of a ϖ bond (unsaturation) creates this type of compound especially susceptible to addition reactions. The addition of halogens and hydrohalic acid to a double bond can occur even at room temperature. Alkenes are also sensitive to reducing and oxidizing agents.

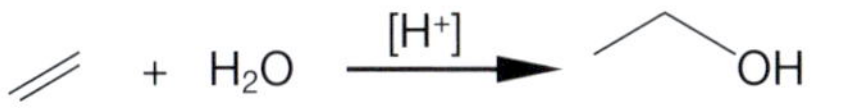

Figure 3.9 Addition reaction of water to ethylene catalyzed by a strong acid generating ethanol as product.

The presence of a strong acid can catalyze electrophilic addition reactions with reagents like water, alcohols, and carboxylic acids. The presence of a strong acid can also catalyze polymerization reactions. In fact, the main application of alkenes is the synthesis of polymers, for example, like polyethylene and polypropylene. .

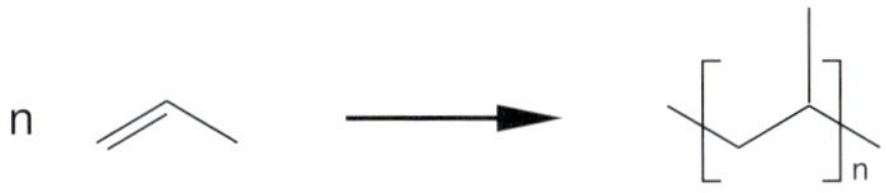

Figure 3.10. Polypropylene synthesis starting from propylene.

3.1.3. Main Production Methods

The main source of hydrocarbons is oil. Alkanes, among them n-hexane, are obtained directly by crude oil distillation while alkenes and alkynes are mostly obtained from alkanes by cracking, reforming, and isomerization reactions. Heavy hydrocarbons (longer chains) are converted into lower molar mass hydrocarbons by cracking reactions. Unsaturated hydrocarbons are produced from alkanes by catalytic reforming reactions and the by-product of this reaction is hydrogen. Isomerization reactions change neither the molar mass of a molecule nor its molecular formula. Only its structure is modified, for example, by changing an unsaturation position.

Figure 3.11. Production of n-butane starting from n-octane using hydrocracking reaction.

Figure 3.12. Production of 2-butene starting from butane using reforming reaction.

Figure 3.13. Production of 1-butene starting from 2-butene using isomerization reaction.

Another less important source of hydrocarbons is Fischer-Tropsch Synthesis. Currently, there are very few industrial sites that use this technology. In very simple terms, this process consists of the synthesis of hydrocarbons starting from carbon monoxide and hydrogen. Carbon monoxide for this process comes from abundant sources like charcoal or natural gas.

Some hydrocarbons can be obtained from alternative sources. Pristane is a component in shark liver oil. Limonene is an abundant component in the oil found in some citric fruit peels. Heptane is a component present in Jeffrey pine resin. Some bicyclic alkenes known as terpenes (basically α e β pinene) are obtained by the distillation of resin from some pine species. This terpene mixture is known as pine oil or turpentine.

Figure 3.14. α-pinene e β-pinene.

3.2. Aromatic Solvents

3.2.1. Definition

Aromatic hydrocarbons, such as the aliphatics, are formed exclusively by carbon and hydrogen, but in the case of aromatics, a special structure, called the aromatic ring or the benzene ring, is present.

In this planar structure, the electrons that constitute the π bonds are delocalized over the six σ bonds of the ring, conferring extra stability to the molecule. This characteristic is also known as resonance. The distance between the six carbon atoms that form the ring is the same and its value is intermediate between the distance of a single bond and a double bond, thus indicating that they are not distinct single and double bonds. That is why the benzene ring is also represented by a hexagon with a concentric circle representing the three delocalized π bonds.

Figure 3.15. Representations of the benzene ring.

3.2.2. Characteristics and Reactivity

Like all hydrocarbons, aromatics are also exclusively formed by carbon and hydrogen. There are no strongly polar bonds, which give aromatic hydrocarbons physical properties similar to the other hydrocarbons.

Aromatic hydrocarbons are also practically immiscible in water. The mixture of aromatic hydrocarbons and water tend, therefore, to form two distinct phases of which the upper phase is organic and the lower phase is aqueous. Aromatic hydrocarbons tend to be soluble with each other in any proportion, with other hydrocarbons, and tend to solubilize other nonpolar compounds. Obviously all compounds that have a benzene ring have at least six carbon atoms, and therefore, are liquids or solids at room temperature and pressure.

In terms of reactivity, aromatic hydrocarbons fall between alkanes and alkenes. Despite having a double bond like alkenes, aromatic compounds do not show the same reactivity. Aromatic hydrocarbons that have no other function in the molecule are more resistant to reduction and oxidation than alkenes. They do not undergo addition reaction as easily as alkenes, and polymerization practically does not occur with this type of molecule. They are resistant to bases including strong bases.

The most common reaction for this type of compound is electrophilic substitution in which one hydrogen atom on the aromatic ring is substituted, generating a chemical function. Thus, molecules with an aromatic ring are relatively sensitive to the presence of a strong acid. This reaction occurs more commonly with concentrated inorganic acids, like sulfuric acid and nitric acid

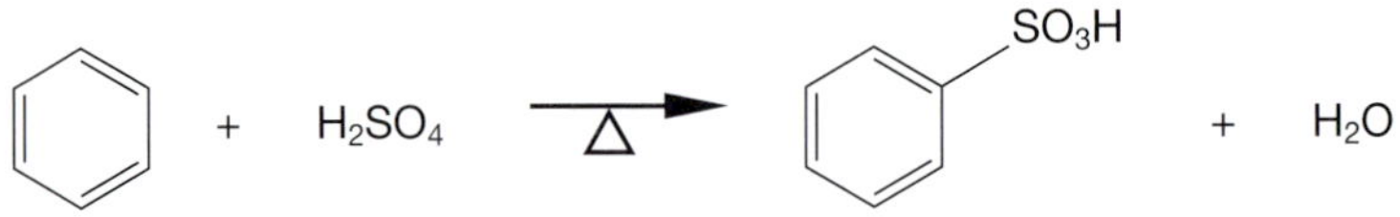

Figure 3.16. Electrophilic substitution of benzene using sulfuric acid creating benzene sulfonic.

Other reactions like acylation or addition of halogen, though well known, by rule need special conditions, like the presence of a Lewis acid catalyst and an anhydrous medium.

Obviously, there are compounds that, in addition to the aromatic ring, have other chemical functions. In these cases, molecule reactivity should also be evaluated, taking into account the other chemical functions present.

3.2.3. Main Production Methods

The larger source of aromatic hydrocarbons, like aliphatic hydrocarbons, is oil. The most common route for aromatic alkane synthesis is catalytic reforming, for example, of n-heptane or methylcyclohexane, resulting in toluene and hydrogen as by-products. From toluene it is possible to produce benzene using the hydro-dealkylation reaction which consumes hydrogen and generates methane as a by-product. It is also possible to produce benzene and xylene (an isomer mixture) by disproportionation of toluene. Furthermore, benzene can be produced directly by the catalytic reforming process, using n-hexane or cyclohexane, for example. These reactions are typical in the petrochemical industry and are usually carried out at temperatures above 500 °C with specific catalysts.

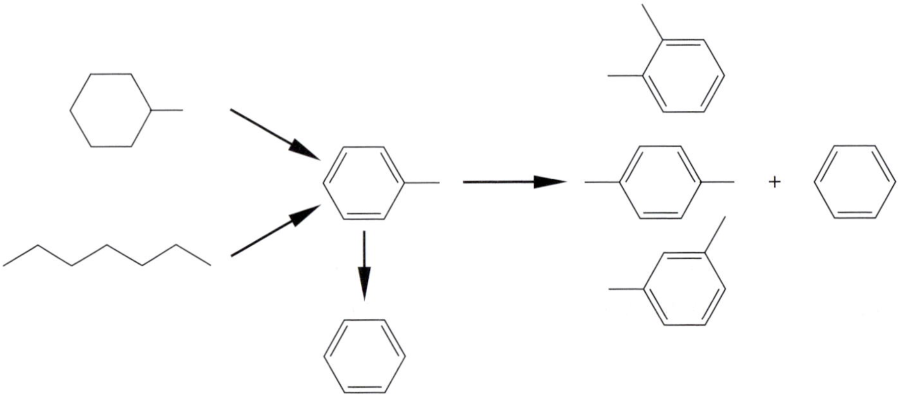

Figure 3.17. Toluene production from methylcyclohexane or heptane, followed by benzene and xylene production.

3.3. Oxygenated Solvents

3.3.1. Definition

Oxygenated compounds can be considered as all those formed by carbon and hydrogen and have in its structure at least one oxygen atom. This very general classification includes a large series of chemical functions that differ among themselves by the way in which the oxygen atom is bonded to the molecular structure. Molecules that have other elements in addition to carbon, hydrogen, and oxygen will be considered later in this chapter. In terms of a carbon chain structure, the same classification can be adopted as the one used for alkanes, or in other words, an oxygenated compound can be aliphatic or aromatic; and being aliphatic, it can be branched or linear, cyclic or acyclic, and saturated or unsaturated.

The presence of an atom with high electronegativity (oxygen) makes the molecule more susceptible to interactions like permanent dipole-induced dipole and dipole-dipole including the hydrogen bonding. Such effect does not occur in hydrocarbons. Chemical properties, like molecule reactivity, are basically determined by the function to which a molecule belongs. On the other hand, physicochemical characteristics, like miscibility and polarity, are determined, in addition to the chemical function to which a molecule belongs, by the number of functionalities and the carbon chain size. In general terms, molecules of high molar mass but with few oxygen atoms have physical-chemical characteristics closer to hydrocarbons.

3.3.2. Characteristics and Reactivity

The way an oxygen atom bonds to the carbon chain determines the function to which the molecule belongs, and therefore, what the chemical properties of this molecule will be. An oxygen atom can typically form two bonds. Thus, an oxygen atom can bond simultaneously to two distinct carbon atoms, to a single carbon atom using a double bond, or to a carbon atom and any other atom, such as hydrogen.

3.3.2.1. Alcohols, Ethers and Acetals

Alcohols are compounds that have a hydroxyl group connected to the aliphatic carbon chain, i.e., an oxygen atom present on the molecule is simultaneously connected to a hydrogen atom and an aliphatic carbon atom. Alcohols are classified as primaries, secondaries, and tertiaries, depending on the carbon to which the hydroxyl group is bonded. Alcohols with two hydroxyl groups connected to an aliphatic chain are called diols and more commonly called glycols.

The O-H bond polarity makes the dipole-dipole intermolecular forces so intense that hydrogen bonding is formed. Thus, all alcohols are liquids or solids; even methanol which has a molar mass of 32 is liquid at room temperature and 1 atmospheric pressure.

Lower molar mass alcohols, like methanol and ethanol, are miscible in water in any proportion. Monoalcohols (having only one hydroxyl group) with a high molar mass tend not to be completely soluble in water since physical-chemical properties tend to be more like nonpolar compounds.

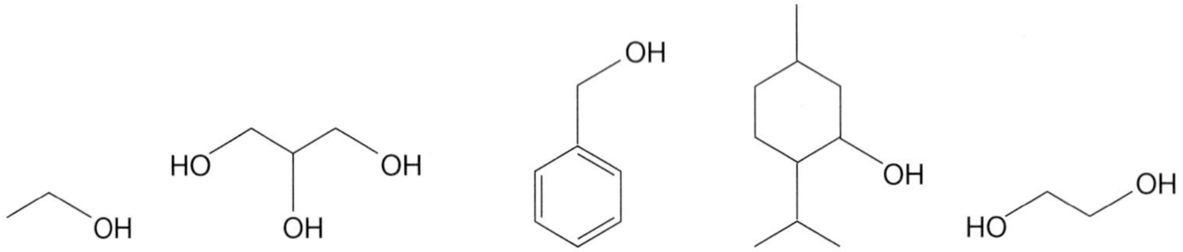

Figure 3.18. Examples of Alcohols.

Alcohols are resistant to the presence of bases and reducing agents. Only extremely strong reducing agents, like alkali metal hydrides, react with alcohols releasing hydrogen. On the other had, alcohols are relatively sensitive to the presence of strong oxidizing agents and strong acids. They can be oxidized to aldehydes, carboxylic acids, or ketones, for example, in the presence of potassium permanganate. The presence of strong acids can catalyze dehydration of alcohols, thus generating alkenes or ethers, depending on reaction conditions and the structure of alcohol in question. Primary alcohols, like methanol and ethanol, are more resistant to dehydration than secondary alcohols (isopropanol) and tertiary alcohols (tert-butanol). Alcohols do not react with water under room temperature and pressure conditions.

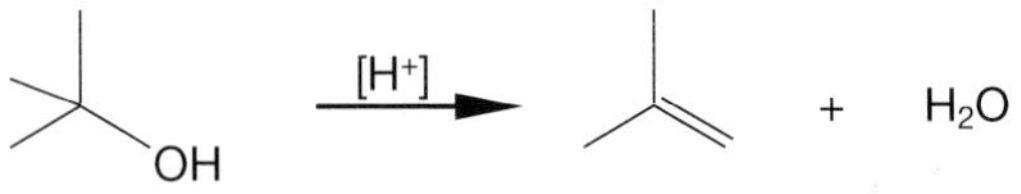

Figure 3.19. Dehydration of tert-butanol resulting in methylpropene.

Ethers have an oxygen atom bonded to two distinct aliphatic saturated carbon atoms. Since there is no O-H bond, ethers tend to be very nonpolar compounds. The miscibility of ethers in water, even for short chain ethers, is low and the melting point of ethers is notably lower when compared to alcohols with the same molar mass.

Ethers are remarkably inert when compared to alcohols. They do not react with water and are not sensitive to strong acids and bases. They are also resistant to oxidizing and reducing agents. Actually, among oxygenated compounds, ethers are most similar to alkanes. Also analogous to alkanes, only cyclic ethers with highly strained rings are reactive, as is ethylene oxide, for example.

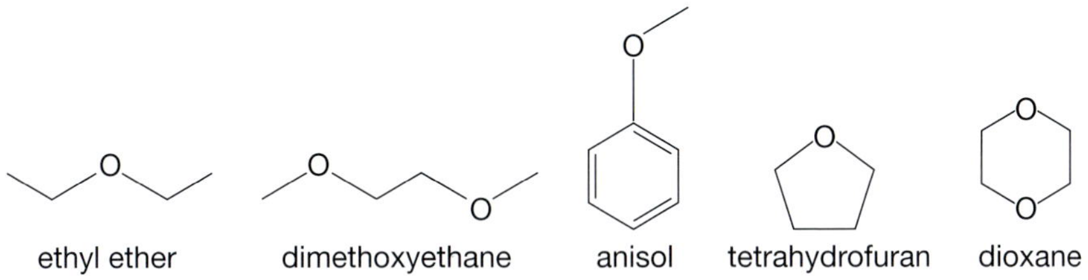

Figure 3.20. Examples of Ethers.

Acetals are molecules that have two oxygen atoms simultaneously connected by a single bond to the same carbon. Each oxygen atom can still be connected to another carbon atom or to a hydrogen, and in this case the molecule can be called hemicetal.

Acetals, as well as ethers, are not capable of forming hydrogen bonds so miscibility of these compounds in water tend to be lower than alcohols with the same molar mass. Acetals are liquids at room temperature and 1 atmospheric pressure and can be solids if the molar mass is very high.

In terms of reactivity, acetals are resistant to reducing and oxidizing agents, and are also resistant to bases, even strong bases. The most common reaction for acetals is hydrolysis in an acid medium. Acetals can react with water if an acid is present, even in catalytic amounts. The reaction results in alcohol and aldehyde or ketone.

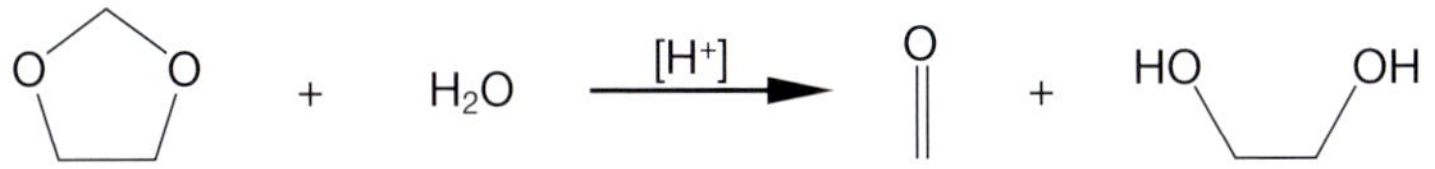

Figure 3.21. Hydrolysis of 1,3-dioxolane in an acid medium generating formaldehyde and ethylene glycol.

3.3.2.2. Main Production Methods

The main alcohols used as solvents are methanol, ethanol, and isopropanol. On a lesser scale, n-butanol, isobutanol, cyclohexanol, and hexylene glycol are also used.

Methanol is produced starting from synthesis gas (Syngas). Methane undergoes steam reform resulting in carbon monoxide and hydrogen, and these two compounds react with each other under appropriate conditions generating methanol.

The olefin hydration process is used in the synthesis of alcohols like isopropanol (propene hydration) and 2-butanol (2-butene hydration) while 1-butanol is obtained by hydroformilation (carbonylation) of propene followed by hydrogenation in one step. Ethanol, as an exception, is mostly produced by fermentative processes of renewable sources, notably, corn starch and sugar cane saccharose. The process of obtaining ethanol using ethene hydration has given way to processes starting with biomass, which is currently the main production method.

$$CO + 2H_2 \longrightarrow MeOH$$

Figure 3.22. Methanol synthesis starting with CO and H_2.

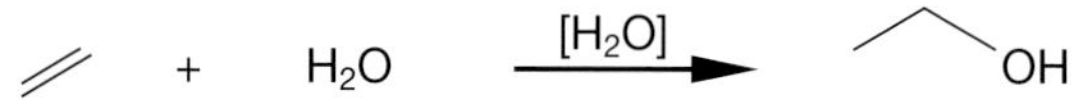

Figure 3.23. Ethanol synthesis starting with ethene hydration.

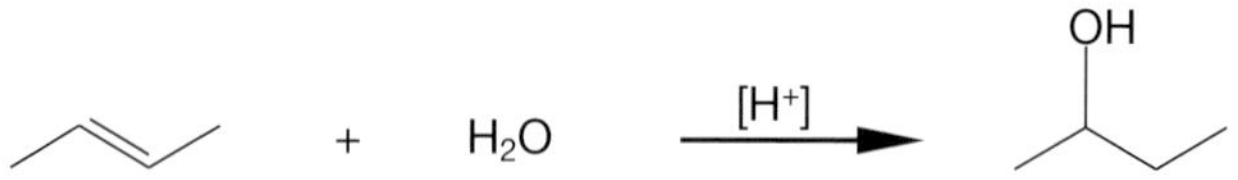

Figure 3.24. 2-butanol synthesis (sec-butanol) starting with 2-butene hydration.

Figure 3.25. n-butanol synthesis starting with propene.

Among glycols, the largest consumption is based on ethylene glycol (MEG) and diethylene glycol (DEG) that, in addition to hydroxyl groups, has an ether function. Ethylene glycol is produced by the hydration of ethylene oxide which is produced by selective oxidation of ethene. Diethylene glycol is obtained through reaction between ethylene glycol and ethylene oxide. This reaction can be performed for that specific intention but it also occurs during ethylene glycol production as a consecutive reaction.

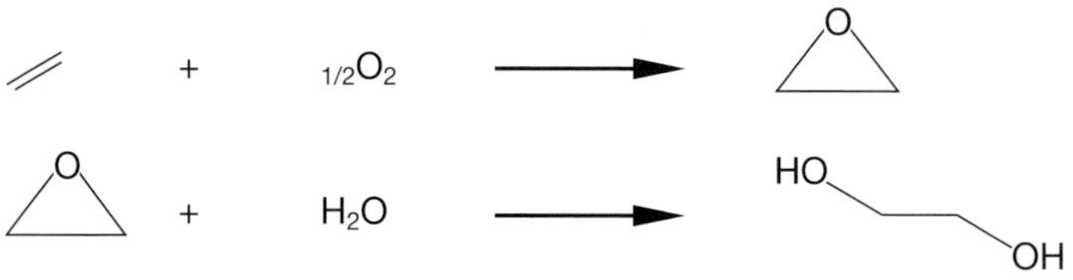

Figure 3.26. Ethylene oxide synthesis followed by hydration and ethylene glycol synthesis.

Figure 3.27. Diethylene glycol synthesis from ethylene glycol and ethylene oxide.

With regard to ethers, ethyl ether production used to be important, however, as a solvent, it has been substituted. Currently, tetrahydrofuran or THF (a cyclic ether) and dioxane occupy some space in the solvent market. Ethyl ether is obtained as a by-product in ethene hydration during the production of ethanol. It can also be obtained through ethanol dehydration over alumina under controlled conditions. THF is obtained from furan

Figure 3.28. Ethanol dehydration and production of ethyl ether.

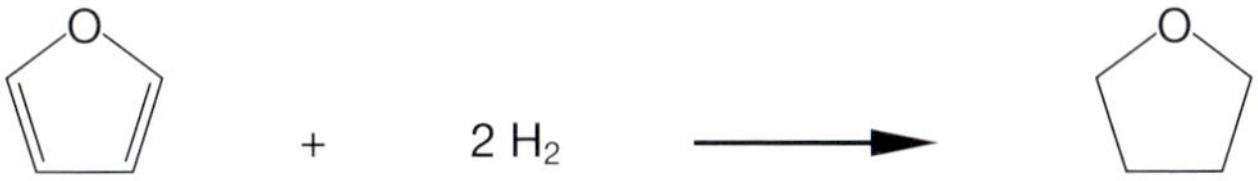

Figure 3.29. Furan hydrogenation and production of tetrahydrofuran.

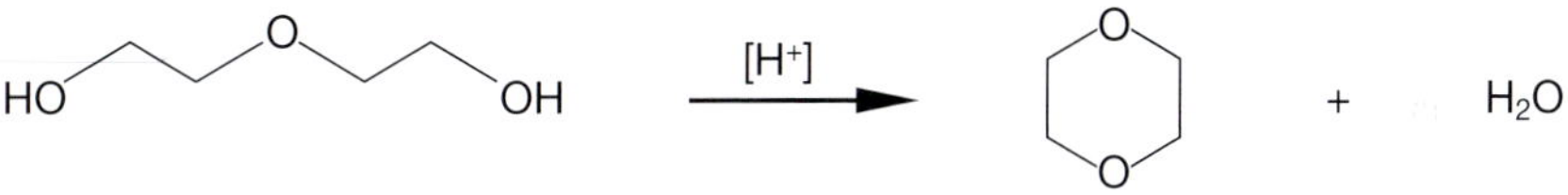

Figure 3.30. Diethylene glycol dehydration and production of dioxane.

3.3.2.3. Carbonyl Compounds

Molecules that have at least one oxygen atom doubled bonded to a carbon atom are called carbonyl compounds. The C=O structure is called a carbonyl group. There are several chemical functions which present carbonyl groups, and among the more common are ketones, aldehydes, esters, carboxylic acids and acid anhydrides.

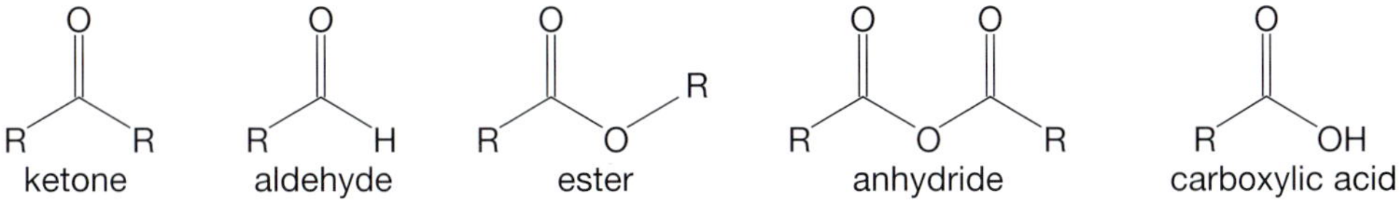

Figure 3.31. Examples of carbonyl compounds.

Some carbonyl compounds are very reactive, and therefore, their use as a solvent is impractical. Anhydrides, for example undergo hydrolysis reaction generating carboxylic acids. This reaction is exothermic and occurs even through air moisture absorption. In fact, anhydrides need to be stored in hermetically-sealed flasks to avoid a hydrolysis reaction. Carboxylic acids also have a high reactivity especially in relation to bases. Furthermore, short chain carboxylic acids have a very pronounced odor.

Aldehydes also make up a class of compounds not used as solvents because of their high reactivity. They are sensitive to oxidizing and reducing agents as well as to acids and bases. They can also undergo polymerization reaction during storage, changing their original characteristics.

Among the carbonyl compounds extensively used as solvents are esters and ketones. The simplest ketone is propanone, known as acetone. Acetone is liquid at room temperature and pressure, contrary to the hydrocarbon with the same molar mass (butane) which is a gas. This happens because of the molecular polarity of the acetone that allows dipole-dipole interactions to exist. Acetone is miscible in water in any proportion. For monoketones, which have only one carbonyl group, water solubility decreases with the increase of carbon atoms.

Ketones are relatively sensitive to reducing agents, but are very resistant to oxidizing agents including potassium permanganate and they practically do not react with water. The formation of a hemiacetal between ketones and water is reversible and equilibrium is shifted toward reagents. Ketones are sensitive to the presence of acids or bases, even in catalytic quantities because they undergo an aldolic coupling reaction. This reaction is reversible but the dehydration that can follow aldolic coupling generates enones, and in this case, the equilibrium is shifted toward the products.

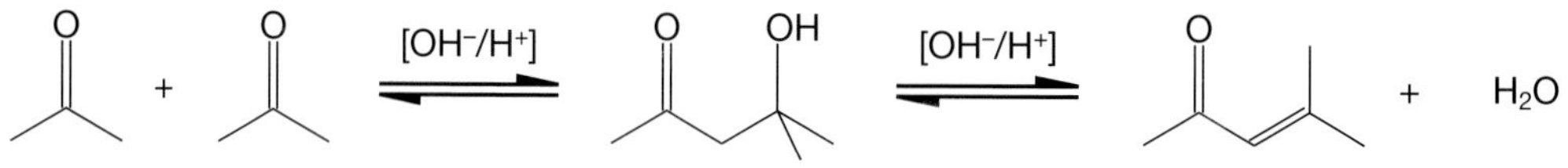

Figure 3.32. Example of an aldolic coupling followed by dehydration. Formation of diacetone alcohol starting from acetone, followed by dehydration with the formation of mesityl oxide.

Esters are another class of compounds largely used as solvents. No simple ester is completely miscible in water at room temperature, not even methyl methanoate (methyl formate) that corresponds to the lowest molar mass ester. Methyl formate, despite having a low boiling point, is liquid at room temperature and pressure. Therefore, all esters are liquids or solids. The immiscibility of esters with water becomes more accentuated with the increase in the length of the carbon chain it is associated with.

Esters are very resistant to reducing and oxidizing agents but are relatively sensitive to the presence of water. Esters undergo hydrolysis reaction generating a carboxylic acid and an alcohol since the formation reaction of esters starting from these components is reversible. This hydrolysis reaction is catalyzed mostly by strong acids. Esters are also sensitive to strong bases due to the saponification reaction.

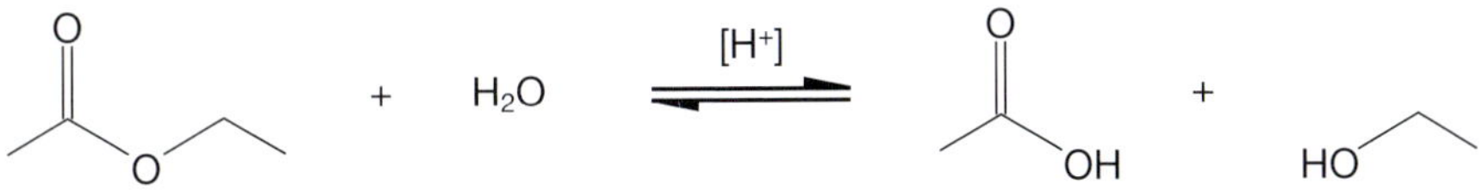

Figure 3.33. Hydrolysis of ethyl acetate catalyzed by acid resulting in acetic acid and ethanol. The reaction is reversible and the final mixture composition tends to reach a thermodynamic equilibrium.

3.3.2.4. Main Production Methods

Among the main ketones used as solvents are acetone, methyl isobutyl ketone (MIBK), methyl ethyl ketone (MEK), and on a lesser scale, diacetone alcohol (which also has an alcohol function), mesityl oxide (enone), and diisobutyl ketone.

The main source of raw material for ketone production is oil. Acetone is basically generated as a by-product in phenol production. Industrial synthesis of phenol begins in the manufacturing of cumene starting from benzene and propene. Cumene is oxidized to cumene hydroperoxide that is cleavaged with an acid medium, generating phenol and acetone. Acetone itself is a raw material used for the production of diacetone alcohol via aldolic condensation. Dehydration of diacetone alcohol generates mesityl oxide and selective hydrogenation of this generates methyl isobutyl ketone. Methyl ethyl ketone is mostly produced using a 2-butanol dehydrogenation reaction and this is produced using the hydration of 2-butene. A diagram of these reactions is shown below.

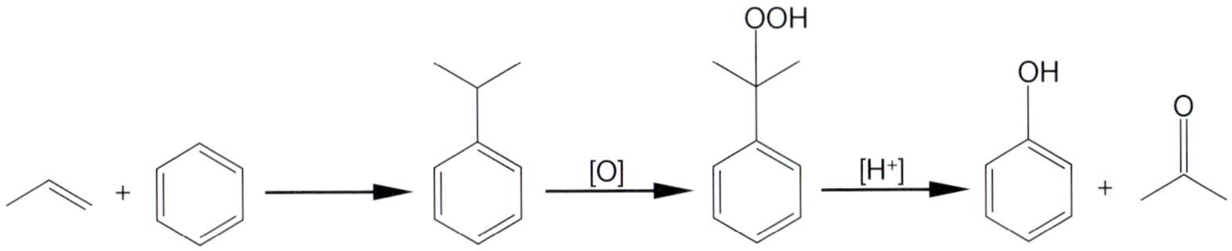

Figure 3.34. Reaction for acetone and phenol production starting from propene and benzene.

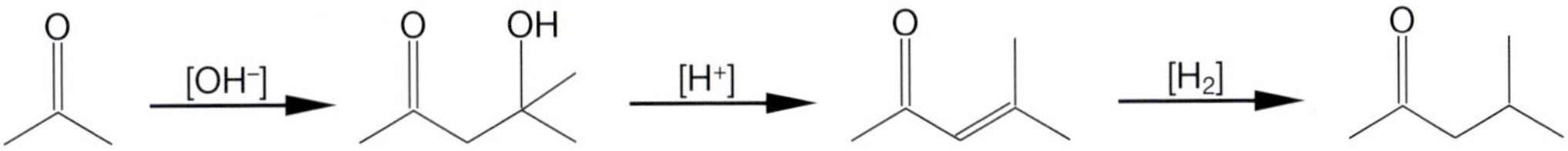

Figure 3.35. Reaction for diacetone alcohol, mesityl oxide and methyl isobutyl ketone production starting from acetoney.

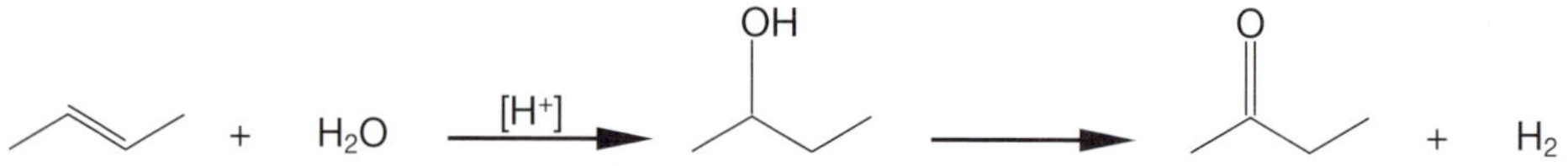

Figure 3.36. Reaction for methyl ethyl ketone production starting from 2-butene.

The utilization of esters as solvents is based mostly on the use of acetates. Acetates are obtained through esterification of different alcohols with acetic acid. The reaction is reversible and occurs with the release of water. Most of the acetic acid for manufacturing acetates is produced using methanol carbonylation. The reaction is catalyzed by rhodium or iridium complexes and takes place in the presence of iodomethane. In Brazil, with its vast production of ethanol from sugar cane, acetic acid is also produced through its oxidation. The main acetates used are ethyl acetate and n-butyl acetate.

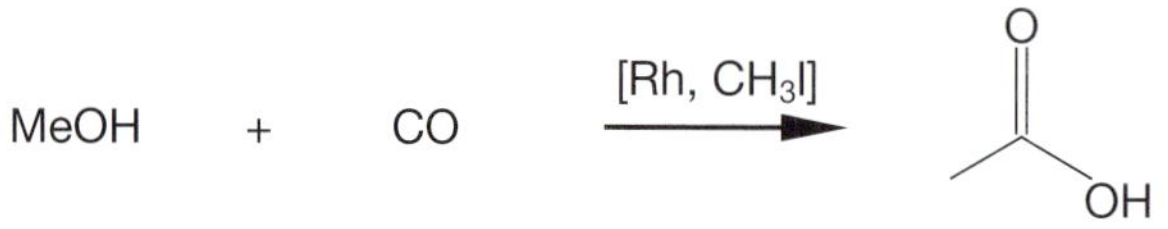

Figure 3.37. Carbonylation of methanol and acetic acid production.

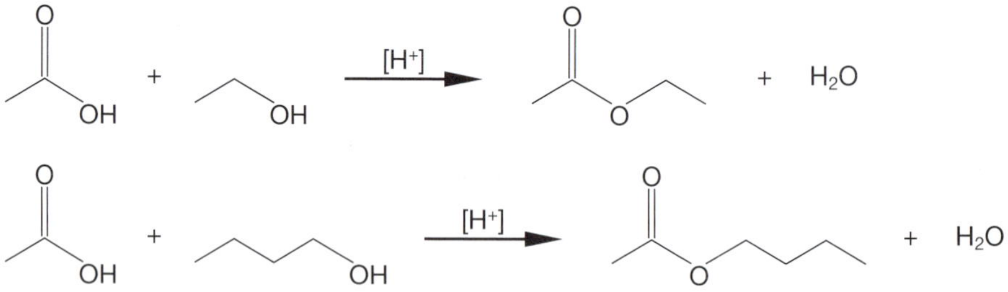

Figure 3.38. Ethyl acetate and butyl acetate production starting from acetic acid.

3.4. Halogenated Solvents

Organohalogen compounds are constituted of carbon, hydrogen, and at least one halogen atom bonded to the carbon chain structure. Despite having very electronegative elements, there is no direct bonding between the hydrogen and halogen atoms in this type of compound, and therefore, no hydrogen bonding. Molecular interactions are dipole-dipole types with the exception of completely symmetrical molecules like carbon tetrachloride.

Organohalogen compounds with low molar mass are gases. This is the case, for example, for some known CFC's (chlorofluorocarbons). Organohalogen compounds with a higher molar mass are usually liquids with very low miscibility in water. The mixture between organohalogen liquids and water tend to form two distinct phases with the lower phase being organic and the upper phase being aqueous since the organohalogens are denser than water.

Reactivity of organohalogens is basically a function of the halogen connected to the carbon chain. Organoiodines and organobromines are very reactive and are used quite frequently as intermediates in organic synthesis. Thus, the use of these compounds as solvents is very rare.

Contrary to this, organofluorines are extremely inert. From the perspective of chemical reactivity, under normal temperature and pressure conditions, organofluorines behave similarly to alkanes. Organochlorides are also relatively inert and are relatively resistant to oxidizing agents and acids but are sensitive to reducing agents and principally bases.

One differentiated property of totally substituted organohalogens (with no hydrogen present in the molecule), like carbon tetrachloride or tetrachloroethylene, is their resistance to combustion. Some compounds like bromotrifluoromethane and bromochlorodifluromethane act as fire extinguishing gases.

3.4.1. Main Production Methods

Currently, the main organohalogen solvents used are organochlorines, more specifically, dichloromethane, tetrachloroethylene, and trichloroethylene. Dichloromethane production is based on the halogenation of methane with chlorine gas. The reaction occurs in harsh temperatures (above 400 °C) in the presence of a catalyst; and at the same time, carbon tetrachloride, trichloromethane (chloroform), and chloromethane are produced. Hydrochloric acid is a by-product of this reaction.

Tetrachloroethylene and trichloroethylene are produced using the halogenation reaction of dichloroethane (also under severe conditions) and the latter is produced using the halogenation reaction with ethene.

$$CH_4 \quad + \quad Cl_2 \longrightarrow HCl \quad + \quad CCl_4 \quad CHCl_3 \quad CH_2Cl_2 \quad CH_3Cl$$

Figure 3.39. Manufacturing of the mixture carbon tetrachloride, trichloromethane (chloroform), dichloromethane, and chloromethane mixture using the halogenation reaction of methane with chlorine in gas phase.

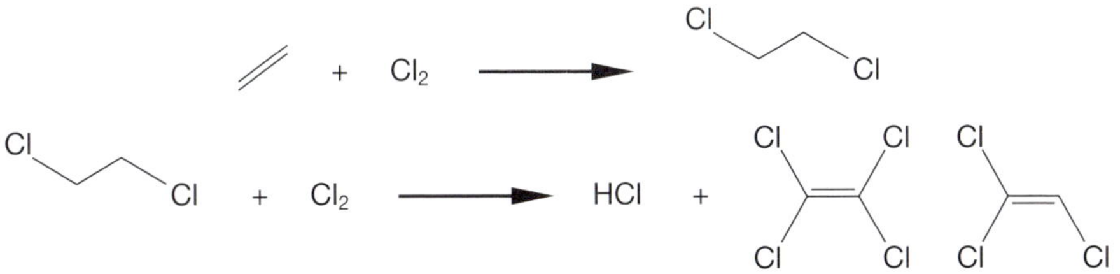

Figure 3.40. Manufacturing of dichloroethane using the halogenation reaction of ethene, followed by the production of tetrachloroethylene and trichloroethylene using the halogenation reaction of dichloroethane.

3.5. Nitrogen and Sulfur-Containing Solvents

Nitrogen compounds and sulfur compounds, as the name implies, have a nitrogen or sulfur atom, respectively, in their structure. The use of this type of compound as a solvent, in comparison to hydrocarbons and oxygenated compounds, is much more restricted. For this reason, only compounds that can have some significant use as a solvent will be mentioned here.

Nitrogen compounds used as solvents can be amines, amides nitriles, or nitro compounds. Amines are based on ammonia structure and have at least one hydrogen atom replaced by carbon chain. In amides, nitrogen is linked to a carbonyl group. Nitriles have at least one nitrogen atom bonded exclusively to a carbon atom by a triple bond. Nitroalkanes have a nitrogen atom connected to a carbon atom and simultaneously to two oxygen atoms. This chemical structure is called a nitro group.

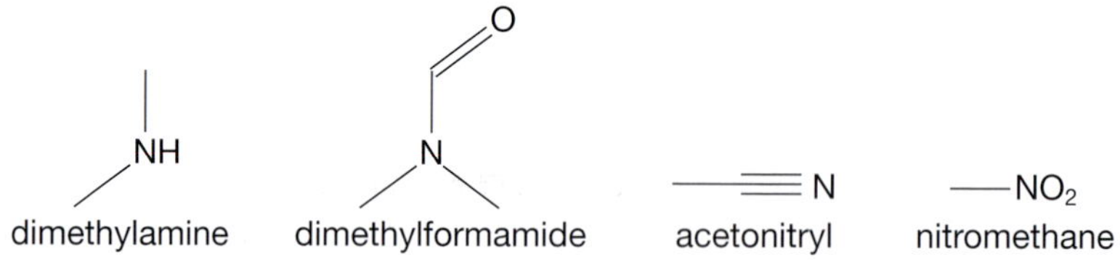

Figure 3.41. Examples of amine, amide, nitrile, and nitroalkane.

Among sulfur compounds used as solvents, sulfoxides, specifically dimethyl sulfoxide, and sulfones, specifically sulfolane, can be mentioned. Sulfoxides are molecules that have one sulfur atom connected to two distinct carbon atoms and a double bonded oxygen atom. Sulfones have the same structure as sulfoxides with an extra oxygen double bonded to sulfur.

Figure 3.42. Example of sulfone and sulfoxide.

Nitrogen and sulfur compounds generally present some reactivity, which means that their use as solvents is restricted to some specific applications that require special properties. Notably, dimethylsulfoxide, sulfolane, dimethylformamide, and dimethylacetamide are polar aprotic solvents.

Amines, for example, are resistant to the presence of bases and reducing agents, but they are sensitive to the presence of acids, oxidizing agents, and water. Primary amines are sensitive to air, which can cause carbonation because of the CO_2. Nitriles are sensitive to the presence of acids, bases, and reducing agents and oxidizing agents. Nitriles can undergo a hydrolysis reaction catalyzed by acids or bases. Nitroalkanes are sensitive to bases and are relatively sensitive to acids and reducing and oxidizing agents. Nitroalkanes can also undergo a hydrolysis reaction. Amides are sensitive to acids and bases and relatively sensitive to reducing agents. Amides are susceptible to hydrolysis in an acid or alkaline medium.

Sulfoxides are relatively resistant to oxidizing and reducing agents. They are relatively sensitive to the presence of bases. Sulfones are resistant to oxidizing agents, but relatively sensitive to reducing agents. They are sensitive to strong bases.

3.5.1. Main Production Methods

The use of amines as solvents is restricted to dimethylamine and methylamine. Both are produced using a reaction between methanol and ammonia in specific conditions. The resulting reaction mixture is subsequently separated. The same strategy is used in the manufacturing of nitro compounds. Nitromethane, nitroethane, 1-nitropropane, and 2-nitropropane are produced in parallel in the same reactor using a reaction between nitric acid and propane. The mixture is then separated for commercialization.

Dimethylformamide and dimethylacetamide are the amides largely used as solvents. Dimethylformamide is produced by reaction between dimethylamine and carbon monoxide. Dimethylacetamide is produced starting from dimethylamine and acetic acid.

The nitrile most used as a solvent is acetonitrile. Acetonitrile is mostly obtained as a by-product during the manufacturing of acrylonitrile (raw material for polymers) using a reaction between propene, ammonia, and oxygen.

Dimethylsulfoxide, the sulfoxide most used as a solvent is obtained from oxidation of dimethylsulfide and this is obtained by reaction between methanol and hydrosulfuric acid. The sulfone most used as a solvent is sulfolane which is manufactured through a reaction between butadiene and sulfur dioxide followed by reduction.

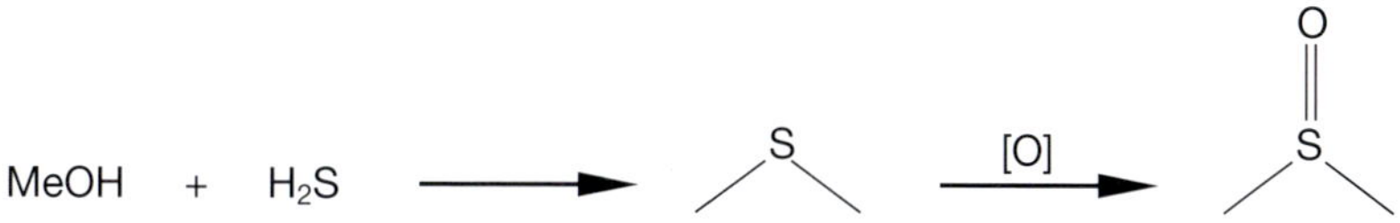

Figure 3.42. Production of dimethylsulfoxide starting from methanol and hydrosulfuric acid.

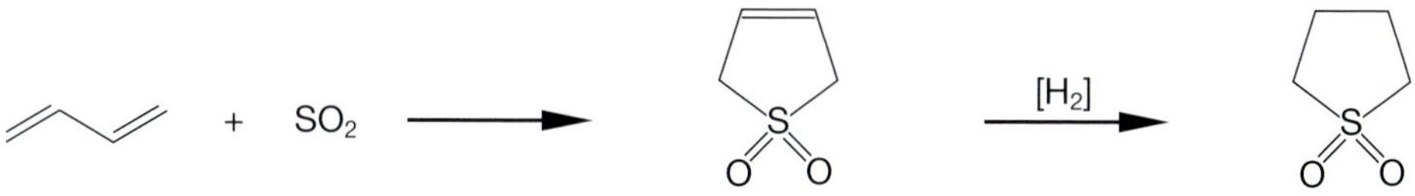

Figure 3.43. Production of sulfolane starting from butadiene and sulfur dioxide.

Table of Incompatibilities

The following table shows the main incompatibilities, from the point of reactivity for some classes of solvents already cited. This is not a definitive guide for chemical reactivity. In practice, each case should be assessed separately, since particular steric and electronic factors of each molecule are determiners of the properties of the compound in question. Furthermore, reactions and interactions between functional groups in a solvent mixture that may significantly change the result of the final formulation were not considered.

Table 3.1. Incompatibilities

Chemical Function	Water	Acids	Bases	Oxidizing Agents	Reducing Agents
Saturated Hydrocarbons	Severe conditions/slow	Severe conditions/slow	Severe conditions/slow	Severe conditions/slow	Severe conditions/slow
Unsaturated Hydrocarbons	Possible	Very probable	Severe conditions/slow	Very probable	Very probable
Aromatic Hydrocarbons	Severe conditions/slow	Very probable	Severe conditions/slow	Possible	Possible
Alcohols	Severe conditions/slow	Possible	Severe conditions/slow	Very probable	Severe conditions/slow
Ethers	Severe conditions/slow	Severe conditions/slow	Severe conditions/slow	Severe conditions/slow	Severe conditions/slow
Acetals	Possible	Very probable	Severe conditions/slow	Severe conditions/slow	Severe conditions/slow
Ketones	Severe conditions/slow	Very probable	Very probable	Severe conditions/slow	Very probable
Esters	Possible	Possible	Very probable	Severe conditions/slow	Severe conditions/slow
Organochlorines	Severe conditions/slow	Severe conditions/slow	Very probable	Severe conditions/slow	Very probable
Organofluorines	Severe conditions/slow	Severe conditions/slow	Severe conditions/slow	Severe conditions/slow	Severe conditions/slow
Amines	Very probable	Very probable	Severe conditions/slow	Very probable	Severe conditions/slow
Amides	Possible	Very probable	Very probable	Severe conditions/slow	Possible
Nitroalkanes	Possible	Possible	Very probable	Possible	Possible
Nitriles	Possible	Very probable	Very probable	Possible	Very probable
Sulfoxides	Severe conditions/slow	Possible	Possible	Possible	Possible
Sulfones	Severe conditions/slow	Possible	Possible	Severe conditions/slow	Possible

Legend:

- (dark blue) Chemical reaction occurs under severe conditions or very slowly
- (grey) Possibility of chemical reaction
- (light blue) Very probable of a chemical reaction

Bibliographical References

Grayson, M. Eds.; Kirk-Othmer Encyclopedia of Chemical Technology. 3rd ed., John Wiley and Sons, New York, 1983.

Wolfgang, G. Eds.; Ullmann's Encyclopedia of Industrial Chemistry, 4th ed., Wiley-VCH, Weinheim, 1985.

March, J.; Smith, M. Advanced Organic Chemistry: reactions, mechanisms and structure, 5th ed., John Wiley and Sons, New York, 2001.

Carey, F.; Sundberg, R. J. Advanced Organic Chemistry, 4th ed., Kluwer Academic/ Plenum Publishers, New York, 2000.

Morrison, R. T.; Boyde, R. N. Química Orgânica, 5 ed., Lisboa: Fundação Calouste Gulbenkian, 1996.

Solomons, T. W. G. Fundamentals of Organic Chemistry, 6th ed., Jonh Wiley and Sons, New York, 1996.

Weast, R. C. Eds.; CRC Handbook of Chemistry and Physics, 61th ed., CRC Press, Boca Raton, 1980.

4 Green Solvents

*(Green Chemistry – a sustainable solution
Rainer Hofer)*

The objective of this chapter is to define principles for solvent classification in relation to the green concept, mainly taking into consideration application performance, toxicity, and environmental impacts.

Sérgio Martins

4.1. Introduction

The purpose of this chapter is to offer a view of the principal aspects that should be taken into consideration for classifying a solvent or a solvent system as green. Along with technological developments and worldwide population growth throughout the 20th century, there has been an important demand for raw materials and energy to support those changes characterized by:

- the use of products with non-renewable resources, such as oil as a main source of energy, fuel, and raw material for industry;
- a focus mainly on processes and products, with little worry on industrial plant emissions;
- unsustainable exploration of valuable natural resources, like minerals, precious metals, and extravagant use of wood, etc.

This type of behavior is not exclusive to contemporary society based on several examples in history that confirm little preoccupation with sustainability on the part of man, which lead to environmental impacts. For example, intensive agricultural activities during the Roman Empire allied with the strong energy demand by Roman society at that time, due to the popular steam baths, the forging of metals and weapons, and the demand for wood used in heating homes or constructing ships are associated with the origin of the first known ecological disasters. These include deforesting in the Apennine region of Italy and later around the Mediterranean Sea, resulting in enormous erosions, with the loss of soil humus and dry periods for some rivers.

As always, nature demands her part, and we are starting to feel the effects of increased unsustainable sources in the last century. The extent of these impacts are greater than can be supported and still it is feared they will be greater than what is currently projected. Below are some examples we are seeing:

- global warming leading to important climate changes on Earth;
- polar icecaps melting, resulting in ocean levels rising which could lead to the loss of islands and cities, like Venice;
- poor air quality in large urban concentrations associated with automobile exhaust emissions and photochemical fog pollution;
- destruction of the stratosphere ozone, the principal filter for UV rays, which when in contact with human beings, cause skin problems, including cancer;
- water pollution, compromising available drinking water;
- forest desertification, as seen in some countries in Africa;[9]

In other words, economic growth and the increase of the standard of living have their price:

- natural resources from exploration are reaching their limits;
- environmental contamination is increasing.

With this reality, there has been a significant change in the technological development concept, with sustainability being practically indispensable in any area of the economy.

From a business point of view, sustainable development is generally separated into three independent areas:

- economics: efficient management of resources
- environment: intolerance of emissions damaging the ecosphere and maintenance of basic living needs
- social: human beings are the center of all preoccupations.[9]

Generally, industry is an important part of the sustainability implementation process, and more specifically, the chemical industry because of its main activities and impact on society and the environment. The counterpart for this process comes from population awareness; it has become vital for companies to align their economic development plans with environmental and social sustainability to guarantee national and international market competitiveness.

One of the results from the sustainable development concept in the chemical industry is the constant concern for processes and their improvement or replacement of processes that are more environmentally acceptable. Concepts used in these changes are a part of what is called Green Chemistry or Sustainable Technology.

Likewise, the concern that product development follows sustainability concepts also guides the main lines of research for companies that have a social and environmental commitment.

Solvents in particular are considered extremely important raw material in some industry segments such as fine chemistry, pharmaceutical industries, and coatings where they are used in large quantities, and therefore are important in the environmental performance of these processes, as well as in cost, safety, and occupational health.

Given their importance, solvents are also targets for sustainability assessment of a process, and today, in addition to application performance and cost, aspects related to environmental and human toxicity are a part of the selection criteria for the most appropriate process. The use of solvents that are more benign to health and the environment is a part of the Green Chemistry movement. The Green Solvent concept expresses the objective of minimizing environmental and human health impact and maintaining performance demanded during different application uses.

Since there is no universal definition of what a Green Solvent is, the objective in this chapter is to present the main concepts that need to be integrated in the criteria when choosing solvents and/or solvent systems with more 'green' characteristics to substitute the current ones in existing diverse applications or in the development of new products.

4.2. Green Solvents

Intrinsic global criteria in the Green Solvent concept are based on four foundations:

- Good application performance;
- Minimizing toxicity to human health;
- Minimizing toxicity in the environment;
- Minimizing environmental impacts.

Therefore, from the perspective of Green Solvents, in addition to performance, toxicity and environmental impacts as key elements in the selection process also need to be considered for choosing those that have less impact on human health and the environment.

In the assessment of the possible substitution of a solvent or solvent system currently being used in a determined application or chemical process, the following considerations should be included when the objective is to add a greener characteristic to a system:

- the possibility of redesigning the chemical nature of a solvent via synthesis;
- identification of alternative solvents to replace traditional ones used in that chemical formulation or process;
- the possibility of reducing or eliminating the use of the solvent for the application in question.

When the objective is the development of a new product, Green Chemistry has fundamental concepts that should be integrated in planning the conception of this new specie. In the next section 12 principal concepts of Green Chemistry are presented.

4.2.1. Green Chemistry

From the perspective of solvents, green chemistry can be seen as a number of techniques to design new solvents, solvent systems, and new roads to using them to reduce or eliminate intrinsic hazards associated with them.

In some cases, new substances are being designed and developed for use as a solvent, while in other cases, already-known substances in other applications are being assessed as solvents.

Anastas, et.al. listed 12 green chemistry principles in the book, Green Chemistry: Theory and Practices[2]:

1. **Prevent waste**: Design chemical syntheses to prevent waste, leaving no waste to treat or clean up.
2. **Design safer chemicals and products**: Design chemical products to be fully effective, yet have little or no toxicity.
3. **Design less hazardous chemical syntheses**: Design syntheses to use and generate substances with little or no toxicity to humans and the environment.
4. **Use renewable feedstocks**: Use raw materials and feedstocks that are renewable rather than depleting. Renewable feedstocks are often made from agricultural products or are the wastes of other processes; depleting feedstocks are made from fossil fuels (petroleum, natural gas, or coal) or are mined.
5. **Use catalysts, not stoichiometric reagents**: Minimize waste by using catalytic reactions. Catalysts are used in small amounts and can carry out a single reaction many times. They are preferable to stoichiometric reagents, which are used in excess and work only once.
6. **Avoid chemical derivatives**: Avoid using blocking or protecting groups or any temporary modifications if possible. Derivatives use additional reagents and generate waste.
7. **Maximize atom economy**: Design syntheses so that the final product contains the maximum proportion of the starting materials. There should be few, if any, wasted atoms.
8. **Use safer solvents and reaction conditions**: Avoid using solvents, separation agents, or other auxiliary chemicals. If these chemicals are necessary, use innocuous chemicals.
9. **Increase energy efficiency**: Run chemical reactions at ambient temperature and pressure whenever possible.
10. **Design chemicals and products to degrade after use**: Design chemical products to break down to innocuous substances after use so that they do not accumulate in the environment.
11. **Analyze in real time to prevent pollution**: Include in-process real-time monitoring and control during syntheses to minimize or eliminate the formation of byproducts.
12. **Minimize the potential for accidents**: Design chemicals and their forms (solid, liquid, or gas) to minimize the potential for chemical accidents including explosions, fires, and releases to the environment.

Originally published by Paul Anastas and John Warner in **Green Chemistry: Theory and Practice** (Oxford University Press: New York, 1998).

4.2.2. Fundamental Concepts

The main characteristics that a solvent or solvent system should have in order to be considered 'green' are:

i. good application performance;
ii. the least amount of toxicity to human health;
iii. the least toxicity for the environment;
iv. the lowest environmental impact.

Chapter 6 has a series of applications and solvents or solvent classes most appropriate for each of them. Clearly, each solvent or solvent system to be assessed as green, should have, more than anything, an appropriate application performance for which it is being used.

Concepts referring to toxicity for human health and the environment are discussed in Chapter 10.

4.2.3. Environmental Impact

The use of certain substances can cause changes in the environment, and the effects are noted below. An example of these effects is acid rain. As is well known, many by-products from combustion processes, like nitrogen oxides and sulfur, are 'washed' in the atmosphere, causing pluviometric acid precipitation, responsible for the mortality of aquatic and vegetation species.

Following are 2 effects that impact Earth's environment:

- The destruction of the ozone layer in the stratosphere;
- Formation of ozone *via* photochemical reactions in the troposphere. .

4.2.3.1. Destruction of the Stratospheric Ozone

In September 1987, the U.S. along with 26 other countries signed a treaty to first limit, and subsequently, through reviews, eliminate the production of all substances causing the destruction to the ozone layer.[14]

This treaty was the result of the verification by many researchers of the solar radiation increase reaching Earth because of the destruction of the protective ozone in the stratosphere, attributed to the halogen compounds, and specifically, chlorofluorocarbons (CFCs). The treaty became known as the Montreal Protocol on Substances that Deplete the Ozone Layer.

These compounds were used often in many applications, mainly associated with heat transfer operations, such as:

- Refrigerating Agents
- Coolants

- Synthetic Foams
- Aerosol Sprays
- Cleaning Solvents

As seen in Figure 4.1, Earth's atmosphere consists of four layers listed respectively starting with the closest to Earth's surface:

- Troposphere
- Stratosphere
- Mesosphere
- Thermosphere

The areas between each of these layers maintain a constant temperature that favors the reduction of them being mixed. However, as the troposphere temperature (where our climate occurs) normally decreases with the increase of the distance from Earth's surface, there is a large degree of turbulence.

This turbulence and instability carry substances that destroy the ozone layer to air currents. The ozone found in the troposphere is a main component of smog, making up part of the air we breathe. The ozone layer near the earth's surface can damage lungs, trees, and other plants; and that is why there is such a large effort towards reducing it. This ozone is made up of the reaction of certain volatile organic compounds (VOCs) with nitrous oxides catalyzed by solar light (photochemical reaction). A detailed explanation of VOCs is addressed in the next item.

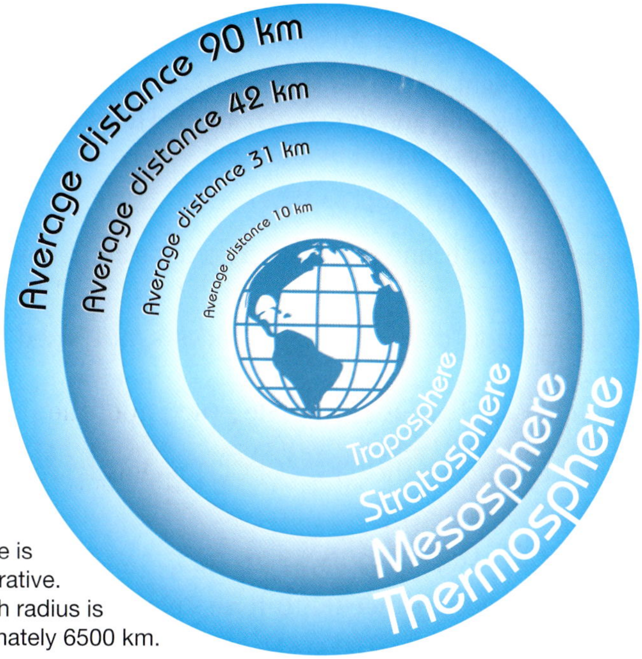

The scale is only figurative.
The Earth radius is approximately 6500 km.

Figure 4.1. Layers that divide Earth's atmosphere[14].

The next layer is the stratosphere, which is more stable than the troposphere, and has little influence on the climate. However, the stratosphere has high concentrations of ozone, referred to in literature as the ozone layer that absorbs a large quantity of the ultraviolet radiation, including ultraviolet B. This is a type of solar radiation linked to skin cancer, eye diseases, immunological system problems, and has several marine and terrestrial ecosystem impacts.

The two nearest layers have low concentrations of ozone and other atmospheric components, and due to this factor, they have low effectiveness in filtering ultraviolet radiation.

Ozone molecules are continuously created and destroyed in our atmosphere, maintaining a constant balance of oxygen (O_2) and ozone (O_3).

Shown below is the ultraviolet radiation action mechanism in the balance between oxygen and ozone. Natural ozone production of the stratosphere follows this mechanism:

$$O_2 + \text{radiation UV} \rightarrow 2\ O^{\bullet} \qquad (4.1)$$

The sun's ultraviolet radiation causes a homolithic break of the oxygen (O_2) molecule in the air into 2 oxygen O• atoms. Then:

$$O^{\bullet} + O_2 \rightarrow O_3 \qquad (4.2)$$

The free oxygen atoms react with (O_2) oxygen molecules to form the ozone (O_3).

Natural oxigen production of the stratosphere follows this mechanism:

$$O_3 + \text{radiation UV} \rightarrow O_2 + O^{\bullet} \qquad (4.3)$$

The ozone absorbs ultraviolet light in the range of 290-320 nanometers, which causes its break into an oxygen molecule and an oxygen atom. This stage is responsible for the filtration of ultraviolet rays emitted by the sun.

$$O^{\bullet} + O_3 \rightarrow 2\ O_2 \qquad (4.4)$$

Free oxygen atoms react with an ozone molecule, regenerating 2 new oxygen molecules.

Now the ozone layer destruction caused by substances synthesized by man follow the following mechanism:

$$CFCl_3 + \text{radiation UV} \rightarrow CFCl_2^{\bullet} + Cl^{\bullet} \qquad (4.5)$$

The sun's ultraviolet radiation breaks the CFC (CFC13) molecule, releasing an extremely reactive chlorine atom.

$$Cl^{\bullet} + O_3 \rightarrow OCl + O_2 \qquad (4.6)$$

Chlorine atoms react with an ozone molecule to form chlorine monoxide and molecular oxygen.

$$O^{\bullet} + OCl \rightarrow O_2 + Cl^{\bullet} \qquad (4.7)$$ [17]

When an oxygen atom reacts with a chlorine monoxide molecule there is a formation of an oxygen molecule and a release of a chlorine atom that destroys more ozone. As can be seen in the mechanism of the ozone destruction by CFC, there is no regeneration of the destroyed ozone, making this stage responsible for the formation of the holes in the ozone layer due to the main consequence of the decreased filtration of solar radiation like those from ultraviolet B rays. The increase in the size of the hole in the ozone layer is a consequence of this chain reaction.

4.2.3.2. Photochemical Smog Formation

The ozone next to the earth's surface is the main component in urban smog and is the yellowish haze in many urban areas, especially in the summer. It is formed by the photochemical reactions of volatile organic compounds (VOCs) with nitrous oxides (NOx) in the atmosphere.

This problem is aggravated in large urban concentrations that emit large quantities of hydrocarbons into the atmosphere from automobiles, for example, creating a climate for this type of smog, with sunny days and hot temperatures.

There are several significant VOC emission sources:

- Natural or Biogenic;
- Automobile Emissions;
- Oil Refinery;
- Combustion Sources.

Tests in smog chambers show that some solvents are more reactive than others, contributing more or less to the formation of ozone.

With changing times, legal defense processes have been ruled in favor of the ozone layer and good air quality

A) U.S. Legislation

A1) CLEAN AIR ACT (CAA)

Due to continual concerns with inadequate air quality in the United States, the American congress passed the Clean Air Act (CAA) with the purpose of defining a set of measures to improve air quality as well as enforcing their application.

Among the different programs developed by the CAA, are the following:

- NAAQS (National Ambient Air Quality Standards)
- SIPs (Station Implementation Plans)
- NSPS (New Sources Performances Standards)
- NESHAPS (National Emission Standards for Hazardous Air Pollutants)

Subsequently, the U.S. government established the Environmental Protection Agency (EPA).[20]

With the creation of the EPA, definitions for VOCs (Volatile Organic Compounds) were outlined. Initially, these definitions were based on the compound mass present in a given formulation, but some compounds produce more ozone than others, leading legislation to consider substituting those with greater photochemical reactivity with those that have less reactivity.

The initial definition generated the first VOC legislation in 1971. In subsequent updates in the legislation, approaching the topic of VOCs in relation to reactivity was always questioned, mainly for compounds on the list of exclusions that present a low tendency to generate ozone. The term used for this characteristic is negligible reactivity.[21]

An example of some compounds considered negligible reactivity and excluded by the VOC legislation are presented below.

Table 4.1. EPA list of photochemically reactive compounds considered negligible. [22]

Compound
Methylene Chloride (Dichloromethane)
1,1,2-Trichloro-1,2,2-Trifluoroethane (CFC-113)
Trichlorofluoromethane (CFC-11)
Trifluoromethane (HFC-23)
1,2-Dichloro 1,1,2,2-Tetrafluoroethane (CFC-114)
Chloropentafluoroethane (CFC-115)
Parachlorobenzotrifluoride (PCBTF)
AcetonePerchloroethylene (Tetrachloroethylene)1,3-Dichloro-
1,3-Dichloro-1,1,2,2,3-Pentafluoropropane (HCFC-225cb)
Methyl Acetate
Methyl Formate ($HCOOCH_3$)
1,1,1,3,3-Pentafluoropropane (HFC-245fa)
Methane
Ethane
t-Butyl Acetate
Ethylfluoride (HFC-161)
Ethylfluoride (HFC-161)
2-(Difluoromethoxymethyl)-1,1,1,2,3,3,3-Heptafluoropropane ($(CF_3)_2CFCF_2OCH_3$)

The table above gives some compounds found on the EPA exclusion list because of their low photochemical reactivity.

A2) TITLE III – Hazardous Air Pollutants (HAP)

In addition to other dangers associated with air quality and the tropospheric ozone, there is more U.S. legislation that deals with the question of intrinsic dangers from compounds that affect human beings and the environment and is known as HAP (Hazardous Air Pollutants).

In 1990, the North American Congress changed the CAA to include a large number of compound air pollutants that can be harmful to humans as well as the environment. The EPA developed regulations applied to main HAP emission sources. A main source is defined as that institution that has the capacity to emit 10 or more tons of an individual HAP or 25 tons of various combined HAPs on an annual basis. The EPA also created a list of 188 substances classified as HAPs. Regulations for the main sources are described in the NESHAP – National Emission Standards for Hazardous Air Pollutants. [18]

In relation to HAP emissions, a large part of them is created by man, with them being differentiated by the EPA according to three principal sources:

- furniture: for vehicles in general
- stationary: factories, refineries, and energy plants in general
- indoor activities: cleaning activities.[18]

These rules do not prohibit the use of HAP solvents, but control HAP quantity that can be emitted into the air. For this reason, it is possible for equipment installation users that control emitted quantities to remain within the regulation.

A3) Other U.S. Legislation

Several classes of volatile organic compounds (VOCs) are emitted into the atmosphere with each of them having different forms of reactivity for the formation of ozone, mainly related to:

- The different reaction mechanisms between them and NOx in function of each chemical compound class;
- The different kinetic reactions that each chemical compound has.

These particular characteristics of each chemical compound that make up the VOC significantly determine different quantities of formed ozone in relation to the reactivity of each of them.

Therefore, an efficient way to decrease ozone formation is by using a VOC emission control strategy based on the reactivity of each compound in its formation, thus decreasing the emission of compounds with high reactivity. This is considered to be the most efficient way rather than considering all VOCs as equally contributing to the formation of ozone in the troposphere layer.

This control strategy based on reactivity was already implemented by agencies in California in the US. Two examples are the California Clean Fuel/Low Emission Vehicle (CF/LV) regulations and the California Air Resources Board (CARB) that developed an option for coatings, allowing them to contain a greater quantity of VOCs, as long as the global photochemical reactivity of the products does not exceed a relative specific reactivity. In June 2002, CARB adopted the Maximum Incremental Reactivity (MIR) concept for aerosol coatings, and in June 2005, the EPA approved the rule as part of the State Implementation Plan. The MIR concept is being considered by the EPA and CARB for use in future regulations for other types of coatings.

The main aspects that determine VOC reactivity are:

- The reaction velocity for ozone formation;
- Quantity of NO molecules oxidized per VOC molecule;
- Quantity of radicals formed per VOC molecule that can participate in the ozone formation route of other VOCs;
- Effect of the NOx removal rate from the atmosphere since ozone formation decreases when the NOx concentration decreases;
- Reactivity of principal VOC oxidation products.

A useful measure of the VOC effect in ozone formation is its incremental reactivity that can be measured by the following equation:

$$\begin{bmatrix} \text{Incremental} \\ \text{Reativity} \\ \text{VOC} \\ \text{Pollution} \\ \text{Situation} \end{bmatrix} = \lim[\text{VOC} \rightarrow 0] \; \frac{\left[g_{O3} \text{ formed by VOC}\right] - \left[g_{O3} \text{ formed without VOC}\right]}{\left[g_{VOC}\right]}$$

This reactivity can be experimentally measured in specific smog chambers that simulate atmospheric conditions or calculated using specific reaction mechanism models.

The main aspects that determine incremental reactivity are:

- NOx availability
    - it is the most important factor because VOCs play an important part in O3 formation when the quantity of NOx is high. When NOx is absent, there is no O3 formation.
- Nature of other VOCs present
    - VOCs that are sources for radical formation decrease the importance of other sources.

- Amount of sunlight and temperature
  - affect the kinetic reaction rate.

The Maximum Incremental Reactivity (MIR) scale represents an average of incremental reactivities in situations were available NOx levels are adjusted for maximum use in ozone formation. MIRs for common solvents are shown in the table below:

Table 4.2. Maximum Incremental Reactivity (MIR) of several solvents

Solvent	MIR (Gram of Formed Ozone/VOC gram)
m-Xilene	10.61
o-Xilene	7.49
Xilenes	7.37
p-Xilene	4.25
1,2,4-Trimethyl Benzene	7.18
Methyl Isobutyl Ketone	4.31
Toluene	3.97
Vinyl Acetate	3.26
Methyl Propyl Acetate	3.07
Methyl Amyl Ketone	2.80
Ethyl Benzene	2.79
Methyl Isoamyl Ketone	2.10
Propylene Glycol Methyl Ether Acetate	1.71
Ethanol	1.69
2-Etoxiethyl Acetate	1.50
Methyl Ethyl Ketone	1.49
Isoamyl Acetate	1.18
n-Butyl Acetate	0.89
Texanol TM	0.89
Propyl Acetate	0.87
Diacetone Alcohol	0.68
Isobutyl Acetate	0.67
Ethyl Acetate	0.64
Acetone	0.40
t-Butyl Acetate	0.20
4-chlorotrifluorbenzene (PCBTF)	0.11
Methyl Acetate	0.07
Methylene Chlorate	0.07
Perchloroethylene	0.04
Methane	0.01
1,1,1-Trichloroethane	0.00

An example of MIR values used for reducing tropospheric ozone formation was the regulation adopted by CARB aerosol base coatings. Emission limits for categories classified as generic entered in vigor in 2002 and those specifically classified entered in vigor in 2003. To finally establish equivalent limits, the reduction calculation for ozone formation was based on the reduction of VOC mass; and based on this, reactivity limits were proposed to ensure the same air quality benefits. These changes allowed coating formulators to create more efficient formulations based on this in relation to ozone formation by VOCs.

Section 94522(a)(3) of the ARB aerosol based coating regulation contains emission limits based on component reactivity. The total reactivity of a solvent formulation is given by the MIR total of each component multiplied by its mass fraction in the formulation. The table below gives reactivity limits for several market segments affected by the regulation.

Table 4.3. Reactivity Limits for Specific Market Segments

	Segments	Average Reactivity Limit (O_3 g/product g)
Generic	Varnishes	1.54
	Fluorescent Coatings	1.77
	Metal Coatings	1.93
	Paints	1.40
	Primers	1.11
Specialized	Sealants	1.80
	Aircraft Primers	1.98
	Corrosion Resistant Coatings	1.78
	Glass Coatings	1.42
	Road Coatings	1.18
	High Temperature Resistant Coatings	1.83
	Varnishes for Boats and Ships	0.87
	Photograph Coatings	0.99
	Aircraft Finishing Primers	1.05
	Aircraft Varnishes	0.59
	Coatings for Leather, Fabric, Polycarbons	1.54
	Wood Coatings	1.38
	Wood Restoration Coatings	1.49

[9]

With Table 4.3 the average total calculated from MIRs for each constituent in automotive varnishes cannot exceed 1.54.

B) European Legislation

European legislation is quite different from American legislation, with an important difference being found in the definition of VOC. The European VOC definition is presented below:

"VOCs are defined as organic compounds with a vapor pressure of less than or equal to 0.01 kPa to 293.15 K, or those that have a volatility equivalent when under usage."[19]

Based on legislation, VOCs are defined as any compound containing at least one carbon atom and one or more hydrogen atoms, in addition to other heteroatoms like halogens, oxygen, phosphorous, sulfur, etc.

It is important to include in the above definitions that at no time is the question of photochemical reactivity of compounds addressed as it is stipulated in the American legislation. The result is that there is no difference between VOCs in European legislation, with vapor pressure the only limiting factor.

In looking at VOC legislation resulting from some industry activities, the directive explicitly presents its main objective as to reduce pollution from tropospheric ozone by preventing or reducing VOC emission limits.

Directive 1999/13/EC of March 11, 1999 specifically limits VOC emissions from organic solvent use in certain activities and industry settings. For dangerous substances, VOCs classified as carcinogenic, mutagenic, or toxic in reproduction, and is described in Directive 67/548/EEC must be substituted with less dangerous substances or formulations. Emission limit for these compounds when used in an application is 2 mg/Nm^3.

The principal industrial categories involved in the Directive are described below:

- Adhesives: with the exception of printing application activities.
- Coating Industry: despite this being a wide area, this refers to the application activity of a film coating on a determined surface, such as:
  - Automotive:
    OEM
    Trucks
    Buses
  - Aircraft:
    Planes
    Trains
    Ships
  - Plastic Surfaces
  - Wood
  - Cloth, Textiles, and Paper
  - Leather
- Coil Coating
- Cleaning

- Pavement Coatings
- Preparation of Coatings, Varnishes, and Adhesives
- Production of Pharmaceutical Products
- Printing Inks:
  - Flexography
    Flexography Plastification
  - Offset
  - Rotogravure
- Rubber Processing
- Cleaning Surfaces
- Animal and Vegetable Oil Extraction, and their refining
- Automotive Refinishing
- Plastic and Wood Plastification

Below is a table giving the emission limits for each of these sectors, with the limits varying according to the total consumed solvents in a unit.

Table 4.4. VOC Emission Limits in an Industrial Activity

Indoor Activities	Annual Consumption of Solvents (ton/year)	Annual Consumption of Solvents (ton/year)
Offset	15 – 25 > 25	100 20
Rotogravure	> 25	75
Flexography Plastification	15 – 25 >25 > 30	100 100 100
Cleaning Surfaces	1 – 5 > 5	20 20
Automotive Refinishing	> 0.5	50
Coil Coating	> 25	50
Plastics	5 – 15 > 15	100 75
Wood Surfaces	156 – 25 > 25	100 50 (drying) and 75 (application)
Leather	10 – 25 > 25	85 75
Adhesives	5 – 15 > 15	50 50
Production of Coatings, Varnishes, and Adhesives	100 – 1000 > 1000	150 150
Production of Pharmaceutical Products	> 50	20

[19]

In terms of OEM (Original Equipment Maintenance), limits are expressed in VOC quantity per m^2 of a total product surface area to be coated. The table below gives some of these values.

Table 4.5. VOC Emission Limits in OEM

Segment	Annual Consumption of Solvents (ton/year)	Production (number of units produced annually)	Emission Limit (g/m^2)
New Cars	> 15	> 5,000	45
Truck Cabins	> 15	≤ 5,000 > 5,000	65 55
Vans	> 15	≤ 2,500 > 2,500	90 70
Buses	> 15	≤ 2,000 > 2,000	210 150

[19]

C) "Natural" Formation and Destruction of Ozone in the Troposphere[16]

In the troposphere, close to Earth's surface, the ozone is formed by a homolithic reaction catalyzed photochemically by sunlight. However, nitrogen dioxide (NO2) in the troposphere is the primary source of oxygen atom production responsible for its formation.

i) Ozone Formation Stage

Sunlight causes an NO_2 homolithic reaction resulting in nitric oxide (NO) and an oxygen atom.

$$NO_2 + \text{sunlight} \rightarrow NO + O^\bullet \tag{4.8}$$

The oxygen atom reacts with the oxygen molecule to form ozone.

$$O^\bullet + O_2 \rightarrow O_3 \tag{4.9}$$

ii) Ozone Destruction Stage

The ozone formed in the previous stage quickly reacts with nitric oxide (NO) again generating NO_2 and O_2.

$$NO + O_3 \rightarrow NO_2 + O_2 \tag{4.10}$$

As shown in the above process, there is no net increase in ozone concentration.

C1) Troposphere Ozone Formation

However, in 1950, some chemists discovered two additional chemical constituents in the troposphere that also contributed to the formation of ozone: NOx and VOCs.

Therefore, troposphere ozone formation requires the presence of NOx and VOCs. A very simplified way of showing this reaction is shown below.

$$\text{NOx} + \text{VOC} + \text{sunlight} \rightarrow O_3 + \text{other products} \quad (4.11)$$

The above equation shows a simplification of what happens in several chemical reactions in series or in parallel, increasing net ozone concentration. They involve:

- VOC oxidation;
- Oxide reduction in NOx ;
- Hydroxyl group catalysts (key reaction for several stages);
- A series of other reactions.

The result of these reactions is the formation of:

- Ozone;
- NO_2 – causing the formation of more ozone;
- The regeneration of hydroxyl groups that will catalyze more ozone formation.

C2) Ozone Formation starting with VOCs and NOx in the Troposphere – Methane Example

Ozone formation caused by methane is a general standard example that the majority of chemical reactions follow. However, the majority of ozone formed in the troposphere involves other hydrocarbons in addition to methane (*non-methane hydrocarbons*). Chemical formation of ozone with other organic compound classes follow a general standard reaction of methane, as shown below, but in a much more complex fashion.

i) Route 1 – Destruction and Formation of O_3 starting with NOx and VOC

There are five stages in this formation of ozone.

a) Stage 1

Sunlight divides the ozone into an oxygen molecule and an oxygen atom ($O^\bullet$).

$$O_3 + \text{sunlight} \rightarrow O_2 + O^\bullet \quad (4.12)$$

The electronically excited oxygen atom reacts with the water vapor creating hydroxyl radicals. Oxygen and water are abundant in the troposphere and take part in the formation process of ozone.

$$O^{\bullet} + H_2O \rightarrow OH^{-} + OH^{-} \qquad (4.13)$$

b) Stage 2

The hydroxyl radicals ($OH^{\bullet}$) react quickly with other chemical compounds and begin another sequence of reactions. One of these reactions is with methane (CH4), creating water and a methyl radical ($CH_3^{\bullet}$).

$$CH_4 + OH^{\bullet} \rightarrow CH_3^{\bullet} + H_2O \qquad (4.14)$$

c) Stage 3

Methyl radicals react with oxygen to produce methyl peroxy radicals ($CH_3O_2^{\bullet}$).

$$CH_3^{\bullet} + O_2 \rightarrow CH_3O_2^{\bullet} \qquad (4.15)$$

d) Stage 4

Methyl peroxy radicals react with nitric oxide (NO), resulting from, for example, the combustion of fossil fuels, to produce a methyloxy radical ($CH_3O^{\bullet}$) and nitrogen dioxide (NO_2).

$$CH_3O_2^{\bullet} + NO \rightarrow CH_3O^{\bullet} + NO_2 \qquad (4.16)$$

e) Stage 5

Sunlight divides nitrogen dioxide into nitric oxide and atomic oxygen that reacts with molecular oxygen forming ozone, as shown in equations (8) and (9).

$$NO_2 + \text{sunlight} \rightarrow NO + O^{\bullet} \qquad (4.8)$$

The oxygen atom reacts with the oxygen molecule to form ozone:

$$O^{\bullet} + O_2 \rightarrow O_3 \qquad (4.9)$$

In addition to this formation there are two more possible routes.

ii) Route 2 – Formation of "Liquid" O_3 by VOCs and NOx

Methyl oxy radicals formed in stage 4 can also participate in other reactions that will also result in more nitrogen dioxide and therefore in a greater formation of ozone. This second way consists of a methyl oxy radical reaction with oxygen to produce formaldehyde and a hyperoxy radical ($HO_2^{\bullet}$).

$$CH_3O^{\bullet} + O_2 \rightarrow CH_2O + HO_2^{\bullet} \qquad (4.17)$$

The hyperoxy radical reacts with nitric oxide forming a hydroxyl radical and nitrogen dioxide.

$$HO_2^{\bullet} + NO \rightarrow OH^{\bullet} + NO_2 \qquad (4.18)$$

Again, the formation stage of ozone starting with sunlight dividing nitrogen dioxide, as shown in reactions (8) and (9).

$$NO_2 + \text{sunlight} \rightarrow NO + O^{\bullet} \quad (4.8)$$

$$O^{\bullet} + O_2 \rightarrow O_3 \quad (4.9)$$

iii) Route 3 – Formation of "Liquid" O_3 by VOCs and NOx

A third way of forming ozone is starting from the formaldehyde resulting from the reaction (17) that increases the concentration of nitrogen dioxide, and consequently leads to more ozone. This reaction sequence consists of sunlight dividing formaldehyde and forming a formyl radical and atomic hydrogen.

$$CH_2O + \text{sunlight} \rightarrow HCO^{\bullet} + H^{\bullet} \quad (4.19)$$

These two species are extremely reactive, and therefore, short lived, reacting almost instantaneously with molecular oxygen forming hyperoxy radicals.

$$HCO^{\bullet} + O_2 \rightarrow CO + HO_2^{\bullet} \quad (4.20)$$

$$H^{\bullet} + O_2 \rightarrow HO_2^{\bullet} \quad (4.21)$$

The hyperoxy radicals react with nitric oxide forming a hydroxyl radicals and nitrogen dioxide, which again forms ozone with the interaction of sunlight.

$$HO_2^{\bullet} + NO \rightarrow OH^{\bullet} + NO_2 \quad (4.22)$$

$$NO_2 + \text{sunlight} \rightarrow NO + O^{\bullet} \quad (4.8)$$

$$O^{\bullet} + O_2 \rightarrow O_3 \quad (4.9)$$

4.3. Criteria for Evaluating Green Solvents

Based on concepts presented above, for a solvent to be considered green, it should perform according to its intended purpose and affect the environment and man as little as possible as well as not contribute to pollution.

In the criteria that follow, it is important that the solvent or solvent system being assessed be the most efficient considering the available technologies. The criteria that must be considered are:

i. Application performance;
ii. Toxicity for humans;
iii. Reproductive carcinogens and toxins;
iv. Skin and eye irritation;

v. Envirnomental toxicity;
vi. Biodegradability;
vii. Smog formation via photochemical reaction;
viii. Destruction of the ozone layer;
ix. Renewable raw material resources: Impact on global warming;
x. Disposal.

Topics related to toxicity for humans, carcinogens, skin and eye irritation, environmental toxicity, and biodegradability will be addressed in detail in Chapter 10.

4.3.1. Application Performance

This topic will be addressed in Chapter 8. However, it is important to note that a profound knowledge of application is needed in order to avoid some criteria being poorly positioned due to alternative formulations substituting a solvent or solvent system. Following is an example of a substitution.

Cleaning solvents are used in many industrial sectors: electronics, metal, textile, etc. Historically, chlorine solvents were preferred for these applications because they were nonflammable; an important requirement in high-temperature cleaning processes, as in steam cleaning. Among those, trichloroethane (TCA) was the most used, in addition to CFC-113 (1,1,2-trichloro-1,2,2-trifluoroethane), trichloroethylene (TCE), perchloroethylene (PERC) and methylene chloride. However, with the Montreal Protocol, and the CAA, the majority of these solvents had restrictions and therefore more environmentally friendly alternatives were needed to substitute them in the processes. As a result, many alternatives appeared in the form of hydrochlorofluorocarbons (HCFCs), hydrofluorocarbans (HFCs), oxygenated solvents, and supercritical fluids.

Another example that has gained space in the market is the substitution of toluene in adhesive formulations by less toxic solvents formulated with oxygenated and aliphatic solvents. There is a law in Chile that does not allow toluene in contact adhesive formulations that are sold in retail stores. Companies use a mixture of oxygenated and aliphatic solvents in its place.

4.3.2. Photochemical Smog and Oxidant Formation

A solvent cannot significantly contribute to the formation of smog, a photochemical formation of ozone in the troposphere. The use of the photochemical reactivity concept is important for the definition of solvent systems with the least possible amount of impact leading to the formation of ozone.

4.3.3. Destruction of the Ozone Layer

A solvent system cannot contain products that destroy the ozone layer. As was shown in item 2.2.3.1, which deals with the destruction of ozone, CFCs are stable in the troposphere and can reach the stratosphere. In the stratosphere, a cleavage reaction is catalyzed by UV radiation, producing chlorine monoxide radicals that interfere in the catalytic cycle of ozone formation and destruction.

4.3.4. Renewable Resources for Raw Material: Impact on Global Warming

For green system evaluation proposals, it is important to include raw material resources of the solvents used. In these terms, resources can be classified as:

- Renewable
- Non-renewable

CO_2 is a gas that contributes to global warming and when any organic substance is burned, it is released into the atmosphere. However, CO_2 is also a carbon source used by plants.

Renewable resources are those that use CO_2 from the atmosphere to produce raw materials used in solvent production. When the solvent degrades or is burned, it creates CO_2, and the net result from this cycle is that there is no additional generation of CO_2 in the atmosphere. This is the case for raw materials from plants, for example, like ethanol from sugar cane.

Non-renewable resources are those that, despite directly or indirectly using CO_2 for raw material production, the time scale for this consumption is extremely long, taking up to thousands of years. In these cases, CO_2 is considered practically non-consumable, and when the substance degrades, CO_2 is generated with the net result being an increase in its concentration in the atmosphere, and consequently contributing to the increase in global warming. Oil is a big source of these raw materials.

With solvent system evaluations, from the point of view of being green, the greater number of solvents from renewable resources should be chosen.

4.3.5. Disposal

The main forms of solvent disposal are incineration and biological treatment. Solvent property assessment is highly important to disposal. In terms of incineration, some important properties include:

- combustion heat influencing energy for incineration;
- emissions from incineration, especially HCl, dioxins, and NOx;
- water solubility.

Here is an example of a property impacting on another. Suppose there is a solvent from a renewable resource but it has a low combustion heat. In this case, there will be a need for more energy to incinerate it, and since the energy source is normally fossil fuel, the CO_2 balance can be unfavorable, creating a net increase in its generation.

The main properties with biological treatment are:

- aeration basin treatment;
- air emission of solvents that have been stripped;
- water solubility.

4.4. Examples of Green Solvent Classes

The main objective of this chapter is to create a greater awareness of the need to look at solvent systems currently used, and if there is an existing alternative with a better classification in one or more criteria, systems that have more green characteristics should be developed.

Below are some compound class examples demonstrating the use of criteria presented for purposes of a greener system:

- substitution of hazardous solvents or solvent systems by others with better properties in terms of security, occupational health, and the environment;
- use of water systems as a solvent;
- replacement of organic solvents by supercritical fluids;
- use of ionic liquids- research in this solvent class is usually on determining human and environmental toxicity;
- solventless system.

4.4.1. Classification Example of a Solvent as a Green Solvent

According to the Globally Harmonized System (GHS), hazards presented by a given chemical substance are related to intrinsic properties that can cause health problems and damage the environment. An example of this is interference in normal biological processes through its ability to cause burns, ignite, or cause damage to the environment. The concept of risk or likelihood of harmful effects is presented when exposure is considered along with hazards of the substance. The basic approach for analyzing a risk is characterized by the formula:

$$\text{Risk} = \text{Hazard} \times \text{Exposure}$$

Therefore, if hazards or exposure are decreased, risk or likelihood of harm will be decreased.

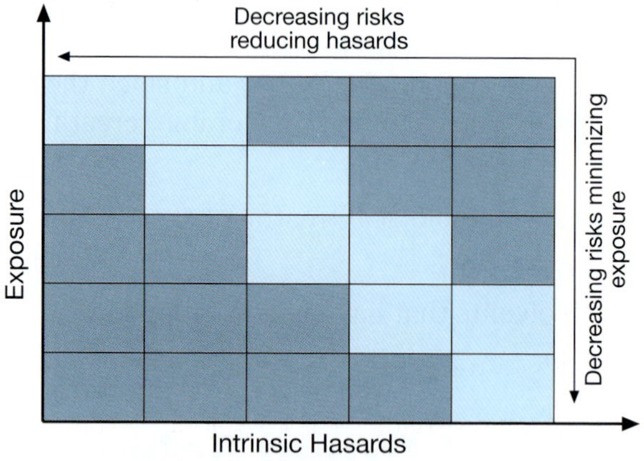

Figure 4.2. Intrinsic Hazards.

In this chapter, the proposal presented for a solvent to be classified as a green solvent is based on comparable criteria in a given end application such that if an alternative solvent compared to the current one shows global gains by minimizing impacts on health and/or the environment, the alternative solvent could be classified as a Green Solvent.

In practice, global gains that minimize impacts on health and/or environment making it possible to create a cycle of continuous improvement in the reduction of hazards for a given application should be considered since it is very difficult to find substitute solvents technically equivalent in performance and that also present fewer hazards in all assessment parameters.

If a comparative assessment is conducted for the same end product in which there is a possibility of using two different solvent systems, or in other words, the exposure factor for health and/or the environment is consistently being neglected, the assessment will be able to be done comparing intrinsic hazards of each solvent system based on the parameters described in the tables below.

- **Physical Hazards**: solvent B has a higher flash point and boiling point than solvent A. However, both are classified as flammable and in relation to minimizing hazards, there are no advantages in substituting one for the other.
- **Environmental hazards**: (note 1 in table) – solvent A vapor pressure is approximately 5 times greater than that of solvent B. However, the photochemical reactivity of the latter is approximately 10 times greater. If both solvents are used in equivalent amounts in the formulation, even with a greater vapor pressure (more volatile) than solvent B, solvent A can represent more important gains for the environment because of its lower photochemical reactivity.

Comparatively, it can be concluded that for a given formulation in which solvents A and B are being used in equivalent proportions, solvent A has greater global gains in reducing health and environmental hazards.

Table 4.6. Intrinsic Hazard Parameters

	Parameter	Why?
Physical Hazards	Flash Point	Flash point and boiling point are necessary to classify substance and mixture flammability.
	Boiling Point	
	Vapor Pressure	In several regulations, vapor pressure is used as classification criteria for VOCs (*Volatile Organic Compounds*)
Health Hazards	Acute Toxicity	Short or long-term systemic effects from inhalation or dermal exposure to a product
	Irritation	Short or long-term local effects from liquid or vapor exposure
	CMR	In several regulations there are restrictions or the prohibition of using classified substances like CMR (carcinogens, mutagens or toxic to reproduction or reprotoxins)
	TOST	In GHS, nauseous or toxic effects on the Central Nervous System (transitory or irreversible) are classified as TOST (*Specific Target Organ Systemic Toxicity*).
Environmental Hazards	Aquatic Toxicity	Potential impact on aquatic organisms (fish, algae, crustaceans)
	Degradability	Identification of degradability potential (biotic or abiotic) of a substance or mixture in the environment
	Bioaccumulation	Accumulation potential in the environment
	Photochemical Reactivity	See 4.2.3.2, section A4

Example:

	Parameter	Solvent A	Solvent B
Physical Hazards	Flash Point	– 22 °C	4.4 °C
	Boiling Point	55.6 °C	114.1 °C
	Vapor Pressure	188 mm Hg (20 °C)	38 mm Hg (20 °C)
Health Hazards	Acute Toxicity (inhalation)	50,100 mg/m^3	10,640 mg/m^3
	Irritation (Respiratory)	Negative	Negative
	CMR	Negative	Negative
	TOST	Positive – Transitory Harm/Damage	Positive – Permanent Harm/Damage
Environmental Hazards	Aquatic Toxicity	CE50 (daphnia): 6,400 mg/L	CE50 (daphnia): 39.2 mg/L
	Degradability	Easily Biodegradable	Easily Biodegradable
	Bioaccumulation	Non-bioaccumulative	Non-bioaccumulative
	Photochemical Reactivity (MIR)	0.40	3.97

Applying these GHS classification criteria for the above data, we have:

	Parameter	Solvent A	Solvent B
Physical Hazards	Flash Point	Flammable Category 1	Flammable Category 2
	Boiling Point		
	Vapor Pressure	Note 1	Note 1
Health Hazards	Acute Toxicity (inhalation)	Category 5	Category 4
	Irritation (Respiratory)	Not Classified	Not Classified
	CMR	Not Classified	Not Classified
	TOST	TOST – category 2	TOST – category 1
Environmental Hazards	Aquatic Toxicity	Not Classified	Toxicity Acute 3
	Degradability		
	Bioaccumulation		
	Photochemical Reactivity (MIR)	Note 1	Note 1

Bibliographical References

1. Air Quality: Revision to definition of volatile organic compounds – exclusion of acetone; junho 1995. Disponível em: www.epa.gov/EPA-AIR/1995/june/day-16/pr-752.html.
2. Anastas, P. T.; Warner, J. C.; Green chemistry: theory and practice, New York: Oxford University Press, 2000.
3. Archer, W. L.; Industrial solvents handbook, New York: Marcel Dekker, 1996.
4. Capello, C; Fischer, U; Hungerbühler, K; What is a green solvent? A comprehensive framework for the environmental assessment of solvents, Green Chemistry, 2007, 9, 927-934.
5. Clark, J. H.; Green chemistry: challenges and opportunities, Green Chemistry, 1999, 1, 2.
6. Clark, J. H.; Green chemistry: today (and tomorrow), Green Chemistry, 2006, 8, 17-21.
7. Curzons, A. D.; Constable, D. C.; Cunninghan, V. L.; Solvent selection guide: a guide to the integration of environmental, health and safety criteria into the selection of solvents. Clean Products and processes, 1999, 82-90.
8. Höfer, R.; Bigorra, J; Green chemistry – a sustainable solution for industrial speciaties applications, Green Chemistry, 2007, 9, 203-212.
9. Linak, E; Global solvent report: the green impact, Specialty chemicals – Sri Consulting, 2006.
10. Nelson, W. M.; Green solvents for chemistry: perspectives and practice, New York: Oxford University Press, 2003.
11. Pianofort, K; Low and Zero VOC coatings, maio 2007. Disponível em: www.coatingsworld.com.
12. Sherldon, R. A.; Green solvents for sustainable organic synthesis: state of the art, Green Chemistry, 2005, 7, 267-278.
13. Shermann, J; Chin, B; Huibers, P. D. T.; Garcias-Valls, Ricard; Hatton, T. A.; Solvent Replacement for Green Processing; EHP, 1997.
14. Stratospheric Ozone Depletion: a focus on EPA'S research, Washington, EPA, março 1995.
15. Tess, R. W.; Solvents Theory and Practice, Washington: American Chemical Society, 1971.
16. NASA; Chemistry in the Sunlight, Earth Observatory; site – http://earthobservatory.nasa.gov/Features/ChemistrySunlight/chemistry_sunlight3.php ---- sunlight3a

17. NASA; Chemistry in the Sunlight, Earth Observatory; site: http://earthobservatory.nasa.gov/Features/Ozone/ozone_2.php

18. EPA, Office of Air Quality, Planning & Standards. Section 112 Hazardous Air Pollutants List.

19. Oficial Journal of the European Communities – Council Directive 1999/13/EC – 11/03/1999 – Annex IIA.

20. Environmental Protection Agency (EPA). (http://www.epa.gov/air/caa/caa_history.html#caa70).

21. Federal Register / Vol. 70, No. 5 / Friday, January 7, 2005 / Proposed Rules.

22. www.epa.gov/ttn/naaqs/ozone/ozonetech/def_voc.htm – Definition of Volatile Organic Compounds (VOC) – As of 2/9/2007 – 40 CFR 51.100(s).

5 Physical Chemical Properties of Solvents

This chapter presents physical-chemical aspects of variations and forms of energy, its relation to covalent bonds and their polarity, as well as existing molecular interactions that influence melting and boiling points, viscosity, surface tension, dielectric constant, and other properties.

Alessandro Rizzato
Léo Santos

An overview of the general industrial solvent market and how Rhodia fits into this context was presented in the initial chapters of this book. Next, general aspects and definitions of the main solvent classes, their characteristics, and reactivity, as well as recent discoveries related to green solvents were discussed.

In this chapter, the physicochemical aspects such as variations and forms of energy, polarity, molecular interactions and their influence in melting and boiling points, viscosity, surface tension, dielectric constant, and other properties will be discussed. Our idea is to briefly present some important concepts in order to understand physicochemical properties of solvent systems.

As a starting point of any experimental work, the first aspect to be considered is the state of the substance involved, i. e., solid, liquid, or gas states. This information is easily found in several sources and is directly responsible for optimized formulation performance influencing film formation, charges dispersion, or solubilization of certain solutes.

5.1. Energy Involved

Several important definitions are found on previous pages with respect to characteristics and reactivity of principal solvent classes, but little has been presented on energy variations involved in these systems.

Energy is defined as the ability to perform work. There are two fundamental types of energy: kinetic and potential. Kinetic energy is energy an object uses when it is in motion, and it is defined as half its mass times velocity squared, i.e., $1/2\ m{\cdot}v^2$. Potential energy is stored energy an object has when it is at rest or at the moment it is about to go into motion. It only exists when there are attractive and repulsive forces between two objects.

The potential energy of a system consisting of two objects joined by a spring increases when the spring is stretched or compressed. When the spring is stretched, an attraction force is created, while being compressed it has stored repulsive force. If the spring is stretched or compressed, the stored potential energy is converted into kinetic energy when the spring is released.

In this context, chemical energy is a form of potential energy that exists due to the attractive and repulsive forces between atoms or molecules. Repulsive force exists when there is an interaction between two nuclei as well as between two electrons while attractive force exists when there is an interaction between a nucleus and an electron. It is practically impossible to describe the absolute amount of potential energy in a substance. For this reason, we often think in terms of relative potential energy and say a system has more or less potential energy than another.

Another term commonly used by chemists is stability or relative stability. The chemical stability of a system is inversely proportional to its relative potential energy. The greater a system's potential energy, the less stable it is.

5.2. Potential Energy and Covalent Bonding

Potential energy, or in this case, chemical energy, can be released in the form of heat when atoms and molecules react. Since heat is associated with molecular movement, the release of heat is the result of potential energy converted into kinetic energy.

In terms of covalent bonds, higher potential energy is manifested when atoms are free, i. e., when they are not bound. This is true because a chemical bond formation is always accompanied by a decrease in potential energy of the isolated atoms.

A convenient way to represent a molecule's potential energy is to consider its relative enthalpy, H. The difference between the reagents' relative enthalpy and that of the product's is called reaction enthalpy, represented by $\Delta H°$. The symbol (°) indicates that the measure was performed under standard conditions (25 °C and 1 atm).

Conventionally, $\Delta H°$ of exothermic reactions (those that release heat) is negative, while endothermic reactions (those that absorb heat) have a positive $\Delta H°$. The $\Delta H°$ measures the variation of enthalpy of reagent atoms when they are converted to products. In an exothermic reaction, product atom enthalpy is less than reagent atom enthalpy, while in an endothermic reaction the opposite is true.

5.3. Bond Polarity

The majority of molecules used as solvents are made up of atoms covalently bonded. Due to the difference in electronegativity of these elements, they show a property called polarity. Two atoms in a covalent bond share electrons between each other, and the nuclei of these atoms are interconnected by the electron cloud. When these atoms are the same and the molecule is symmetrical, the electron cloud is uniformly distributed throughout the bond.

However, in the majority of cases the two nuclei do not equally share electrons, thus the electron cloud is denser in one atom than in the other. For this reason, one of the sides of the bond is negative and the other is positive, i.e., it has a negative pole and positive pole. Under these conditions, the bond has a polarity, or in other words, it is a polar bond. Polarity can be expressed using the symbols, δ^+ and δ^-, which represent a partial positive or negative charge. Figure 5.1 is a water molecule showing polarities.

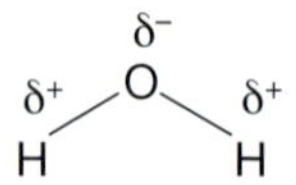

Figure 5.1. Polarity in a water molecule.

Consequently, polarity in a bond exists when the linked atoms with different tendencies to attract electrons have different electronegativity. Bond polarity will be greater when there is a greater difference in the electronegativity between the two atoms.

The elements with higher electronegativity are those in the upper right corner of the periodic table, except for the noble gases making up group 18. Among the elements most frequently found in solvent molecules, fluorine is the most electronegative, followed by oxygen, nitrogen, chlorine, bromine, and lastly carbon. Hydrogen electronegativity is not considerably different from that of carbon.

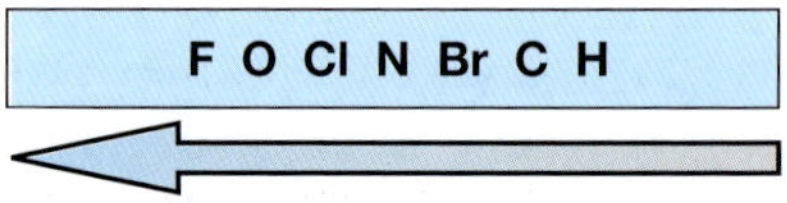

Bond polarity is closely related to physical and chemical properties, influencing melting and boiling points, as well as solubility, dielectric constant, and surface tension, among others.

5.4. Molecular Polarity

Polarity exists as a consequence of different atoms constituting covalent bonds in molecules. Atoms making up a covalent bond in which electronegativity is significant are considered charged centers. Thus, a molecule will be polar if the negative charges do not coincide with the positive charges. In this case, the molecule consists of a dipole: two equal charges with opposite signs separated by space. Dipole moment p_e, is defined by the product of the charge on the negative side, Q, times the distance l between the charged centers (atoms).

$$p_e = Q \cdot l \tag{5.1}$$

The International System (SI) dipole moment unit is the Coulomb meter, C · m, but the Debye, which is $3.33 \cdot 10^{-30}$ C · m, is still used.

5.5. Intermolecular Forces

There are two intermolecular forces responsible for holding neutral molecules together: dipole interaction and van der Waals forces.

The dipole interaction observed in polar molecules results in a mutual attraction between the positive charge of one molecule and the negative charge of another. In hydrogen chloride, for example, the relatively positive hydrogen of one molecule is attracted by the relatively negative chlorine in another molecule. Due to the dipole-dipole interaction, two polar molecules are more aggregated than nonpolar molecules with similar masses. The difference in intensities between intermolecular forces are reflected in physical properties of the respective substances.

An example of a relatively strong dipole-dipole interaction is the hydrogen bond in which the hydrogen atom serves as a bridge between two electronegative atoms establishing a covalent bond with one and a purely electrostatic interaction with the other. When a hydrogen atom bonds to a highly electronegative atom, the electron cloud is displaced considerably in the direction of the more electronegative atom, leaving the hydrogen nucleus uncovered. The positive charge of the hydrogen nucleus is strongly attracted to the negative charge of a more electronegative atom in the second molecule. This attraction has an energy of nearly 21 KJ · mol^{-1}, therefore being weaker than the covalent bond, which is nearly 210 to 420 KJ · mol^{-1}, but much stronger than other dipole-dipole interactions. For the effect of hydrogen bonds to be important, both electronegative atoms must be from the fluorine, oxygen, and nitrogen group.

The other intermolecular force that can exist in nonpolar molecules is the van der Waals interaction. Similar to what happens with dipole-dipole interactions, the average distribution of charges in molecules in a van der Waals interaction is symmetric and the resultant dipole moment is null. However, since electrons move, it is quite probable that in a given moment, charge distribution distorts according to the average symmetric distribution. In this instance, a small electric dipole is produced in which its momentary existence affects electronic distribution in another neighboring compound molecule. The negative pole of the dipole tends to repel electrons while the positive pole tends to attract them; therefore, the first molecule induces an opposite dipole in the second one. Even though momentary dipoles and induced dipoles constantly vary, the final result is the existence of a mutual attraction between the two molecules. These van der Waals interactions show a very short action radius and influence the space very close to other molecules, or in other words, only affect molecule surfaces. Therefore, the relationship between the intensity of van der Waals interactions and molecule surface areas helps us to understand the dependence between the chemical physical properties and the molecular size and shape.

Next, we describe the fundamental thermodynamic aspects that define the physical states of the substances, as well as their relation to important physical-chemical properties like boiling and melting points, dielectric constant, density, and viscosity, among others

5.6. Phase Stability

A substance's phase is defined as the state of the matter in which the chemical composition and physical state are uniform throughout a determined system. There are three well-defined phases: solid, liquid, and gas. A phase transition is a spontaneous conversion from one phase to another. They occur in well-defined temperatures and pressures and are exclusive for each substance. For example, at 1 atm of pressure and a temperature below 0 °C, the solid is water in its stable phase. In a given fixed pressure, transition temperature is that where two phases are in equilibrium.

5.7. Phase Limit

A phase diagram of a pure substance shows pressure and temperature regions where substance phases are thermodynamically stable (Figure 5.2.). The lines separating the different regions, known as phase limits, show pressure and temperature values of where phases can simultaneously coexist in equilibrium. The lines divide the diagram in three distinct regions of solid, liquid, and gas. If we take a representative point in the system located in the solid phase region, we can say the substance is in a solid state. If a point is located in the liquid phase region, the substance will be found in a liquid state. If the point is at the limit of two phases, or in other words, on the line separating regions, we can say the substance shows two co-existing phases in equilibrium.

Considering a pure solvent in a liquid phase in a closed vessel, vapor pressure in equilibrium with the liquid is called the solvent's vapor pressure (Figure 5.3). Vapour pressure is characteristic to each substance and its behavior can be predicted by the phase limit curve seen in the phase diagram (Figure 5.2.). As molecular temperature increases, they gain enough energy to break existing interactions between a molecule and its neighbors. If a substance is in a liquid state, it will undergo a phase transformation to a gaseous state in which the molecule-molecule interactions are strongly minimized.

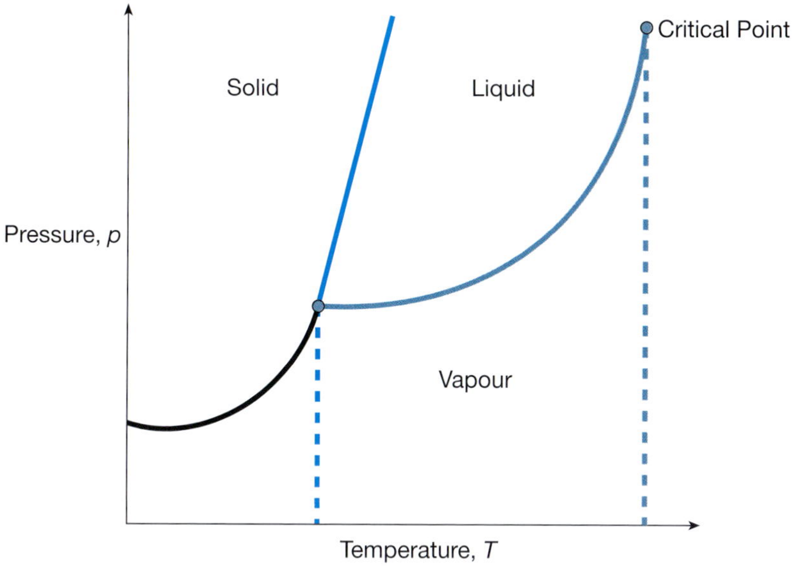

Figure 5.2. Phase diagram of a pure substance showing the coexistence of the three phases: solid, liquid, and gas in relation to temperature and pressure.

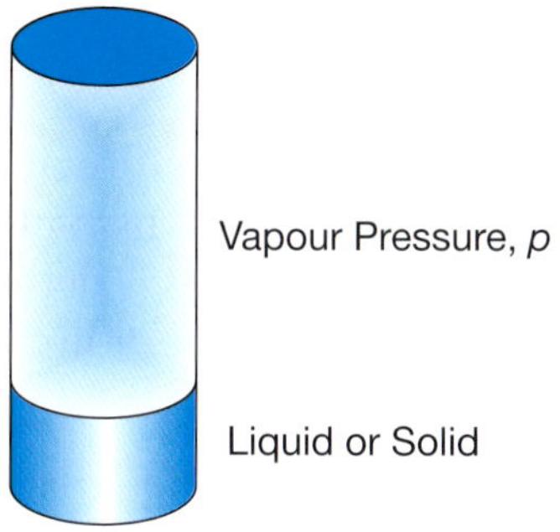

Figure 5.3. Vapour pressure of a liquid or solid maintained in a closed container.

5.8. Critical Point and Boiling Point

When a solvent in a liquid state is heated in an open vessel, the liquid evaporates from the surface. At the temperature where vapour pressure is equal to external pressure, or atmospheric pressure, evaporation can occur from the bulk of the solution, and thus, vapour expands freely to its neighbors. Free evaporation of a liquid phase in an open vessel is said to be boiling and the temperature at which vapour pressure of a liquid is equal to external pressure is called the boiling temperature. When external pressure is at 1 atm, boiling temperature is defined as boiling point.

On the other hand, if a liquid is heated in a closed container (Figure 5.4), it does not boil. Instead, it apears that pressure continuously increases with system temperature. At the same time, liquid density decreases, causing the boundary between liquid and gas states to disappear. The temperature at which the limit between phases disappears is called critical temperature (T_c). Similarly, when vapour pressure reaches critical temperature, this point is defined as critical pressure (p_c). Above this point there is only one phase called supercritical fluid.

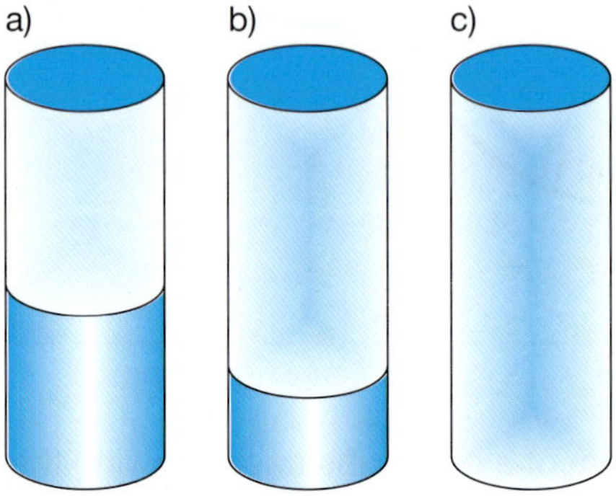

Figure 5.4. (a) Liquid in equilibrium with vapour; (b) When a liquid is heated in a closed cup, the vapor density increase and the liquid decrease reaching a certain stage; (c) Such that the densities of bath phases (liquid and vapor) are equal the interface between the fluids disapear.

The boiling point of a liquid is the temperature at which its vapour pressure is equal to atmospheric pressure, so it can be said that boiling point depends on pressure applied to a given medium, and generally, this is determined at 1 atm. A substance that boils at 150 °C and 1 atm will boil at a temperature significantly lower if pressure is reduced.

Molecule (or ion) transformation from a liquid state to a gas state creates an important separation with the molecules themselves. Due to this, many organic compounds frequently decompose before they reach boiling point. The need for thermal energy to completely separate ions is so high that the majority of the chemical reactions, like decomposition, occur before a substance is made volatile.

Nonpolar compounds, where intermolecular forces are weaker, boil at lower temperatures; even at room temperature. However, this is not always true since molecular weight and size have a large amount of influence. Very heavy molecules need a quite a bit of thermal energy to acquire enough velocity to escape the liquid's surface. Additionally, since molecule surface area is quite large, van der Waals intermolecular attraction will also be very strong, causing the boiling point to increase.

At this time, more specific parameter aspects that influence boiling temperature of some organic compound classes will be discussed. Boiling point of linear alkanes, for example, increases regularly with the increase of molecular weight. At room temperature 25 °C and 1 atm of pressure, the first 4 members of the homologous series of alkanes are gases. For those carbon chains that have from 5 to 17 atoms they are in a liquid state, and those with more than 18 carbons are solids.

Otherwise, the presence of branching in alkane carbon chains, the boiling point in most cases, decreases. Part of the explanation for these effects are related to van der Waals interactions. Considering the unbranched alkanes, the molecular size increases with molecular weight, rendering the surface area less important. For the branched chain molecules, the surface area is influenced more significantly by van der Waals interaction because the molecules are so close to each other. This leads to increasing interactions and in turn, the boiling temperature decreases.

With alcohols, physical properties can be understood in a more simple way if we recognize the fact that structurally, an alcohol is formed by a portion similar to an alkane and another similar to water. The group similar to alkane (C_nH_{2n}) is lipophilic (affinity for nonpolar molecules), while the hydroxyl group (—OH) is hydrophilic (affinity for water). It is the —OH group provides physical properties of alcohols, while the alkyl group modifies these properties according to its carbon chain length. The —OH is highly polar and capable of forming hydrogen bonds with other molecules of the same alcohol, with other neutral and anion molecules.

The boiling point of ethers is comparable to that of hydrocarbons with the same molecular weight. For example, the boiling point of diethyl ether is 34.6 °C and that of pentane is 36 °C (molecular weight equal to 74 and 72 respectively). On the other hand, alcohols show a boiling point higher than ethers and hydrocarbons. Very strong dipole-dipole attractions occur between hydrogen atoms and more electronegative atoms (O, N, or F) and with any other free electron pair of the electronegative atoms. The hydrogen bond (bond dissociation energy between 4 and 36 $kJ \cdot mol^{-1}$) is weaker than a covalent bond, but stronger than a dipole-dipole interaction occurring in ketones. Generally the fact that ethanol shows a higher boiling point (78.5 °C) than dimethyl ether (–24.9 °C), which has the same molecular weight, is attributed to hydrogen bonds. In other words, in ethanol molecules, the hydrogen atoms are covalently bonded to oxygen atoms, making them stronger. However, in dimethylether molecules, there are no hydrogen atoms bonded to oxygen atoms and this results in only weak dipole-dipole intermolecular forces.

Carbonyl compounds – aldehydes and ketones – as the name itself implies, presents a carbonyl group in its chain, giving it its polarity since the oxygen atom is more electronegative than the carbon. Due to the polar group, these molecules have a higher boiling point than hydrocarbons with the same molecular weight. However, since aldehydes and ketones have no strong hydrogen bonds with molecules, they have a lower boiling point than that of alcohols with the same weight.

Methyl iodide (bp = 42 °C) is the only monohalomethane that is liquid at 1 atm and 25 °C. Bromide (bp = 38 °C) as well as ethyl iodide (bp = 72 °C) are liquids, but ethyl chloride (bp = 12 °C) is a gas. Except in cases cited above, halogenated solvents – alkyl chlorides, bromides, and iodides – generally are liquids at room temperature and pressure, and tend to a have boiling point close to that of alkanes with similar molecular weights. However, polyfluoroalkanes tend to have unconventional boiling points. Hexafluoroethane has a molecular weight of 138 and boils at –79 °C, while decane which has a molecular weight of 144 and boils at 174 °C.

5.9. Melting Point and Triple Point

Crystalline solids, particles, ions, or molecules constituting structural units have absolute regular symmetrical configurations corresponding to the regular repetition of a geometric cell unit. Melting represents the transition from this highly ordered state, which is characterized by the existence of a crystalline network, to another state where molecules are less ordered, that is, a liquid state. Melting is produced when the system reaches the temperature at which the thermal energy of the molecule overcomes the action of intracrystalline forces that keep them in their configuration.

The structural units in ionic crystal compounds are the ions. Sodium chloride, for example, is composed of positive sodium ions, and negative chloride ions, alternating with absolute regularity. Around each positive ion there are six equally-spaced negative ions. Analogously, around each negative ion there are six positive ions. The crystal's rigid structure, which is extremely resistant, is explained by the high intensity of the electrostatic forces that keep the ions in their position. In order to overcome the intense inter-ionic forces, the temperature needs to be considerably increased. The melting point for sodium chloride is 801 °C.

Crystal compounds in which atoms are covalently bonded or in non-ionic compounds, as is the case for solvents, the structural units are the molecules. Forces that must be overcome in order to melt these crystals are those that keep the molecules aggregated. These intermolecular forces in general are weaker than inter-ionic forces.

In order to reach sodium chloride's melting point, enough energy needs to be supplied to destroy ionic bonds between Na^+ and Cl^- ions. However, to reach ethanol's melting point, for example, we do not need to provide energy to destroy covalent bonds in the molecules; we only need enough energy to separate the molecules from each other. At the extreme opposite of sodium chloride, ethanol melts at –115 °C.

As noted above, the force holding molecules together (van der Waals interactions) are weak and have a short range, only acting among neighboring molecular parts they come in contact with, or in other words, between molecule surfaces. It should be expected that in a compound family, the larger the molecule, and therefore its surface area, greater will be the forces that will act on them increasing the melting point. However, beyond molecular size, we do not know how they fit in the crystalline network.

From a thermodynamic view, the melting temperature is that under a given pressure which allows simultaneous coexistence of a pure substance in liquid and solid phases. Since a substance melts exactly at the same temperature that it freezes, the melting and freezing points have the same temperature. The freezing temperature at 1 atm of pressure is called the normal freezing temperature, also known as normal melting point.

There is a set of parameters where solid, liquid, and gas phases can exist simultaneously in equilibrium known as the triple point. Geometrically speaking, the triple point is at the intersection of three lines that define the phase boundary in the phase diagram (Figure 5.4). There is only one point at which temperature and pressure define the triple point, and this property is an intrinsic characteristic of each substance. The triple point of water is 273.16 K and 611 Pa, and its three phases (solid, liquid, and vapor) coexist in equilibrium and at no other temperature and pressure conditions is this possible. It is interesting to note that the triple point defines the lowest pressure at which the liquid phase can exist.

In general, we can say that the phase diagram of a substance, and here we can extend the observation to solvents, allows us to accurately determine temperature and pressure where phase transitions occur. These properties are intrinsic characteristics of each substance, and consequently, they are intimately correlated with bond polarity intensity and intermolecular interaction.

5.10. Dielectric Constant

When two charges q_1 and q_2 are separated by a distance r in a vacuum, potential energy of these interactions is

$$V = \frac{q_1 \cdot q_2}{4 \cdot \pi \cdot \varepsilon_0 \cdot r} \tag{5.2}$$

When these same two charges are immersed in a liquid or gas, their potential energy is reduced to

$$V = \frac{q_1 \cdot q_2}{4 \cdot \pi \cdot \varepsilon \cdot r} \tag{5.3}$$

where ε is the medium permittivity. Permittivity is normally expressed in dimensional terms of relative permittivity, ε_r, or the dielectric constant of the medium.

$$\varepsilon_r = \frac{\varepsilon}{\varepsilon_0} \tag{5.4}$$

The dielectric constant of a substance is high if the molecules are polar or highly polarizable. The quantitative relationship between dielectric constant and molecule electric properties is obtained considering the medium polarization; it is expressed by the Debye equation:

$$\frac{\varepsilon_r - 1}{\varepsilon_r + 2} = \frac{\rho \cdot P_m}{M} \tag{5.5}$$

Where ρ is the density, M is the molecular molar mass and P_m is the polarization:

$$P_m = \frac{N_A}{3 \cdot \varepsilon_0}\left(\alpha + \frac{\mu^2}{3 \cdot k \cdot T}\right) \tag{5.6}$$

The term $\mu^2/3 \cdot k \cdot T$ is derived from the thermal motion of electric dipole moment in the presence of an applied field.

We can then define the dielectric constant (ε) as the relationship between the capacitance of the capacitor filled with a standard liquid and the capacitance of the same capacitor in a vacuum. Since it is a relative constant, it is non-dimensional and frequently measured at 20 °C or 25 °C. The dielectric constant of nonpolar solvents decreases linearly with the increase of temperature mainly because of the decrease in density. For polar solvents, it is non-linear, due to the decrease in density and the increase of the thermal motion of molecules. In practice, knowledge of the dielectric constant is closely correlated with the ability of a solvent or solvent mixture to dissolve or dilute a determined solid or other liquid.

To overcome electrostatic forces that maintain the ionic network in ionic solids or intermolecular interactions in molecular compounds with covalent bonds, a considerable amount of energy is required. Only water or other polar solvents are capable of dissolving/diluting polar molecules. Consequently, there are attractions between positive and negative regions of a polar molecule with a positive and negative side. These attractions are called ion-dipole bonds and although individually they are relatively weak, together they supply enough energy to overcome existing inter-ionic forces in crystalline solids and molecular compounds.

In solutions, each solute ion or molecule is surrounded by several solvent molecules; in other words, ions or molecules are solvated. In specific cases where the solvent is water, we say the ion or molecule is hydrated. Generally, we say an ionic compound is going to be dissolved if the solvent used has a dielectric constant capable of reducing or even isolating the attraction between oppositely charged ions, through solvation. Figure 5.5 is a schematic representation of the solvation effect on ions by the polar molecules of the solvent.

The superiority of water as a solvent for ionic substances is due, in part, to the respective polarity as well as the high dielectric constant. It is interesting to note, however, there are other solvents with high dipole moments and high dielectric constants, but they are inadequate to solubilize ionic compounds. In this case, solvation power is required. For a more detailed study of solvent structures and what determines substance solvation power, water is used as an example.

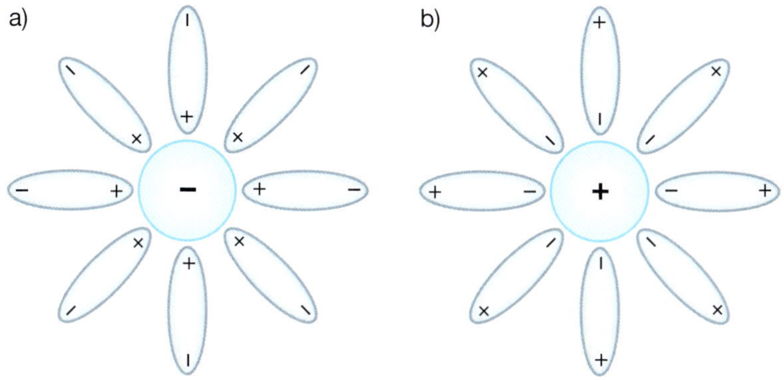

Figure 5.5 Ion-dipole interaction: solvated negative (a) and positive (b) ions by a solvent's polar molecules.

Positive ions or positive regions in molecular compounds are attracted by the polar negative charge of the solvent. In water, the negative pole is, undoubtedly, an oxygen atom, which is a highly electronegative element, and in addition, it has free electron pairs and only two hydrogen atoms bonded to it. However, negative ions or negative regions of molecular compounds are attracted to the positive charges of a polar molecule. In water molecules, these negative regions are found in hydrogen atoms. The ion-dipole bonds that link negative ions to water are hydrogen bonds. Solvents, like water, are called protic solvents; that is, solvents have hydrogen atoms bonded to oxygen or nitrogen atoms. This group of substances solvates ions or molecules as water does: positive ions or positive regions of molecular compounds through unshared electron pairs and negative ions or negative regions of molecular compounds through hydrogen bonds.

In recent years, a high nonprotic solvent has been developed, or in other words, polar solvents presenting moderately high dielectric constants, but without hydrogen atoms. In these solvent classes, the negative charge is concentrated in the most exposed oxygen atom in the molecule, where through free electron pairs, positive ions or positive regions of the molecules will be solvated. The positive pole, on the other hand, is hidden inside the molecule, thus, negatively charged ions or negative regions of the molecules will be weakly solvated. Therefore, nonprotic solvents dissolve or dilute ionic compounds, especially by solvation of positively charged ions.

5.11. Viscosity

The ability of a liquid to flow is measured by its viscosity. In this context, a liquid presenting a high viscosity is a liquid where the ability to cover a surface is lower, a term known as liquid mobility. Glasses and melted polymers are highly viscous because their molecules become overlapped, impeding fluidity. Water is more viscous than benzene;

its molecules interact with more intensity, thus making flow difficult. An increase in temperature is generally accompanied by a decrease in viscosity due to the molecule's increase in motion, thus allowing molecules to escape easier from neighboring molecular interactions. Because a molecule changes from one position to another, it implies a rupture of weak van der Waals interactions, and therefore a portion of the molecules with enough energy to move, follow the Boltzmann distribution. This suggests that the capacity of liquid to flow should behave according to the equation:

$$\text{Fluidity} \propto e^{-\Delta E/RT} \tag{5.7}$$

where E is the energy required to overcome liquid flow, R is a constant and T is the temperature. Viscosity is the inverse of Fluidity.

$$\text{Viscosity} \propto e^{\Delta E/RT} \tag{5.8}$$

Experimental observations show viscosities obey an exponential shape in a limited temperature range and the value of ΔE is similar to interaction energy or molecular bonding (some $kJ \cdot mol^{-1}$, for example, 11 $kJ \cdot mol^{-1}$ to benzene and 3 $kJ \cdot mol^{-1}$ for methane).

5.12. Refractive Index

When light crosses a given medium, the refraction phenomena occurs. To visualize this effect, put a glass, wooden or plastic stick in water (Figure 5.6). If the stick was put in perpendicularly, we see that it will remain vertical in water. However, if we slightly incline the stick and then look at it through the side of the glass, the stick seems to be broken. This is because the beam of light changes direction when it crosses the liquid's surface due to the difference between the velocity of light in both the mediums. This change in direction is known as refraction.

Figure 5.6. Light refraction phenomena ilustration.

This direction change when it crosses two distinct media depends on the speed of light in both media. The relative refractive index (n_{21}) is the physical value related to the speeds in two media. It is defined as being the relationship between the speed of light in the first medium (v_1) and the second medium (v_2).

$$n_{21} = \frac{v_1}{v_2} \tag{5.9}$$

When the first medium is the vacuum, v_1 is equal to the speed of light in a vacuum (c). The relationship between light speed in a vacuum with any other medium is called the absolute refractive index (n).

$$n = \frac{c}{v} \tag{5.10}$$

As light speed in a vacuum is always greater than any other medium, the refractive index value in any medium is always going to be greater than the unit. For example, light speed in glass is 200,000 km $\cdot$ s^{-1}, thus the refractive index for glass will be equal to 1.5, or in other words, 300,000 km $\cdot$ s^{-1} divided by 200,000 km $\cdot$ s^{-1}.

The Refraction Law was discovered by Snell in 1621 and republished in 1637 by Descartes. That is why it is also called the Snell-Descartes Law. It relates the angles of incidence and refraction with the refractive index and is expressed as:

$$n_1 \sin(\theta_1) = n_2 \sin(\theta_2) \tag{5.11}$$

where n_1 and n_2 are respectively refractive index in media 1 and 2, θ_1 is the incidence angle, and θ_2 is the refraction angle.

The refractive index (n) is one of the most used parameters in identifying organic solvents. Only under certain conditions is the refractive index measured at different temperatures of 20 or 25 °C. The most used notation to designate temperature is the exponent after the symbol (ex. n^{20}). To ensure data precision to the fourth decimal place, the temperature should be constant with a ± 0.2 standard deviation. Further, the refractive index depends a lot on the light wavelength used in the measurement; with an increase in wavelength, refractive index decreases (and vice-versa). Refractive index measurements are normally quick and easy using the Abbe refractometer, or modern digital refractometers. Refractometry, the name given to the measurement technique of the refractive index, is very popular and used by several quality control laboratories.

5.13. Surface Tension

The existence of short range van der Waals attraction forces between molecules is a well-known fact responsible for the existence of the liquid state. Surface tension and interface tension phenomena are explainable in terms of these forces. Molecules situated in the liquid bulk are subject to equal attraction forces in all directions, whereas liquid-air surface molecules are under action of unbalanced attraction force resulting in a force that tends to pull molecules to the bulk of the liquid (Figure 5.7). Due to the attraction of these unbalanced forces, a large part of these molecules will tend to move from the surface to the bulk of the liquid; and because of this effect, the surface will tend to spontaneously contract. This phenomenon also explains the fact that droplets tend to acquire a spherical shape.

Surface tension performs a very important physicochemical role. Surface tension of a determined liquid is often defined as the force that acts orthogonally at any unit segment imagined on the liquid surface. However, this definition is more appropriate for liquid films. It is more convenient to define surface tension (γ) and the free surface energy as the work (w) necessary to increase a surface in one area unit (A_{sup}), through an isothermic reversible process. The same considerations are valid for separation surface between two immiscible liquids. In this case, we will also have unbalanced intermolecular forces but with less intensity.

$$w = \gamma \cdot A_{sup} \tag{5.12}$$

This model considers the liquid remaining in static conditions. However, it should be noted that at a molecular level, a liquid surface apparently at rest can, in reality, be found in a state of high perturbation, as a result from the Brownian motion of molecules between the bulk and the surface, and the surface and the atmosphere. Just to have a number in mind, molecule permanence on a liquid surface is on the order of 10^{-6} s.

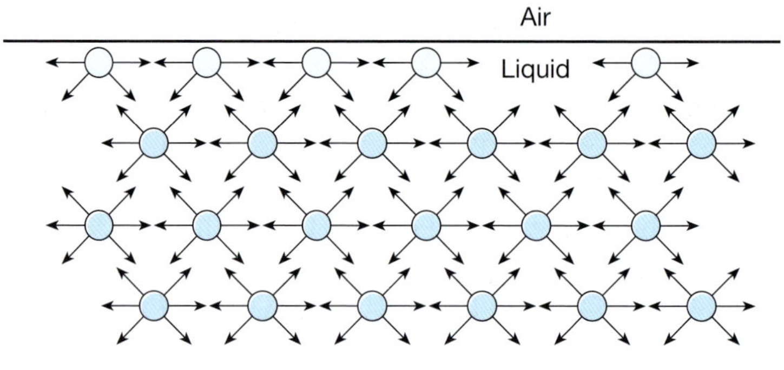

Figure 5.7. Attraction forces between a liquid's molecules.

5.14. Density

Absolute substance density is defined as the ratio between the mass and the volume it occupies:

$$\rho = m/V \tag{5.13}$$

where ρ is density, m is substance mass, and V is volume. Thus, the density of any liquid can be determined by carefully weighing a given quantity and then measuring its volume. Further, care should be taken in controlling the temperature on which the measure is being made since any temperature variation can considerably vary the density. Although on the International System the density unit is kilogram per cubic meter ($kg \cdot m^{-3}$), the better known unit is grams per milliliters ($g \cdot ml^{-1}$).

Density is also influenced by intermolecular forces previously described in the explanation of the physicochemical properties of the substances. As intensity of interactions increases, the attraction tendency increases between molecules, or in other words, the number of molecules per volume unit increases leading to the increase in substance density. On the other hand, as intermolecular forces decrease, the quantity of molecules per volume unit will be lower and thus causing a decrease in substance density.

We can also define relative density; that is, the ratio between substance absolute density and a standard absolute density. Since the relative density is the quotient between these two absolute density values, its value is an adimensional number

$$d = \rho / \rho^{o} \qquad (5.14)$$

The relative density is a physical characteristic property inherent in the substance, which can be used to determine a liquid's degree of purity. Since its value is the result of interactions between molecules of the same substance, if this substance is contaminated, resultant interactions between molecules will be different from those observed in a pure substance. The presence of impurities causes a change in the amount of a substance's molecules per volume unit, thus modifying density value. Therefore, density is a measurement used extensively by quality control laboratories to show whether or not solvents have been adulterated.

Following are several tables giving the physicochemical properties discussed in this chapter related to the most used industrial solvent classes. Despite their being found in literature, our idea was to group some information for a quick consult that can be used in choosing the most appropriate solvent for several applications.

Table 5.1. Physical Properties of Solvents

Units	Mol $g \cdot mol^{-1}$	Structure	b.p. (°C)	m.p. (°C)	p (mmHg)	ρ $g \cdot cm^{-3}$ (20 °C)	n_d^{20}
ALCOHOLS							
Methanol	32.04	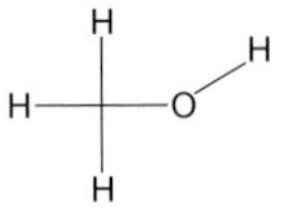	64.5	–97.8	100 (21 °C)	0.791	1.326
Ethanol	46.07	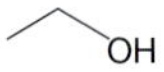	78.3	–114.1	40 (19 °C)	0.790	1.359
n-Propanol	60.09		97.2	–127.0	20.8 (25 °C)	0.803	1.383
Isopropanol	60.11	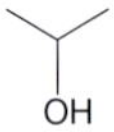	82.5	–88.9	44.0 (25 °C)	0.785	1.375
n-Butanol	74.12	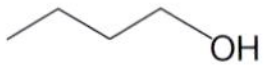	117.7	–89.0	5.5	0.809	1.397
Isobutanol	74.12	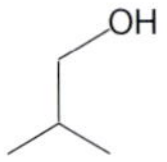	107.8	–108.0	10.0 (22 °C)	0.806	1.394
Methyl Isobutyl Carbinol	102.18	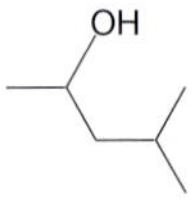	131.8	–90.0	2.8	0.808	1.409
2-ethyl-hexanol	130.23	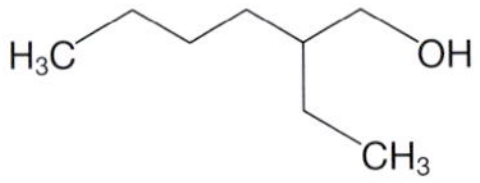	184.8	–76.0	30.0 (98 °C)	0.830	1.429
Cyclohexanol	100.16	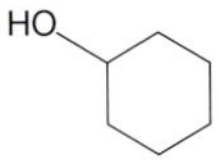	161.0	24.0	80.0 (25 °C)	0.968	1.464

η cP (20 °C)	γ dina·cm^{-1} (20°C)	Flash point open vase (°C)	Flash point close vase (°C)	Solubility in water % mass (20 °C)	Water solubility in solvent % mass (20 °C)	Auto ignition (°C)	Evaporation rate n-Butyl Acetate = 100	Distilation range (°C)
0.58	22.5	15.6	11.1	complete	complete	464.0	181	64.0-65.0
1.20	22.39	15.6	8.9	complete	complete	423.0	150	78.3-78.5
2.26	23.71	29.4	15.0	complete	complete	412.0	89	96.0-98.0
2.41	21.32	21.0	12.0	complete	complete	399.0	135	81.5-83.0
3.0	24.52	37.2	35.0	7.9	20.1	365.0	46	116.5-118.5
4.0	23.0	31.1	27.8	9.5	16.9	415.0	62	106.9-108.9
5.2	22.8	55.0	41.1	1.64	6.35	–	29	130.0-133.0
9.8	26.7	–	77.3	0.1	2.6	231.0	1.9	183.0-185.0
41.07 (30 °C)	33.4	67.8	67.7	0.13	11.78	300.0	5.8	160.0-162.0

Continued…

Table 5.1. (continued)

Units	Mol $g \cdot mol^{-1}$	Structure	b.p. (°C)	m.p. (°C)	p (mmHg)	ρ $g \cdot cm^{-3}$ (20 °C)	n_d^{20}
KETONES							
Acetone	58.08	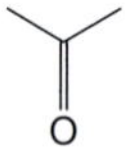	56.2	–94.6	184.5 (20 °C)	0.790	1.359
Methyl Ethyl Ketone	72.11	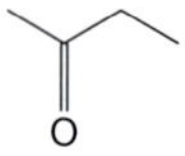	79.6	–86.7	90.96 (25 °C)	0.805	1.379
Methyl isobutyl Ketone	100.18	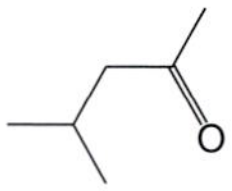	115.9	–83.9	15.7 (20 °C)	0.800	1.396
Diisobutyl ketone	142.24	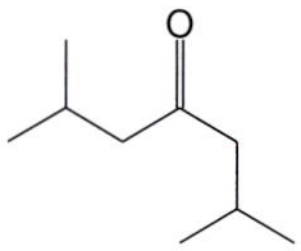	169.3	–41.5	4.0 (30 °C)	0.810	1.413
Diacetone alcohol	116.16	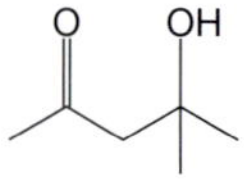	167.9	–42.8	1.23 (20 °C)	0.940	1.415
Isophorone	138.20	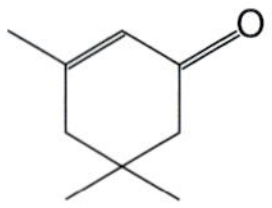	215.2	–8.1	0.43 (25 °C)	0.923	1.476
Cyclohexanone	98.14	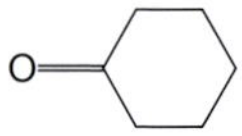	156.7	–47.0	4.6 (25 °C)	0.948	1.450
Methyl n-amyl ketone	114.18	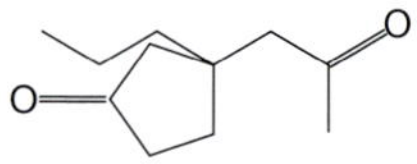	151.4	–26.9	7.0 (30 °C)	–	1.408
Methyl n-propyl-ketone	86.13	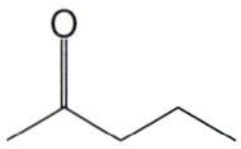	102.3	–77.5	16.0 (25 °C)	0.809	1.391
Isoamyl methyl ketone	114.18	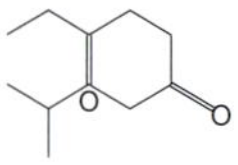	144.9	–73.9	4.5 (20 °C)	0.812	1.407
Acetophenone	120.15	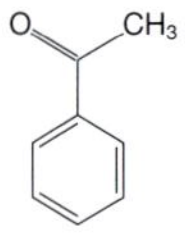	201.6	19.7	0.28 (20 °C)	1.03	1.532

η cP (20 °C)	γ dina·cm^{-1} (20 °C)	Flash point open vase (°C)	Flash point close vase (°C)	Solubility in water % mass (20 °C)	Water solubility in solvent % mass (20 °C)	Auto ignition (°C)	Evaporation rate n-Butyl Acetate = 100	Distilation range (°C)
0.33	23.38 (22 °C)	–15.5	–18.0	complete	complete	538.0	520	55.6-56.6
0.4	24.49 (25 °C)	–5.6	–3.3	27.0	12.5	516.0	340	78.0-81.0
0.59	23.6 (20 °C)	23.0	13.3	1.7	1.9	460.0	155	114.0-117.0
1	23.92 (22 °C)	48.9	48.9	0.05	0.75	396.0	21	163.0-173.0
3.2	24.6 (20 °C)	62.2	54.0	complete	complete	602.0	12	155.0-175.0
2.6	32.2 (20 °C)	96.1	84.4	1.2	4.3	460.0	2.5	215.0-220.0
2.2	32.32 (20 °C)	54.4	46.7	2.3	8.0	420.0	31	153.2-157.2
0.7	26.17 (25 °C)	48.8	47.2	0.43	1.5	393.0	40	150.0-154.0
0.5	23.26 (25 °C)	14.4	12.2	3.1	4.2	504.0	88	101.0-105.0
0.8	25.03 (28 °C)	42.2	41.1	0.5	1.2	191.0	53	141.0-148.0
1.8	39.5 (20 °C)	93.3	82.2	0.55	1.65	465.0	3	196.0-202.0

Continued…

Table 5.1. (continued)

Units	Mol $g \cdot mol^{-1}$	Structure	b.p. (°C)	m.p. (°C)	p (mmHg)	ρ $g \cdot cm^{-3}$ (20 °C)	n_d^{20}
ESTHERS							
Ethyl Acetate	88.12	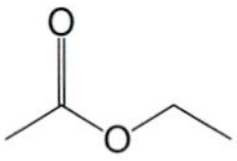	77.0	–83.6	100.0 (27 °C)	0.901	1.370
n-Butyl Acetate	116.18	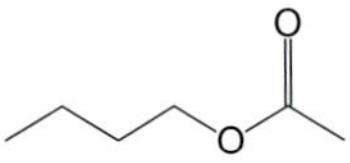	126.5	–76.8	15.0 (25 °C)	0.883	1.392
n-Propyl Acetate	102.13	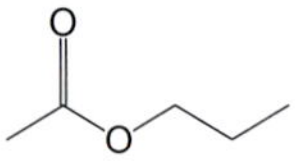	101.6	–92.5	25.21 (20 °C)	0.888	1.383
Isopropyl Acetate	102.13	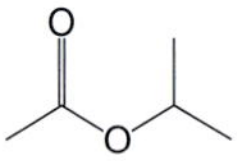	88.7	–73.1	60.59 (25 °C)	0.870	1.375
Isobutyl Acetate	116.16	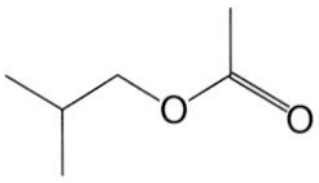	117.2	–99.8	13.0 (20 °C)	0.871	1.390
n-Pentyl Acetate	130.19	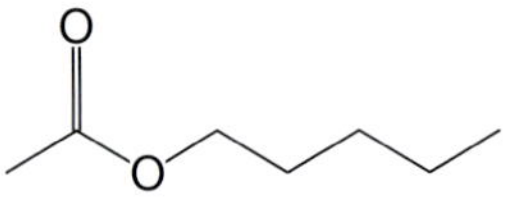	146.0	–100.0	28.5 (20 °C)	0.875	1.401
Methyl Acetate	74.08	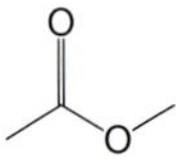	57.1	–98.1	400.0 (40 °C)	0.933	1.358
2-Ethylhexyl Acetate	172.27	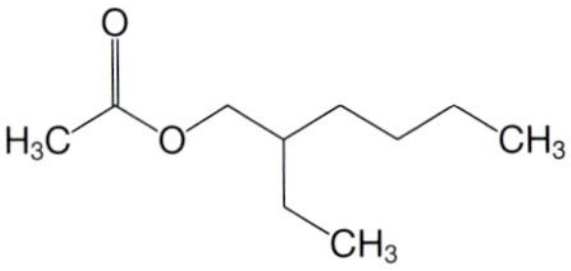	199.0	–80.0	0.4 (20 °C)	0.871	1.417

η cP (20 °C)	γ dina·cm^{-1} (20 °C)	Flash point open vase (°C)	Flash point close vase (°C)	Solubility in water % mass (20 °C)	Water solubility in solvent % mass (20 °C)	Auto ignitionn (°C)	Evaporation rate n-Butyl Acetate = 100	Distilation range (°C)
0.45	23.9 (20 °C)	13.3	–3.3	8.7	3.3	427.0	430	76.0-78.0
0.73	14.5 (25 °C)	32.2	22.2	0.7	1.6	421.0	100	124.0-127.0
0.59	24.28 (20 °C)	18.3	–	2.3	2.6	450.0	226	99.0-103.0
0.6	21.12 (20 °C)	16.6	5.5	2.9	1.8	460.0	355	84.0-90.0
0.700	23.70 (20 °C)	28.3	25.0	0.75	1.64	423.0	145	116.0-119.0
0.45	4.0 (20 °C)	41.1	25	0.2	0.9	–	45	140.0-150.0
–	25.37 (20 °C)	–5.56	–1.11	–9.5	8.2	502.0	660	53.0-59.0
1.500	26.91 (25 °C)	87.8	–	0.03	0.55	268.0	3.7	192.0-205.0

Continued…

Table 5.1. (continued)

Units	Mol $g \cdot mol^{-1}$	Structure	b.p. (°C)	b.p. (°C)	p (mmHg)	ρ $g \cdot cm^{-3}$ (20 °C)	n_d^{20}
Cyclohexyl Acetate	142.19	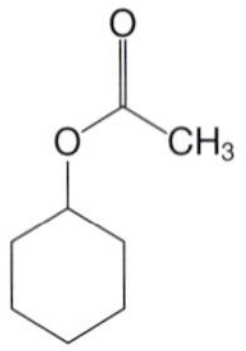	177.0	–65.0	7.0 (30 °C)	0.969	1.439
P.G.M.M.E. Acetate	132.16	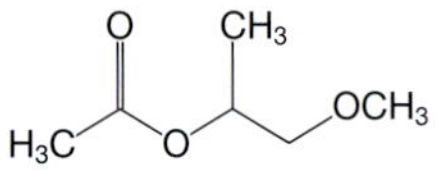	145.8	<–67.0	27.4 (20 °C)	0.966	1.399
Butylglycol Acetate	160.22	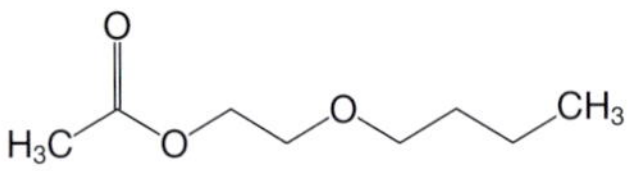	191.6	–64.6	0.35 (20 °C)	0.940	1.413
Ethyl Glycol Acetate	132.16	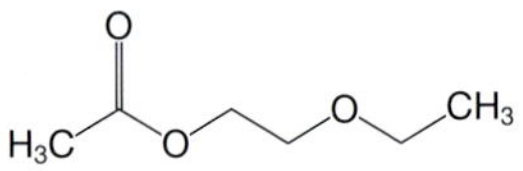	156.3	–61.7	2.0 (20 °C)	0.974	1.402
Diglycol Monobutyl Ether Acetate	204.27	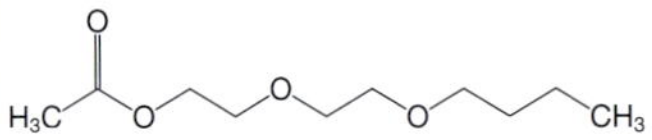	246.0	–32.2	–	0.985	1.423
Diglycol Monoethyl Ether Acetate	176.22	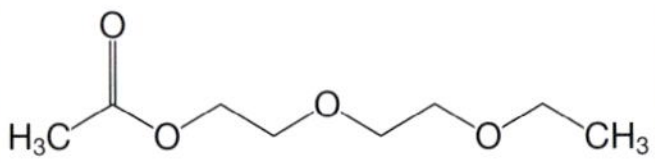	217.4	–25.0	0.10 (20 °C)	1.	1.423
Ethyl Lactate	118.14	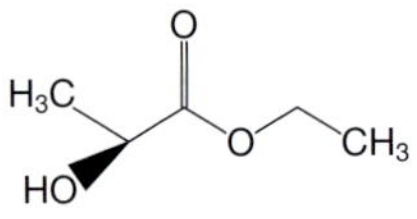	154.0	–25.0	1.03	50.0 (100 °C)	1.412

η cP (20 °C)	γ dina·cm^{-1} (20 °C)	Flash point open vase (°C)	Flash point close vase (°C)	Solubility in water % mass (20 °C)	Water solubility in solvent % mass (20 °C)	Auto ignition (°C)	Evaporation rate n-Butyl Acetate = 100	Distilation range (°C)
3.000	31.31 (20 °C)	64.00	57.80	1.44	0.20	334.0	15	174.0-178.0
1.140	3.7 (20 °C)	49.40	–	19.80	3.21	–	35	140.0-150.0
1.800	–	87.80	76.10	1.50	1.70	340.0	3.7	188.0-192.0
1.300	31.8 (25 °C)	55.00	–	23.80	6.50	379.0	20	150.0-160.0
3.600	–	115.60	112.70	6.50	3.70	–	0.14	235.0-250.0
2.800	29.31 (40 °C)	110.00	95.00	completa	completa	360.0	0.63	214.0-221.0
2.610	29.2 (20 °C)	54.40	–	completa	completa	400.00	21	140.0-163.0

Continued…

Table 5.1. (continued)

Units	Mol $g \cdot mol^{-1}$	Structure	b.p. (°C)	b.p. (°C)	p (mmHg)	ρ $g \cdot cm^{-3}$ (20 °C)	n_d^{20}
GLYCOLS							
Diethylene Glycol	106.1	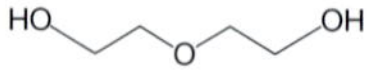	245.8	–7.8	1.0 (92 °C)	1.118	1.446
Propylene Glycol	76.1	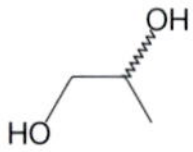	187.3	–60.0	0.07 (20 °C)	1.038	1.432
Ethylene Glycol	62.1	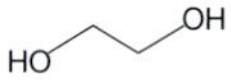	197.6	–12.7	0.06 (20 °C)	1.115	1.430
Hexylene Glycol	118.2	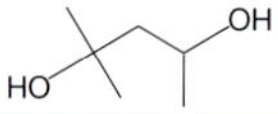	197.0	–40.0	0.05 (20 °C)	0.922	1.426
Dipropylene Glycol	134.2	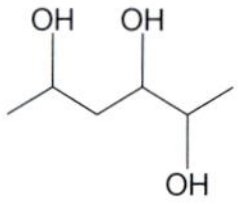	232.8	–4.4	0.01 (20 °C)	1.023	1.440
GLYCOL ESTERS							
Ethylene Glycol t-Butyl Ether	118.2	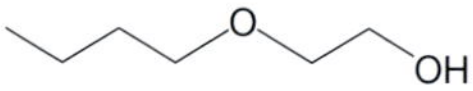	171.2	–70.0	0.76 (20 °C)	0.901	1.418
Ethylene Glycol Ethyl Ether	90.1	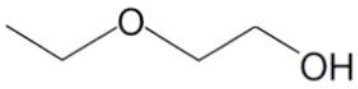	135.1	–76.0	5.29 (25 °C)	0.931	1.405
Ethylene Glycol Methyl Ether	76.1		124.5	–85.1	6.2 (20 °C)	0.966	1.400
Diethylene Glycol Monobutyl Ether	162.2	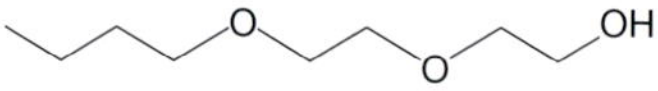	230.6	–68.1	0.02 (20 °C)	0.955	1.423
Diethylene Glycol Monoethyl Ether	134.2		202.7	–76.0	0.13 (25 °C)	0.989	1.425
Diethylene Glycol Monometyl Ether	120.2	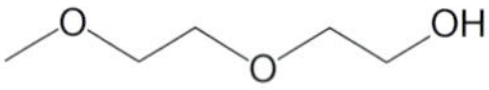	194.2	–85.0	0.25 (25 °C)	1.021	1.424
Ethylene Glycol i-Butyl Ether	118.2	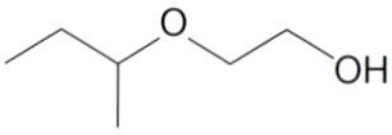	160.5	–87.0	26.0 (71 °C)	0.893	1.416
P.G.M.M.E.	90.1	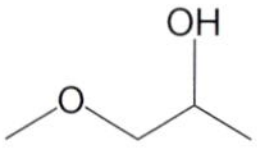	120.1	–96.6	11.8 (25 °C)	0.919	1.402
D.P.G.M.M.E.	148.2	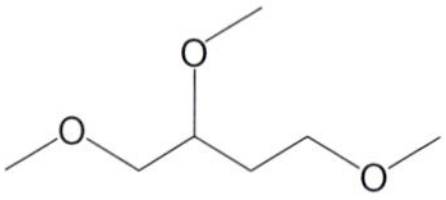	188.3	–82.0	0.40 (26 °C)	0.951	1.422
T.P.G.M.M.E.	206.6	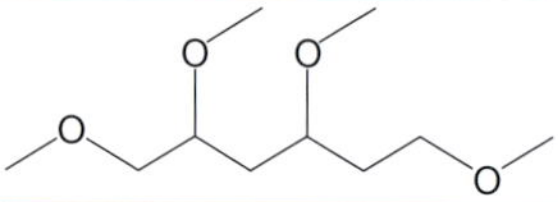	242.4	–78.9	2.0 (100 °C)	0.965	1.427

η cP (20 °C)	γ dina·cm^{-1} (20 °C)	Flash point open vase (°C)	Flash point close vase (°C)	Solubility in water % mass (20 °C)	Water solubility in solvent % mass (20 °C)	Auto ignition (°C)	Evaporation rate	Distilation range (°C)
							n-Butyl Acetate = 100	
36.000	48.43 (20 °C)	143.0	143.0	complete	complete	229.0	<0.1	242.0-250.0
60.500	35.46 (30 °C)	–	124.0	complete	complete	427.0	<1	185.0-189.0
21.000	48.43 (20 °C)	115.5	112.7	complete	complete	400.0	<1	193.0-201.5
34.400	33.1 (20 °C)	102.0	93.8	complete	complete	270.0	<1	196.0-199.0
107.000	32.8 (25 °C)	137.7	–	complete	complete	–	<0.1	228.0-236.0
6.400	27.4 (25 °C)	73.9	60	complete	complete	238.0	6.8	169.0-173.0
2.100	28.2 (25 °C)	54.4	48.4	complete	complete	235.0	39	132.0-136.0
1.700	33.0 (20 °C)	46.1	41.6	complete	complete	285.0	58	124.0-125.0
6.500	34.0 (20 °C)	115.6	100	complete	complete	228.0	0.35	220.0-235.0
4.500	31.18 (25 °C)	96.1	95.1	complete	complete	204.0	1.3	198.0-204.0
3.900	34.84 (101 °C)	93.3	83.1	complete	complete	193.0	2	188.0-198.0
5.500	–	59.4	–	complete	complete	–	11	157.0-162.0
1.700	27.7 (20 °C)	37.7	33.9	complete	complete	–	71	117.0-125.0
3.400	28.8 (20 °C)	85	74.4	complete	complete	–	3	184.0-193.0
5.600	30.0 (25 °C)	126.6	112.7	complete	complete	–	<1	236.0-251.0

Continued...

Table 5.1. (continued)

Units	Mol $g \cdot mol^{-1}$	Structure	b.p. (°C)	b.p. (°C)	p (mmHg)	ρ $g \cdot cm^{-3}$ (20 °C)	n_d^{20}
HALOGENATED							
Trichloroethylene	131.4	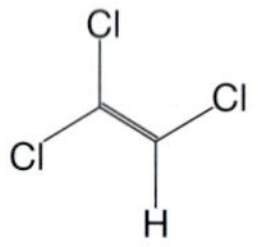	86.7	–	57.8 (20 °C)	1.464	1.475
Methylene Chloride	84.9		40.4	–	400.0 (21 °C)	1.326	1.421
1,1,1 Ethane Chloride	113.4	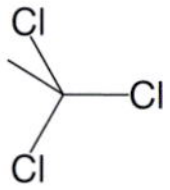	74.0	–	100.0 (25 °C)	1.338	1.431
Perchloroethylene	165.8	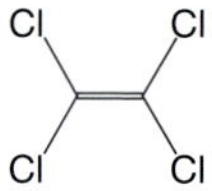	121.2	–22.3	184.7 (25 °C)	1.622	1.505
AROMATIC HIDRO CARBONS							
Toluene	92.1		110.5	–95.1	36.7 (30 °C)	0.870	1.493
Xilene	106.2	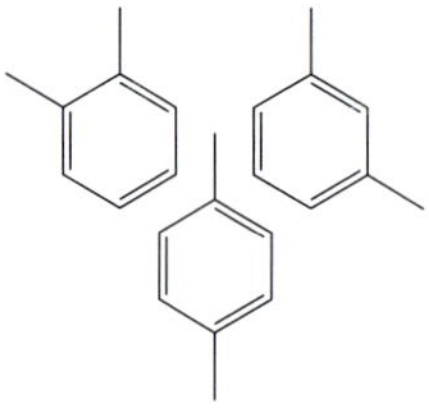	140.0	-45.0	6.72 (21 °C)	0.870	1.497
ALIPHATIC HIDRO CARBONS							
Hexane	86.2		68.7	–	150.0 (20 °C)	0.659	1.372
Heptane	100.2		98.4	–	40.0 (22 °C)	0.684	1.385
Cyclohexane	84.2		80.7	–	10.0 (61 °C)	0.779	1.423

η cP (20 °C)	γ dina·cm^{-1} (20 °C)	Flash point open vase (°C)	Flash point close vase (°C)	Solubility in water % mass (20 °C)	Water solubility in solvent % mass (20 °C)	Auto ignition (°C)	Evaporation rate n-Butyl Acetate = 100	Distilation range (°C)
0.580	29.5 (20 °C)	–	–	0.11	0.033	410.0	450	86.0-87.0
0.425	28.10 (20 °C)	–	–	1.3	0.198	615.0	990	39.0-40.0
0.725 (30 °C)	26.4 (25 °C)	–	–	0.44	0.034	537.0	530	74.0-76.0
0.880	32.22 (20 °C)	–	–	0.015	0.01	–	–	120.0-122.0
0.600	28.52 (20 °C)	8.9	4.4	0.06	0.05	536.0	190	109.0-110.0
0.800	29.48 (20 °C)	31.6	–	0.04	0.05	–	60	136.0-144.0
0.298	18.4 (20 °C)	–	–23.3	0.011	0.001	234.0	620	—
0.396	20.14 (20 °C)	–1.0	–1.1	0.009	0.000	223.0	290	—
0.398	24.6 (20 °C)	–	–18.3	0.006	0.01	260.0	440	—

Bibliographical References

Morrison, R. D.; Boyd, R. N. *Química orgânica*. 8. ed. Lisboa: Fundação Calouste Gulbekian, 1986.

Atkins. *Physical Chemistry*. 6. ed. Oxford University Press, 1998.

Verneret, H. *Solventes industriais* – propriedades e aplicações. Toledo Assessoria Técnica e Editorial, 1984.

Solomons, T. W. G, *Química orgânica*. vol. 1 e 2, 6. ed., 1996.

Shaw, D. J., *Introdução à química de colóides e de superfície*. São Paulo: Blucher, 1975.

Sedivec V.; Flek, J. *Handbook of analysis of organic solvents*. John Wiley & Sons, 1976.

Behring, J. L.; Lucas, M., Machado, C.; Barcellos, I. O. *Adaptation of the drop-weight method for the quantification of surface tension:* a simplified apparatus for the CMC determination in the chemistry classroom. Quím. Nova, 2004, vol. 27, n. 3, p. 492-495.

Gregory W. Kauffman; Peter C. Jurs. *Prediction of surface tension, viscosity, and thermal conductivity for common organic solvents using quantitative structure-property relationships*. J. Chem. Inf. Comput. Sci., 2001, 41 (2), p. 408-418.

6 Solubility Parameters

This chapter deals with the energies involved in the solvent/polymer interaction and the different quantifying methods: Hildebrand, Prausnitz and Blanks, and Hansen solubility parameters.

Denílson José Vicentim
Sérgio Martins

When planning a solution formulation for a resin or resin system, one of the main points to consider is the criteria used to define the solvent system to be used.

Criteria for choosing a solvent system is discussed in Chapter 7. However, one of the fundamental points in determining choice is solvent power in relation to the resin.

Solvent power is the interaction between the polymer(s) and solvent(s).

This chapter presents concepts that govern thermodynamic interactions involved in a system as well as the Hansen solubility theory currently used to predict polymer-solvent interaction in diverse systems.

6.1. Cohesive Energy Density

Considering basic thermodynamics, the change in the energy needed for vaporization of a molecule initially in a liquid state is called enthalpy of vaporization.

When molecules are in a vapor state, and using the perfect gas theory to explain its behavior, i.e., where no intermolecular interaction exists between them, then vaporization energy can be considered a measure of forces between the molecules when in a liquid state.

The correlation between internal energies in a system of polymers and solvents, taking into account vaporization energy as a measurement of intermolecular attraction in a solution, was provided by Hildebrand and Scott in 1950.

They defined Cohesive Energy Density (CED) as the force necessary to maintain molecules together in a liquid per volume unit, determined by vaporization energy.

Using the first law of thermodynamics[1], [4], [6]:

$$dU = \delta U - \delta W \tag{6.1}$$

where:

U – total system energy
Q – heat transferred during a change of state
W – work carried out by the system during change of state.

Entropy of a system is given as:

$$dS = \left(\frac{\delta Q}{T} \right) \tag{6.2}$$

Thus, heat transferred during a change of state is given as:

$$\delta Q = TdS \tag{6.3}$$

The work of expansion against a medium, without change in potential or kinetic energy of a system can be represented by:

$$PdV = \delta W \tag{6.4}$$

Substituting equations (6.3) and (6.4) in equation (6.1), and considering total energy of a system being represented by internal energy E, according to thermodynamics, we arrive at the following correlation:

$$dE = TdS - PdV \tag{6.5}$$

The differentiating equation (6.5) in relation to volume V at constant temperature, and using the Maxwell correlation, gives the thermodynamic equation of state representing the basis of the CED:[6]

$$\left(\frac{\partial E}{\partial V}\right)_T = T\left(\frac{\partial P}{\partial T}\right)_V - P \tag{6.6}$$

6.2. Solubility Parameters

Below is the development of the solubility parameter models leading up to the

6.2.1. Hildebrand Solubility Parameter

The correlation between the CED and solubility was made by Hildebrand and Scatchard, starting with the thermodynamic criteria that two substances are mutually soluble. For this to be true, Gibbs free energy in a system needs to be decreased, i.e.:

$$\Delta_m G \leq 0 \tag{6.7}$$

where:

$\Delta_m G$ – change in Gibbs free energy of a mixture.

However, based on the definition of Gibbs free energy:

$$\Delta_m G = \Delta_m H - T\Delta_m S \tag{6.8}$$

where:

$\Delta_m H$ – change in mixture enthalpy
$\Delta_m S$ – change in mixture entropy

During the mixing of substances, the change in system entropy is generally positive, i.e.:

$$\Delta_m S \geq 0 \tag{6.9}$$

And since absolute temperature (T) is always greater than zero:

$$T > 0 \tag{6.10}$$

The two conditions expressed in equations (6.9) and (6.10) when used in equation (8) show that the term:

$$-T\Delta_m S < 0 \tag{6.11}$$

In other words, this term generally contributes positively to the mixture of the two substances being considered.

Using the same evaluation for the value of $\Delta_m H$ in equation (6.8), and taking into consideration conditions (6.9), (6.10), and (6.11), the conclusion is there are three possible situations for the value of $\Delta_m H$ for there to be solubilization of two of the substances:

a) $\Delta_m H > 0$ – there is a positive limit value for enthalpy, such that the condition expressed in equation (6.7) is satisfied.
b) $\Delta_m H = 0$ – in this case, equation (6.7) will always be satisfied.
b) $\Delta_m H < 0$ – in this case, equation (6.7) will always be satisfied.

Hildebrand and Scatchard (1950) correlated the enthalpy of a mixture ($\Delta_m H$) with the CED using the equation: [6] [7] [8] [9] [10]

$$\frac{\Delta_m H}{V} = \left[\left(\frac{E_1}{V_1}\right)^{1/2} - \left(\frac{E_2}{V_2}\right)^{1/2}\right]^2 \phi_1\phi_2 \tag{6.12}$$

where:

V is the total molar volume of the mixture,
E_1 is the energy of vaporization for species 1 and 2, respectively;
V_1 and V_2 are molar volumes for species 1 and 2, respectively;
E_1/V_1 are CEDs for species 1 and 2;
ϕ_1 and ϕ_2 are volume fractions for species 1 and 2, respectively.

Thus, the Hildebrand solubility parameter is defined as the square root of the CED and is expressed as:

$$\delta = \left(\frac{\Delta E}{V}\right)^{1/2} \tag{6.13}$$

where:

δ represents the Hildebrand solubility parameter.[7] [8] [12]

The main limitation of this theory is the contributions of entropy changes in the mixture are not considered in the solubility evaluation; just those energy changes that can be explained using the definition itself in the equation (6.13).

$$\frac{\Delta_m H}{V} = \left[\delta_1 - \delta_2\right]^2 \phi_1 \phi_2 \qquad (6.14)$$

Upon evaluating equation (6.14), it can be seen that:

δ_1 and δ_2 correspond to solubility parameters of substances 1 and 2.[2] [6]

In order to minimize enthalpy in the mixture of substances 1 and 2, and as a consequence, increase the probability of solubilization, the following conditions need to be satisfied:

- $\Delta_m H \rightarrow 0$: for the enthalpy of a mixture to be zero
- $\delta_1 = \delta_2$: solubility parameters for the two substances need to have close values.

The energy units adopted for δ solubility parameters are expressed as $MPa^{1/2}$.

In evaluating this condition, substances with similar chemical structures present similar solubility parameters and they are miscible with each other. [2]

This thermodynamic definition agrees with the definition popularly used to measure substance solubility of "like dissolves like."

Based on the definitions adopted by Hildebrand, the solubility parameter is associated with the enthalpy of vaporization at a constant temperature. Therefore, energy supplied to a system for the liquid to reach its boiling point is not included in the vaporization enthalpy at a constant temperature.

In fact, the energy considered is the energy that when the liquid reaches boiling point, it is enough to cause the separation of molecules of a substance without leading to any increase in temperature of the system, changing them from a liquid state to a vapor state. Currently, the principal energies keeping molecules of substances close to each other are the van der Waals forces, which are characterized by weak intermolecular interactions mainly for nonpolar bonds present in the molecules.

Below are some molecule structures that have weak intermolecular interactions which characterize the van der Waals interaction.

Table 6.1. Main Intermolecular Interactions Characterized by van der Waals

Compound	Chemical Structure
Hexane	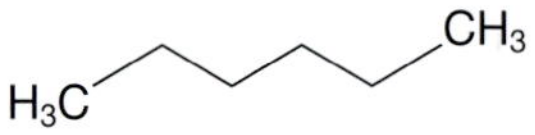
Cyclohexane	
Isooctane	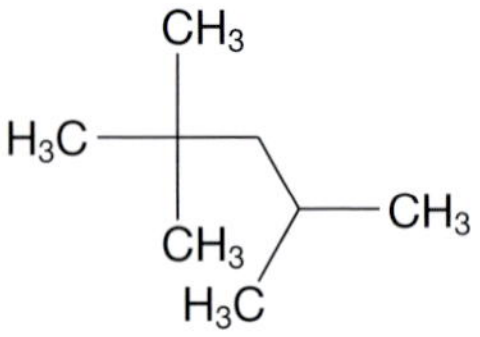
Octane	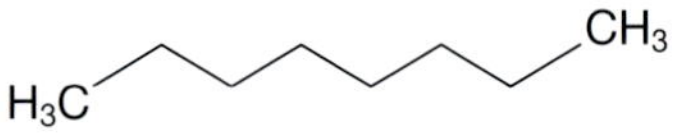

Evaluating chemical structures of these molecules shows there is no important dipole moment in any of them since existing bonds are C—C and C—H, and all carbon atoms are saturated; therefore, all electrons in the outside layer are shared.

6.2.2. Prausnitz and Blanks Parameter Model

It is worth considering the chemical structure of another molecule, acetone:

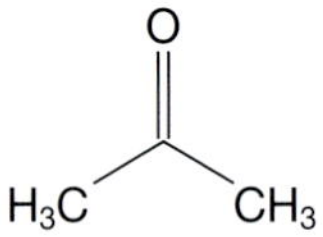

In this molecule, a carbonyl bond is represented by the C=O double bond. In relation to this, there are a couple of properties to be taken into consideration:

- oxygen has 6 electrons in the outside layer
- oxygen presents a greater electronegativity than that of carbon.

These properties determine a charge distribution in the molecule defining a polarization in the structure with the distribution as presented below.

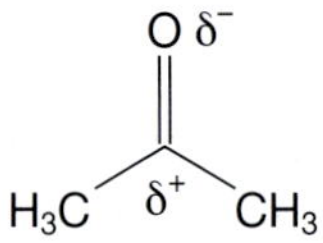

There is a partial negative charge concentrated in the oxygen atom and a partial positive charge located in the carbonyl carbon. The charge distribution in the molecule generates a dipole moment being displaced to the oxygen atom. These distributions directly influence molecules rearranging in a solution since a part of the molecule with the negative charge tends toward positive centers and vice-versa.

These forces correspond to dipole-dipole interactions and are known in literature as Keesom interactions. They are dependent on the following properties of the molecules:

- atomic composition;
- chemical structure;
- difference in electronegativity;
- distribution of electron density.[2]

Below is a diagram of a dipole moment, with an arrow indicating the most electronegative region of the molecule.

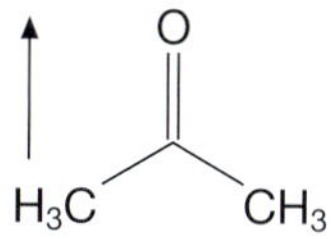

The degree of polarity for each molecule is directly related to the comparison of its chemical structures of the atoms that make up each one, and the geometric form of each polar and nonpolar grouping.

Below are chemical structures for acetone and heptanone.

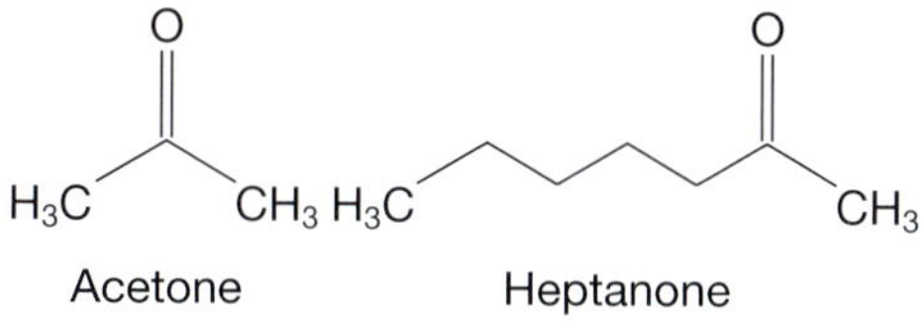

The region of the carbonyl function in the molecules, in the rectangles, are the same, and therefore, the same contribution for polarity of both should be expected, as shown below.

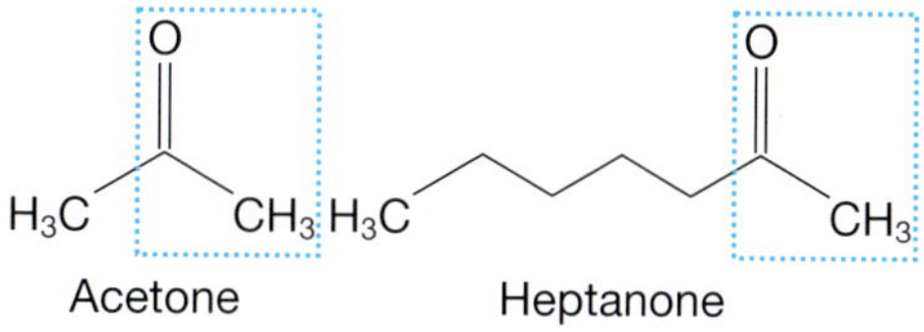

The main difference between these structures are found in the nonpolar region, with this being the aliphatic chain region, where in the acetone there is only one carbon atom, while in the heptanone there are 5 carbon atoms as shown below.

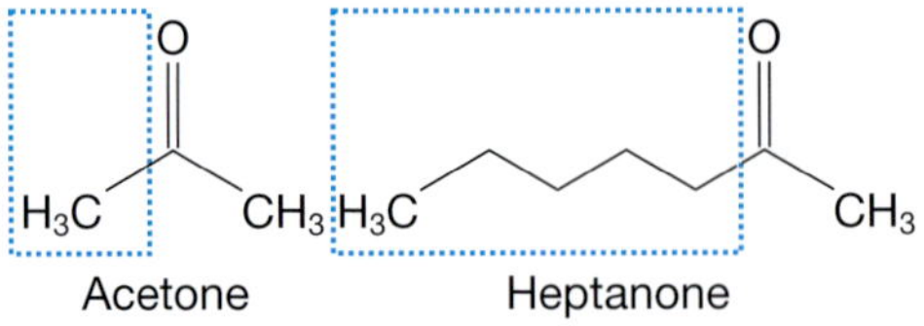

This difference implies there is a greater nonpolar characteristic in the heptanone molecule and the acetone characteristic is more polar.

Another example of the impact of a molecular polarity structure is the CH_2Cl_2 (dichloromethane) and CCl_4 (carbon tetrachloride). Below is a figure of the CH_2Cl_2 molecule.

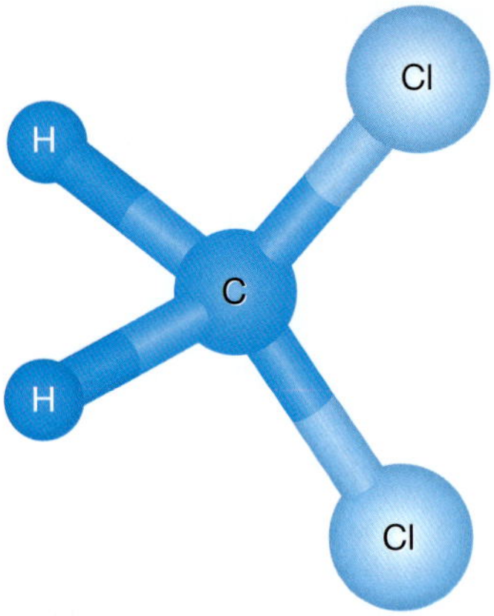

The chlorine atoms, which are more electronegative than the carbon atom, create a dipole moment as in the diagram below.

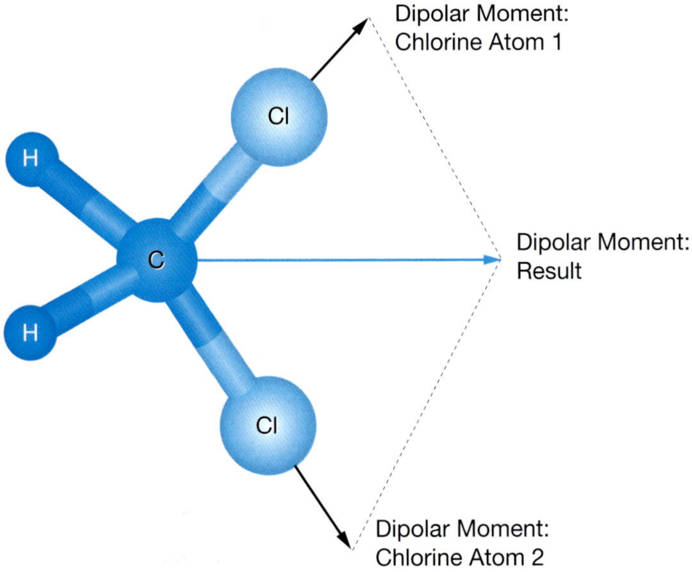

Carrying out the same type of evaluation for the CCl_4, the result is as follows, with the carbon atoms red and the chlorine atoms grey.

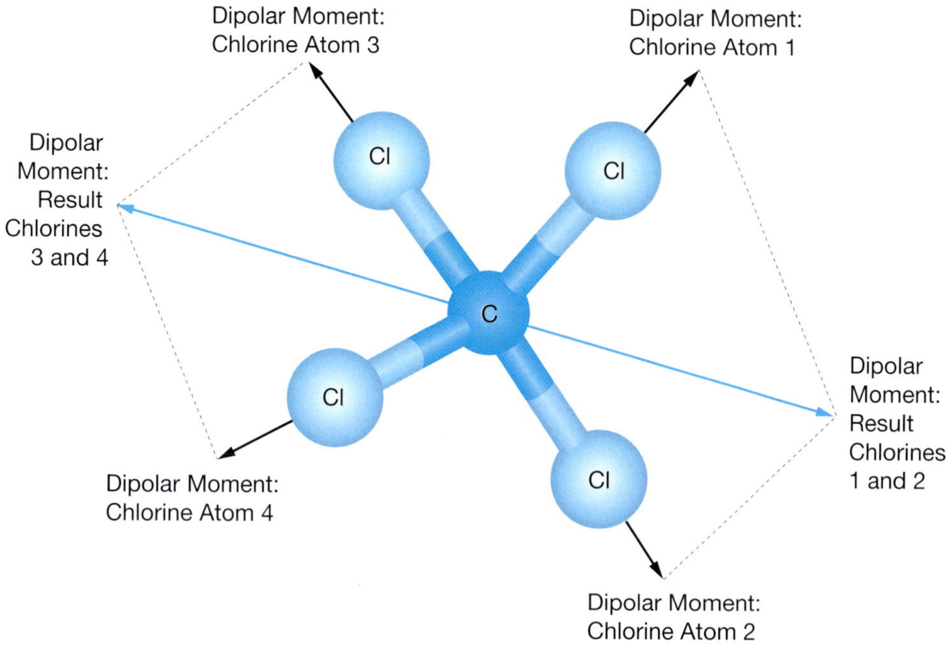

Based on the figure above, the dipole moments resulting in chlorines 1 and 2 are the same as the dipole moment resulting in chlorines 3 and 4. However, they are in exactly opposite directions, and the subtraction of these vectors results in a value of zero. Therefore, the dipole moment resulting from the CCl_4 molecule is zero, explaining its low polarity.

Finally, in comparing CH_2Cl_2 and CCl_4 polarity, it can be concluded that CH_2Cl_2 is more polar than CCl_4 due to the simple influence of the chemical structure, mostly, despite that both have chlorine atoms in their structure that are more electronegative than the carbon and should favor the polarity of the molecules.

These examples show chemical structure influences molecular polarity.

As laid out above, in addition to van der Waals interactions, considered by the Hildebrand solubility model, molecular polarity plays an important role in solution properties.

Prausnitz and Blanks (1964) defined a solubility parameter model that included two polarity dispersion forces influencing a system's solubility.

6.2.3. Hansen Parameter Model

In addition to the interactions previously mentioned, in 1967, Hansen proposed a parameter model that included a third type of energy: hydrogen-bonding energy.[2]

When some chemical functions are present in molecules, like primary and secondary hydroxyls and amines, they are characterized by having hydrogen atoms directly bonded to oxygen and nitrogen atoms.

Due to the difference in electronegativity between these atoms with hydrogen, electrons with this chemical bond are strongly attracted by the heteroatoms, creating a positive partial charge in a hydrogen atom, as shown in the ethanol molecule below.

$$CH_3\text{—}CH_2\text{—}\overset{\delta^-}{O}\text{—}\overset{\delta^+}{H}$$

The positive partial charge created in the hydrogen atom may be attracted by other oxygen atoms, thus producing a hydrogen bond (shown below), where the hydrogen atom is shared between two molecules.

$$\overset{\delta^-}{H}\text{—}\overset{\delta^+}{O}\text{—}CH\backslash_2\text{—}CH_3$$

$$CH_3\text{—}CH_2\text{—}\overset{\delta^-}{O}\text{—}\overset{\delta^+}{H}$$

The hydrogen bond is weaker than a covalent bond, but stronger than a van der Waals interaction. An interesting comparison of the power of this bond can be made at the boiling point for ethanol and dimethyl ether molecules shown in the table below.

Tabla 6.2. Comparison of Properties between Ethanol and Dimethyl Ether

Name	Molecular Formula	Structural Formula	Boiling Point (°C)
Ethanol	C_2H_6O	CH_3-CH_2-OH	78
Dimethyl Ether	C_2H_6O	CH_3-O-CH_3	–25

The above table shows that both present the same molecular formula, but when the boiling points are considered, they are very different, with the difference being approximately 100 °C. One of the main factors determining this difference is the hydrogen bond formation in ethanol, while the dimethyl ether, which has an oxygen atom bonded exclusively to carbon atoms, does not form hydrogen bond. With ethanol, hydrogen bonds increase the difficulty and energy necessary to separate these molecules in order for them to go from a liquid state to a gaseous state.

If ethanol molecules can create these hydrogen bonds among themselves, they can also do this with other molecules; in particular, in solubilization and also with the polymer chains they come into contact with if those chemical functions permit that formation.

Due to this important property, Charles M. Hansen (1967) created a model that included three forces influencing solubilization properties:

- Dispersion Forces (δ_D) – van der Waals Forces
- Polarity Force (δ_P) – Dipole-Dipole Interactions
- Hydrogen Bonding Force (δ_H).

In his model, he proposed that the forces interacted, making additive contributions, as expressed in the equation below.[1] [2] [4] [6]

$$\delta_T^2 = \delta_D^2 + \delta_P^2 + \delta_H^2 \tag{6.15}$$

Below is a table using solubility parameters of some of the more common solvents used in coatings and adhesives.

Table 6.3. Solubility Parameters of Solvents used in Coatings and Adhesives[2]					
Class	Solvent	δ_D ($J^{1/2}/cm^{3/2}$)	δ_D ($J^{1/2}/cm^{3/2}$)	δ_D ($J^{1/2}/cm^{3/2}$)	δ_D ($J^{1/2}/cm^{3/2}$)
	Acetone	15.5	10.4	7.0	19.7
	Acetophenone	19.6	8.6	3.7	21.6
	Cyclohexanone	17.8	6.3	5.1	21.3
	Di-isobutyl Ketone	15.9	3.7	4.1	16.5
	Diacetone Alcohol	15.7	8.2	10.8	20.0
	Isophorone	16.6	8.2	7.4	19.1
	Methyl n-Amyl Ketone	15.1	7.5	7.1	18.3
	Methyl Ethyl Ketone	15.9	9.0	5.1	19.3
	Methyl Isoamyl Ketone	15.9	5.7	4.1	17.7
	Methyl Isobutyl Ketone	15.3	6.1	4.1	17.5
	n-Butanol	15.9	5.7	15.7	23.7
	Cyclohexanol	17.4	4.1	13.5	22.3
	Ethanol	15.7	8.8	19.4	26.1
	2-Ethyl Hexanol	15.9	3.3	11.9	20.8
	Isobutanol	15.1	5.7	15.9	22.7
	Isopropanol	15.7	6.1	16.4	23.4
	Methanol	15.1	12.3	22.3	29.6
	Methyl Isobutyl Carbinol	13.0	7.5	10.4	18.3
	n-Propanol	15.9	6.7	17.4	24.9
	Amyl Acetate	15.3	3.3	6.9	17.7
	n-Butyl Acetate	15.7	3.7	6.3	17.8
	Ethyl Acetate	15.7	5.3	7.2	18.2
	Propylene Glycol Mono-methyl Ether Acetate	14.9	4.7	6.1	15.6
	Ethyl Hexyl Acetate	14.7	6.3	5.3	16.8
	Butyl Glycol Acetate	14.0	8.2	8.6	18.4
	Methyl Acetate	15.5	7.2	7.6	19.3
	Ethyl Glycol Acetate	15.9	4.7	10.6	19.1
	Butyl Glycol	15.9	5.1	12.3	20.2
	Ethyl Glycol	16.2	9.2	14.3	21.9
	Propylene Glycol Mono-methyl Ether	15.43	7.9	13.9	22.1
	Ethylene Glycol	17.0	11.0	26.0	34.9
	Diethylene Glycol	16.2	14.7	20.4	29.1
	Propylene Glycol	11.8	13.3	24.9	30.6
	Hexylene Glycol	15.7	8.4	17.8	25.1
	Toluene	18.0	1.4	2.0	18.3
	Xylene	17.8	1.0	3.1	18.5
	Hexane	14.9	0.0	0.0	14.9
	Cyclohexane	16.8	0.0	0.0	16.8

As previously discussed, qualitatively, there is an important impact of the chemical structure of the compounds on polarity and hydrogen bond properties. The table above presents these differences quantitatively.

In carrying out a correlation between properties and solubility parameter values, it can be shown that:

- low association or interaction between hydrocarbon molecules (ex. hexane) due to a dominance of van der Waals interaction being reflected by an elevated dispersion force (δ_D) value. Since this is practically the only interaction force in this molecule class, solubility parameters related to the polarity and hydrogen bond have a value of zero.
- polar behavior of ketones, like acetone, is due to the partial dipole of the carbonyl group (C=O) present in the structure. This separation of a partial charge gives the polarity of the molecule. Further, acetone will also have a tendency to form a hydrogen bond.
- Alcohols and glycol ethers exhibit strong intermolecular hydrogen bond interaction. The increase in size of carbon atoms present in a chain leads to the decrease of the hydrogen bond characteristic.
- When alcohols and glycol ethers are compared to the ketones and esters, they show a greater tendency to form hydrogen bonds reflected by the higherδ_H values.
- Ketones and esters have solubility parameters that are similar to each other in the three contributions: dispersion force, polarity, and hydrogen bond. This fact explains the possibility that in different applications, one or the other is used with the same performance.
- Hexane is a solvent with only carbon and hydrogen atoms and presents 100% of the nonpolar characteristic, therefore the dispersion force value is high: $\delta_d = 14.9$.
- Acetone has carbon, hydrogen, and a carbonyl group of atoms that give polarity to the structure with the following parameters:
  - $\delta_d = 15.5$
  - $\delta_p = 10.4$
  - $\delta_h = 7.0$
- Methanol has carbon and hydrogen atoms, in addition to a hydroxyl group (OH), which is responsible for the polarity and for the hydrogen bond formation. Its solubility parameters are:
  - $\delta_d = 15.1$
  - $\delta_p = 12.3$
  - $\delta_h = 22.3$
- The percentage of each value of δ in the total solubility parameter for the three solvents discussed above is shown in the table below:

Tabla 6.4. Percentage of Solubility Parameter Total

Solvent	Solubility Parameter		
	Nonpolar (dispersion) (%)	Polar (%)	Hydrogen Bond (%)
Hexane	100	0	0
Acetone	47	32	21
Methanol	30	22	48

Since a solvent mixture is generally used for obtaining a good performance in an application, knowing the solubility parameter and the percentage of each solvent in the mixture, it is possible to determine the solubility parameters of this mixture.

For example, suppose a mixture has the following solvents:

- Solvent 1
- Solvent 2
- Solvent 3
- Solvent 4

These solvents have the following solubility parameters:

- Solvent 1: δ_{D1}, δ_{P1}, δ_{H1}
- Solvent 2: δ_{D2}, δ_{P2}, δ_{H2}
- Solvent 3: δ_{D3}, δ_{P3}, δ_{H3}
- Solvent 4: δ_{D4}, δ_{P4}, δ_{H4}

And they have the following volumetric fractions:

- Solvent 1: χ_1
- Solvent 2: χ_2
- Solvent 3: χ_3
- Solvent 4: χ_4

The mixture's solubility parameters for the four solvents can be calculated as:

$$\delta_{Dm} = \chi_1 \cdot \delta_{D1} + \chi_2 \cdot \delta_{D2} + \chi_3 \cdot \delta_{D3} + \chi_4 \,.\, \delta_{D4}$$

$$\delta_{Pm} = \chi_1 \cdot \delta_{P1} + \chi_2 \cdot \delta_{P2} + \chi_3 \cdot \delta_{P3} + \chi_4 \,.\, \delta_{P4}$$

$$\delta_{Hm} = \chi_1 \cdot \delta_{H1} + \chi_2 \cdot \delta_{H2} + \chi_3 \cdot \delta_{H3} + \chi_4 \,.\, \delta_{H4}$$

where:

m – mixture's solubility parameter

The solubility parameter δ_t, determined by the equation (6.15), of a material is the 3D point in space where the vectors of three solubility parameters meet.

Therefore, it is possible to determine a solvent or solvent system position and that of the polymer in 3D starting from their respective solubility parameters.

The distance in space between the 3D position of the solvent and the polymer can be determined using the interaction radius ${}^{ij}R$ of the sphere.[1] [3]

$$ {}^{ij}R = [4({}^{i}\delta_D - {}^{j}\delta_D) + ({}^{i}\delta_P - {}^{j}\delta_P) + ({}^{i}\delta_H - {}^{j}\delta_H)]^{1/2} \qquad (6.20) $$

where:

i – corresponds to solubility parameters of resin;

j – corresponds to solubility parameters of the solvent.

Using an appropriate group of solvents, with the known solubility parameters and evaluating its interaction with a determined polymer, it is possible to construct the solubility surface of this polymer.

The surface can be represented in a 3D space showing axes as:

- Dispersion Force (δ_D)
- Polarity Force (δ_P)
- Hydrogen Bonding Force/Interaction (δ_H)

Below is a diagram showing the solubility surface of a polymer.

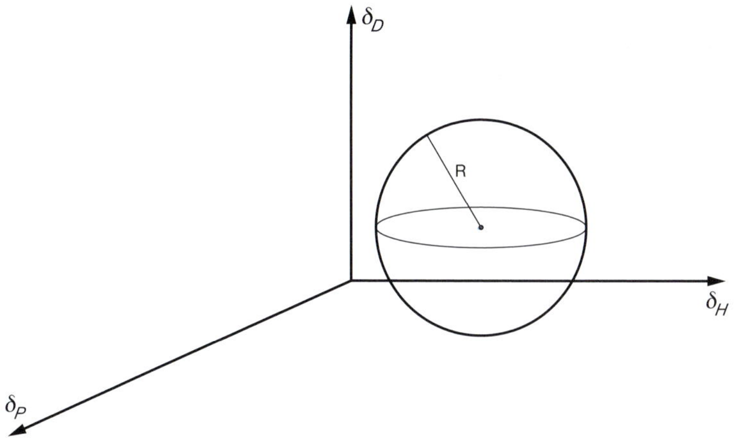

Figure 6.1. Spherical Solubility Surface of a resin. The total solubility parameter (δ_t) of a solvent is a point in 3D space where three vectors, respective of each solubility parameter, meet.

If the interaction radius (${}^{ij}R$) for the solvent's combination and that of the resin is less than the sphere's radius of the resin solubility (R), the solvent probably will dissolve the resin, and the solvent's solubility point will be inside the resin's sphere, as shown in the figure above. On the other hand, a solvent, which has a solubility point outside the resin sphere will be a non-solvent for it.

Below is an example showing the usefulness of the Hansen method. Neither xylene nor methanol is a good solvent for the Epoxy NOVOLAC D.E.N. 438 resin. However, calculations show a xylene and methanol mixture in a 50:50 ratio (per volume) should solubilize it. Below are calculations that show the results of the solubility parameters of the xylene/methanol mixture:

$$^{j}\delta_{D} = 0.5(17.6) + 0.5(15.1) = 16.35$$

$$^{j}\delta_{P} = 0.5(1.0) + 0.5(12.3) = 6.65$$

$$^{j}\delta_{H} = 0.5(3.1) + 0.5(22.3) = 12.7$$

The coordinates of the central point and the solubility radius determined by the epoxy resin are:

$$^{i}\delta_{D} = 20.3$$

$$^{i}\delta_{P} = 15.4$$

$$^{i}\delta_{H} = 5.3$$

$$Sphere\ radius = 15.1$$

Substituting the resin and the values for the solvent mixture solubility parameter in equation (6.20) results in a value, ^{ij}R of 13.9. As long as the calculated distance between the coordinates for the resin and the solvent mixture is less than the resin sphere radius, which has a value of 15.1, the solvent mixture should dissolve the epoxy resin since the total solubility parameter would be inside the resin's sphere of solubility. When laboratory tests were run, a xylene/methanol mixture was confirmed as a good solvent for an epoxy resin.

Bibliographical References

1. Hertz, Daniel L. *Solubility parameter concepts* – A New Look; American Chemical Society – Meeting of the Rubber Division; 1989.
2. Burke, John. *Solubility parameters:* theory and application; August 1984; The Oakland Museum of California.
3. Burke, John. *Solubility parameters:* theory and application; 1984; v. 3;The book and paper Group Annual; The American Institute for Conservation.
4. Wolf, B. A. *Solubility of polymers*; Pure & Appl. Chem. v. 57; n. 2. pp. 323-336, 1985; Printed in Great Britain.
5. Ribar, Travis; Rohit Bhargava; Jack L. Koenig. FT-IR *Imaging of polymer dissolution by solvent mixtures. 1 Solvents*. Macromolecules 2000, 33, pp. 8842-8840.

6. King, J. W. *Determination of the solubility parameter of soybean oil by inverse gas chromatography*. Lebnsm. Wiss. u. Technol, 28, 190-195 (1995).

7. Belmares, M.; M. Blanco; W.A. Goddard, III; R. B. Ross; G. Caldwell; S. H. Chou; J. Pham; P. M. Ololfson; Cristina Thomas. *Hildebrand and hansen solubility parameters from molecular dynamics with applications to electronic nose polymer sensors*; 2004; J. Comput Chem 25: 1814-1826.

8. Vessof, S.; J. Andrieu; P. Laurent; J. Galy, J.F. Gerard. *Curing study and optimization of a polyurethane-based model paint coated on sheet molding compound part II*: Drying defects related to curing conditions; 2000; Drying Technology, 18:1, 219-236.

9. Christodoulou, K. N.; E. J. Lightfoot; R. W. Powell. *Model of stress-induced defect formation in drying polymer films*; July 1998; AICHE Journal v. 44, N. 7 14, N. 7 1484-1498.

10. Huang, Alvin Y.; Donald K. Montrose, John C. Berg. Transient Aggregation during Dry Down of Solvent-Borne Dispersions; Langmuir 2005, 21. p. 9926-9931.

11. Lindvig, Thomas; Michael L. Michelsen; Georgios M. Kontogeorgis; *Thermodynamics of paint-related systems with engineering models* AICHE Journal; November 2001 v. 47 n. 11. pp. 2573-2584.

12. Zellers, Edard T.; Daniel H. Anna; Robert Sulewski, Xiarong Wei. *Improved methods for the determination of hansen's solubility parameters and the estimation of solvent uptake for lighly crosslinked polymers*; Journal of Applied Polymer Science; Volume 62, Issue 12, Pages 2081 2096 Published online 7 Dec 1998.

7 Principal Criteria for Choosing a Solvent

This chapter deals with the main criteria used for choosing a solvent or solvent system – Solvent Power, Evaporation Rate, and Safety, Health and Environment.

Denílson José Vicentim
Sérgio Martins

Resins have an important role in system performance but a poor solvent selection can compromise it significantly. Solvents address in a large part of the coating applicability properties, film formation, leveling, running, degree of reticulation, and film hardness.

7.1. Solvent Power

With the solubilization of any substrate, two pieces of information are essential:

- solvent power – this is included in the solvent effectiveness assessment with a solute. The formulator must be able to qualitatively compare various solvents;
- an assessment as to whether there is an existing solvent for this substance or if a solvent mixture is necessary to dissolve it.

With the need to have a mixture of two, three, or several solvents and dilutants, the problem gets complicated. Consecutive testing can be carried out but is synonymous with a waste of time and cannot guarantee improvement in the results obtained.

However, there is a rational method that allows us to know ahead of time the solvents capable of being used in a given formulation. This method which has presented considerable progress, mainly thanks to research done by Charles M. Hansen, is based on solubility parameters for polymer/solvent groups. This methodology was discussed in detail in Chapter 6 – Solubility Parameter.

After basing a selection on solubility parameters, a selection based on solution viscosity can be used to choose a more appropriate solvent (or mixture) for the desired application.

Product solvent power analysis is important and the viscosity study is the most efficient measurement to use for this information. Three other methods are also used in solvent power analysis: Dilution Rate, Kauri-butanol Index, and the Aniline Point.

Definitions and principal methods for solution viscosity measurement assessment and two viscosity measurement methodologies will be presented in this chapter.

It is known that solvent power is greater when solution viscosity of a constant dry extract is lower.

As discussed in detail in Chapter 6, the theory of solubility parameters shows that a resin solution in a solvent becomes more stable thermodynamically as the distance from the representative point of a solvent to the resin solubility surface center diminishes.

A low-viscosity solvent farther from the sphere's center can give a very fluid solution at the moment of preparation that, in some cases, runs the risk of transforming into a thermo-reversible gel with time.

Solvent power based on viscosity measurements is very important for the definition of a solvent system, but viscosity should be confirmed throughout its shelf life. In this case, the solvent power of a liquid is as high as the resin quantity added to a solution in a given viscosity.

Below are the properties that characterize the choice of a good solvent.

- Solvency Power;
- Evaporation Rate;
- Flash Point;
- Chemical Stability;
- Surface Tension;
- Color;
- Oder;
- Toxicity;
- Biodegradability;
- Cost × Benefit Assessment.

7.1.1. Viscosity

7.1.1.1. Dynamic Viscosity

Viscosity of a fluid, in general, translates molecule resistance into moving some of them in relation to the others. It all happens as if the liquid were made of overlapping layers of molecules.

As described by Hubert Verneret in Industrial Solvents[3], on applying a τ force, called shear stress, tangentially in the direction of movement, this pressure causes a disturbance that spreads to each layer, diminishing as distance increases. There is a resistance with each successive layer from intermolecular friction.

Motion is at V velocity, varying in function of distance.

According to distance $x + dx$, molecules move at a velocity $V + dV$.

The differential relation $D = dx/dV$ is called velocity gradient.

When shear stress τ increases proportionally to velocity gradient D, the fluid is called *Newtonian*.

$$\tau = \eta D$$

The factor of proportionality η is the dynamic viscosity which is expressed in pascal-second (Pa · s) in the international unit system (S.I.).

In this relation:

τ = Force (F), in Newton, applied to the surface unit (S) in square meter.

$$\tau = \frac{F}{S}$$

D = the relation of a velocity (V) at a length (L), in meters.
V = Distance (L), in meter, at a unit of time (s), in seconds.

Where:

$$\eta = \frac{\tau}{D} \therefore \eta = \frac{F}{S} \times \frac{L}{V} = \frac{F}{S} \times \frac{Ls}{L} = \frac{F}{S} \cdot s$$

The F/S relationship is considered to be assimilative with pressure and is expressed in pascal, though rigorously, forces are applied tangentially and not perpendicularly to the surface unit.[3]

Dynamic viscosity of a fluid is a proportionality constant related to the shear stress profile and fluid speed submitted to this stress.

The relationship between the pascal-second and the poise:

1 Pa · s = 10 P
1 mPa · s = 1 Cp

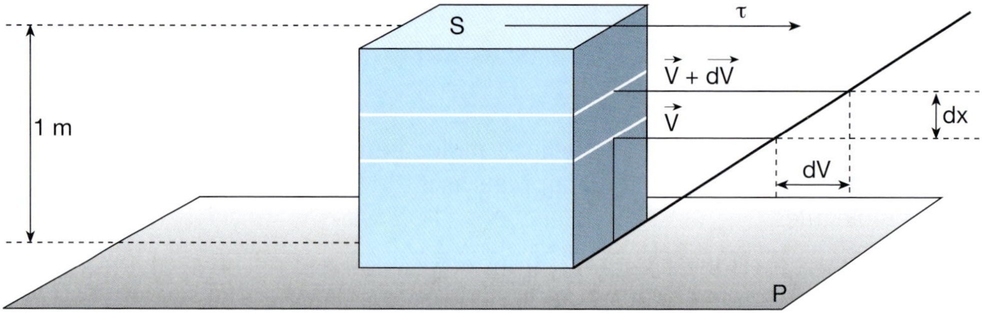

Figure 7.1. Dynamic Viscosity[3].

7.1.1.2. Kinematic Viscosity

Where the gravitational force interferes, it is not possible to directly use dynamic viscosity, so in this case kinematic viscosity v is used and is expressed in square meter per second (m^2/s) in the international system (S.I.), or in stokes (St) and in centistokes (cSt) in the CGS system.[3]

1 m^2/s = 10,000 St
1 mm^2/s = 1 cSt

The relationship between dynamic and kinematic viscosities is given in the formula below:

$$\nu = \frac{\eta}{\rho}$$

in which:

ρ, is the volumetric mass of a fluid at a given temperature;
υ, is kinematic viscosity; and
η, is dynamic viscosity

Therefore, it can be said that one square meter per second (m^2/s) is the kinematic viscosity of a fluid in which its dynamic viscosity is 1 pascal-second (Pa · s) and in which its volumetric mass is 1 kilogram per cubic meter (kg/m^3).

The systems can behave rheologically as a Newtonian fluid or non-Newtonian fluid.[3]

7.1.1.3. Rheology

A Newtonian fluid presents viscosity independent of the rotation speed of the equipment doing the measuring. Water and several coatings and varnishes with low viscosity are classified as being a Newtonian fluid.

$$\frac{\tau}{D} = \eta = \text{constant}$$

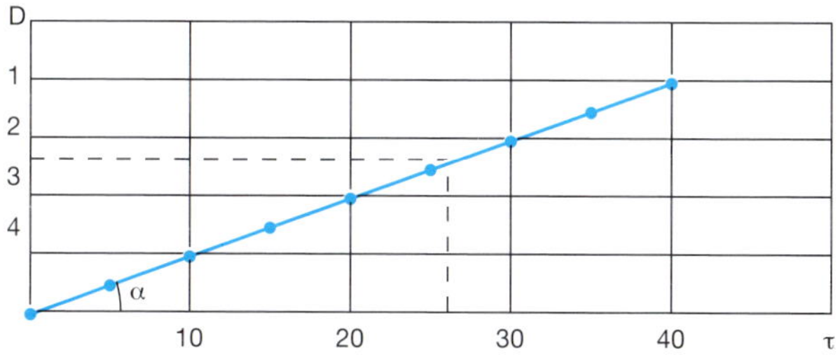

Figure 7.2. Rheological Behavior – Newtonian fluid[3].

With non-Newtonian fluid, viscosity decreases with the increase of velocity gradient D and vice-versa. Several concentrated coatings and varnishes show this behavior.

At rest, molecular chains are in a spiral and rolled up. When a liquid starts flowing, these chains unroll and run parallel to the fluid's movement. Frictional forces decrease until a limit value corresponding to the alignment of all molecules is reached.

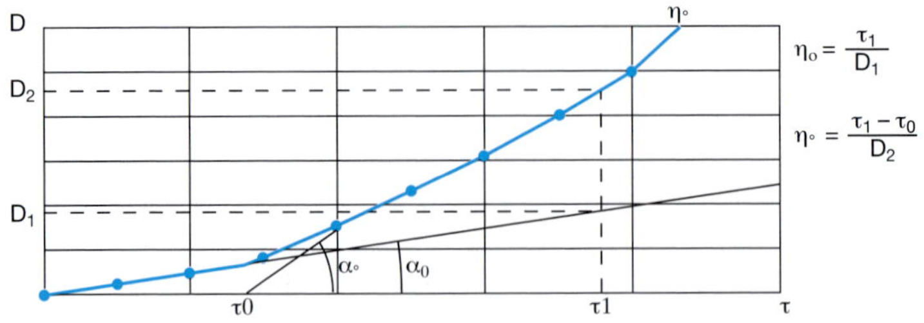

Figure 7.3. Rheological Behavior – non-Newtonian fluid[3].

This behavior is reversible. Macromolecules behave a little like springs that return to their initial position after the traction force relaxes.

$$\eta o = \frac{\tau_1}{D_1}$$

$$\eta^{\circ} = \frac{\tau_1 - \tau_0}{D_2}$$

where

τ_0 = shear stress 0
τ_1 = shear stress 1
D_1 = velocity 1
D_2 = velocity 2
η = viscosity

Coating formulations, when very heavy with solid particles, fall under the category of **thixotropic fluid**.

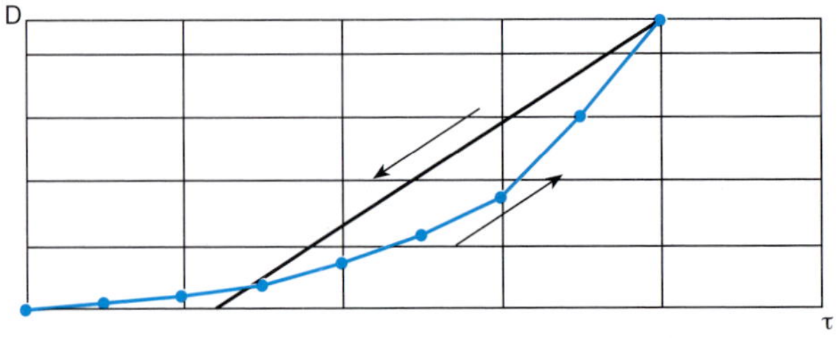

Figure 7.4. Rheological Behavior – Thixotropic Fluid[3].

At a constant speed, viscosity decreases in function with time, but it decreases equally when the velocity gradient increases.

Thus, this type of rheological behavior is more complex than the previous ones, but it is frequently found.

Viscosity can be measured using an efflux viscometer, sphere drop viscometer, glass capillary viscometer, and rotational viscometers.[3]

7.1.1.4. Measurements

Efflux viscometers are used frequently in the industry because of their simplicity and quick results. They are simple cylindrical containers in which the conic-shaped end is a standard diameter orifice, and therefore soft and easy to clean.

The more commonly used cup models are:

Ford n. 04 (Standard ASTM D-1200)
Funnel Diameter: 4.115 mm
Interior Volume: 102.5 cm^3 at 20 °C

AFNOR n. 04 (Standard NFT30-014)
Funnel Diameter: 4.00 mm
Interior Volume: 102 cm^3 at 20 °C

They are easy to handle, but as less precise devices, they only serve to make comparison measurements. Their indications can only be converted into very approximate absolute units due to the turbulence that occurs at the level of the orifice and must be treated as a Newtonian substance.

The falling sphere viscometers are more precise apparati in which the calibrated sphere falls due to gravity in a normalized cylindrical tube. The fall time increases with the medium's viscosity in which the sphere is moving.

Hoppler created a type of apparatus in which a cylindrical tube is inclined a few degrees in a vertical position (80 to 100°) and the sphere rolls around the wall of the tube. Its precision is better because of physical reasons.

These apparati can give absolute viscosity only with Newtonian liquids. However, they need to have a long falling time for good precision, making this their limitation.

One of the easiest models to use is the viscometer built according to Standard DIN 53015. It does not have a heated skirt, but its simplicity explains its frequent use. Nitrocellulose manufacturers use this method to classify their resins.

The distinction is initially made using letters A and E (soluble qualities in alcohol and esters, respectively). Each type is characterized by the time necessary for a metal sphere of 7.95 mm diameter, weighing 2.043 g to fall from a height of 254 mm in a solution of 12.2% weight of nitrocellulose in a standard mixture made of 25% ethanol to 95 GL, 55% toluene and 20% ethyl acetate.

The solution at 25 °C is contained in a tube of 1 pol. diameter (25.4 mm) and 350 mm in length. Then the classification in nitrocellulose 5 s, 1/2 s, 1/4 second, etc.

Capillary viscometers are more precise when measuring, but are more difficult to handle. For absolute measures, they can only be used with Newtonian fluids. The principle includes determining the flow time between two points of a determined liquid in a capillary tube. There are many models, with the Ubbelohde being the most precise and simplest to use with transparent liquids.

Knowing the coefficient K, depending on each viscometer, liquid viscosity in millimeters square per second (or in centistokes) is simply a function of time.

$$\nu = kt$$

With opaque liquids, the Cannon-Fenske viscometer is used with inverted flow (Standard NFT60100).

Rotation viscometers are used for non-Newtonian substances, with coatings, glues, polymers in solution, emulsions, biological liquids, etc.

This type of viscometer is the only one that gives an exact study of rheological behavior of substances, and from them deduces absolute η viscosity. Thus, dynamic viscosity is defined as the relation of shear stress τ, over velocity gradient D, in classic rheological behavior.

$$\eta = \frac{\tau}{D}$$

This relation is not as simple with non-Newtonian substances, but knowing τ and D is extremely necessary for calculating absolute viscosity.

The most commonly adopted principle is based on rotation of an Ri radius cylindrical cup dipped in a liquid, and inside this cup there is an Re radius coaxial cup.

The Re/Ri relation must remain as close as possible to the unit, preferably below 1.1 and should not be greater than 1.5 (Standard ISO 3219).

The cylinder is suspended on a thread by which its twisting increases with the viscosity of the liquid. The degree of twisting is read on an arbitrary scale in which the values allow shear stress τ values to be calculated for the system being used. The velocity gradient D depends on system characteristics for measurement and rotation speed of the apparatus, but it is independent of the substance being studied. Absolute viscosity η is deduced from given conditions.

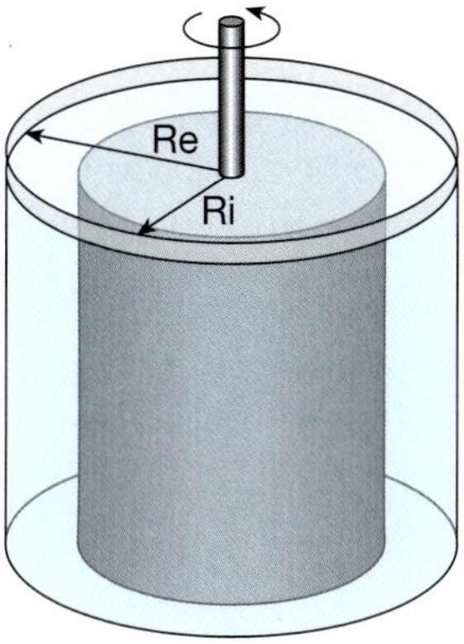

Figure 7.5. Coaxial Cup[3].

7.1.1.5. Other Methods for Determining Solvent Power

Dilution Rate

Dilution rate is related to the amount of dilutant accepted by the formulation; in some cases, amounts are significant, and in others, resin or polymer turbidity and/or precipitation occur rapidly. The market uses the dilution rate as another way of showing solvent power.

Dilution rate is the relation of the dilutant volume and the solvent at the moment cloudiness is observed.

It is known that the higher the dilution rate, the better the solvent power.[3]

American standard ASTM D-1720 defines it, for example, in the case of nitrocellulose, as being the maximum number of dilutant volume units possible to add to a unit volume of solvent to produce the first persistent heterogeneity in a solution of 8 g of cellulose nitrate per 100 ml of solvent and dilutant mixture at 25 °C.

The dilutants (usually hydrocarbons) at times contribute to product cost reduction. Viscosity determination and dilution rate are frequently results that agree with each other.[3]

Kauri-butanol Index *(Standard ASTM D-1133)*

This index deals equally with dilution rate, indicating solvent volume in milliliters in which it is necessary to add 20 g of a kauri gum solution in butanol at 25 °C to get a cloudy solution. Solvent power increases when the Kauri-butanol index increases.

Aniline Point *(Standard ASTM D-611)*

The aniline point determination is rarely used outside oil-based solvents. It is another method that allows solvent power to be assessed in relation to aromatic hydrocarbon content. In principle, it is specific for this type of solvent.

The aniline point is the lowest temperature for which equal volumes of aniline and solvent to be tested are completely miscible, showing the miscibility rupture with the clear appearance of cloudiness.

Hydrocarbon solvent power will be higher as the aniline point goes down and increases as the aromatic content increases.[3]

7.2. Evaporation Rate

Evaporation rate is as important as solvent power because it is essential for the development of a balanced formula and adapted to application conditions.

Exact knowledge of evaporation rate is very important when dealing with balancing coating and varnish drying or getting a dry film at a very restricted time interval, cases of applications in rotation engraving and flexography.

A solvent must evaporate relatively quickly at the start of drying in order to avoid running, however it needs to evaporate slow enough to allow leveling and film adhesion to the substrate.

The evaporation process uses heat exchange between the solvents and the medium which is acquired through ambient cooling, including the substrate in which the coating is being applied.

With more volatile solvents, cooling is still more pronounced, lowering the temperatures to below the dew point, which can provoke humidity condensation on the film's surface. This phenomenon is known by painters as blushing normally through the absorption of water.[3]

We are going to need the meaning of some words for these phenomena which are studied or observed but in which the meaning is not always clear.

7.2.1. Vaporization

Vaporizatin is a general term used for the change of a liquid state to a gas state and that includes the following physical-chemical phenomena.

- *Evaporation*: slow transformation of a liquid to vapor at the gas/liquid interface, with the liquid temperature being lower than its boiling temperature. Without energy being supplied;

- *Boiling Point*: large production of vapor with the formation of bubbles in a liquid with an equal or lower pressure to its vapor tension. With energy being supplied.

7.2.2. Classification of Solvents According to their Boiling Point

Solvents are also classified according to boiling point, delineating their volatility. This definition is used throughout the coating sector.[3]

- Light Solvents: those with a boiling temperature below 100 °C.
- Medium Solvents: those with a boiling temperature between 100 °C and 150 °C.
- Heavy Solvents: those with a boiling temperature above 150 °C.

This classification only serves to orient since there is no simple relation between boiling temperature of a solvent and its evaporation rate.

The Heen formula is more satisfactory. According to this relation, the quantity of an evaporated liquid is proportional to the vapor tension product by the liquid molecular mass.

$$\text{Evaporation Rate} = T_v \; x \; M \; x \; K$$ [3]

T_v = Vapor tension at a measured temperature.
M = Solvent Molar Mass.
K = Proportionality factor, depending on test conditions.

However, accentuated differences are observed, above all, with water and alcohols in which their calculated evaporation rates are shown to be stronger than the measurements.[3]

Finally, only experimental measurements can give a valid solution. The AFNOR proposes two methods. One is simple and quick, but results are approximations (volatility index), while the other needing a more specialized apparatus gives better values.

7.2.2.1. Determining a Solvent Volatility Index

(Standard AFNOR NFT -30301)

This method consists of determining a solvent volatility index in relation to the solvent in reference, using evaporation of small quantities of two products in a filter paper support.

The volatility index (V_e) is the quotient between the evaporation temperature of n-butyl acetate on the evaporation time of the solvent being studied, with these times being measured in precise conditions.

Methodology

The material used and the process of the operation are described in detail in Standard NFT-30301.

With the help of two micropipettes, 0.5 ml of a solvent to be measured and 0.5 mL of n-butyl acetate are poured, successively. in the center of two paper-filter discs kept horizontally. The two discs are simultaneously put in a vertical position, starting two chronometers at the moment in which the papers are put in place. The chronometers are stopped when the filter papers are dry. Five tests are run.

Result Formulation

The mathematical average of evaporation times from the five measurements is calculated and the volatility indices are expressed according to the formula presented previously.

Methodology Deficiencies

The standard itself outlines the numerous causes of errors that can affect the measurements.

- With heavy solvents, the end of evaporation is determined by very weak precision;
- Quick evaporation of light solvents causes an intense cooling, where systematic errors in results are made;
- Viscosity and surface tension differences among solvents are the cause for differences in diffusion speeds in paper fibers, i.e., differences in the evaporation surface.

7.2.2.2. Determining a Solvent Evaporation Curve (Standard AFNOR NFT -30302)

This consists of measuring, according to temperature, the quantity evaporated from the sample solvent or from the solvent mixture, in a ventilated oven maintained at a constant temperature. The evaporation curve gives the relation between evaporation duration and the remaining solvent quantity.

The pure solvent evaporation "curve" is almost always a straight line, which almost never occurs with solvent mixtures.

$$V_e = \frac{\text{n-Butyl Acetate evaporation time}}{\text{Solvente on study evaporation time}}$$

Solvent evaporation rate is defined as being the tangent inclination for an evaporation curve with 50% of evaporated product.

Methodology

The material used and the process are described in detail in the standard.

The solvent for which evaporation conditions are wanted is placed in a recipient at a known mass, with the recipient itself placed inside the ventilated oven. The stainless steel recipient is suspended on a scale with a range of 0 to 15 g, in increments of 0.05 g.

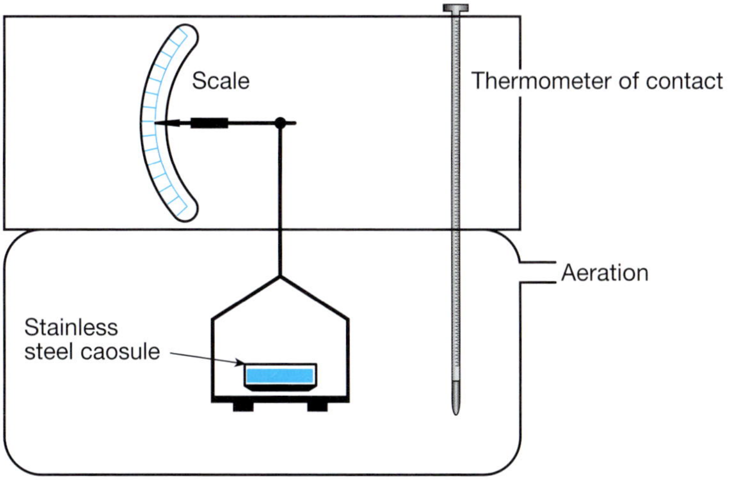

Figure 7.6. Oven for Solvent Evaporation[3].

Two cases must be noted:

1. Solvents lighter than butyl acetate.

- Put a circular paper filter in the bottom of a clean recipient;
- The recipient is placed in a regulated oven at 30 °C;
- Add 5 g of the solvent to be studied;
- The weight is noted minute by minute until the solvent is completely evaporated.

2. Solvents heavier than butyl acetate.

    - The same operation is followed, but the oven's temperature is maintained at 80 °C;
    - Weighing is done every 2 minutes for the first 10 minutes of the test, afterwards, every 5 minutes until solvent evaporation is complete.

7.2.3. Formulation of Solvent Equilibrium

Preparing coatings and varnishes requires a careful balance between light, average and heavy solvents that is called solvent system equilibrium.

Light solvents evaporate very quickly and cannot be used alone owing to intense cooling that occurs on film surfaces. This fact causes water condensation from the relative humidity leading to resin precipitation and surface blush. Hot air drying can avoid this phenomenon.

Medium solvents do not have the inconveniences of light solvents. Their evaporation rate is, however, sufficiently quick for the drying time to be acceptable. They are essential in formulations.

Heavy solvents are used, overall, for good shine qualities and resistance given to the film. They can only be added in small quantities in order to avoid sticky or wet films because of a very slow drying time or a large retention in the thickness of the skin. In addition to their esthetic role, they help conserve the exposed film the most time possible and avoid the precipitation of the resin. They must be true resin solvents, helpful in the formulation.

The dilutants and co-solvents always have their place in formulas because of their low cost. Their evaporation rate must be sufficiently quick to evaporate from the film before or at the same time the other solvents do.

Solvent mixture evaporation rates in coatings, varnishes, and glues are studied using two complementary methods.

Table 7.1. Physicochemical Properties of Solvents

Solvents	Evaporation Rate (Butyl Acetate = 100)	Boiling Point at 760 mmHg (°C)	Vapor Pressure mmHg	°C
Ketones				
Acetone	520	56.2	184.5	20
Acetophenone	3	201.6	0.28	20
Cyclohexanone	31	156.7	4.6	25
Diisobutyl Ketone	21	169.3	4	30
Diacetone Alcohol	12	167.9	1.23	20
Isophorone	2	215.2	0.43	25
Methyl n-Amyl Ketone	40	151.4	7	30
Methyl Ethyl Ketone	340	79.6	90.96	25
Methyl Isoamyl Ketone	53	144.9	4.5	20
Methylisobutylketone	155	115.9	15.7	20
Methyl n-Propyl Ketone	88	102.3	16	25
Mesityl Oxide	–	130.0	8	20
Alcohols				
n-Butanol	46	117.7	5.50	20
Cyclohexanol	5.8	161.0	80	25
Ethanol	150	78.3	40	19
2-ethyl-hexanol	1.9	184.8	30	98
Isobutanol	62	107.8	10	22
Isopropanol	135	82.5	44	25
Methanol	181	64.5	100	21
Methylisobutylcarbinol	29	131.8	2.8	20
n-Propanol	89	97.2	20.8	25

Continued…

Table 7.1. (continued)				
Solvents	Evaporation Rate (Butyl Acetate = 100)	Boiling Point at 760 mmHg (°C)	Vapor Pressure mmHg	°C
Esters				
Pentyl Acetate	45	146	28.5	20
Cycloexyl Acetate	15	177	7	30
Ethyl Acetate	430	77	100	27
M.P.G.M.M.E. Acetate	35	145.8	27.4	20
2-Ethyl Hexyl Acetate	3.7	199	0.4	20
Butyl Glycol Acetate	3.7	191.6	0.35	20
Methyl Acetate	660	57.1	400	40
Ethyl Glycol Acetate	20	156.3	2	20
t-Butyl Acetate	280	96	30.5	20
Butyl Diglycol Acetate	0.14	246	–	–
Ethyl Diglycol Acetate	0.63	217.4	0.1	20
Isobutyl Acetate	145	117.2	13	20
Isopropyl Acetate	355	88.7	60.59	25
n-Propyl Acetate	226	101.6	25.1	20
Ethyl Lactate	21	154	50	100
Glycol Ethers				
Butyl Glycol	6.8	171.2	0.76	20
Ethyl Glycol	39	135.1	5.29	25
Methyl Glycol	58	124.5	6.2	20
Butyl Diglycol	0.35	230.6	0.02	20
Ethyldiglycol	1.3	202.7	0.13	25
Methyl Diglycol	2	194.2	0.25	25
Isobutyl Glycol	11	160.5	26	71
M.P.G.M.M.E. (1)	71	120.1	11.8	25
D.P.G.M.M.E. (2)	3	188.3	0.4	26
T.P.G.M.M.E. (3)	< 1	242.4	2	100

Continued...

Table 7.1. (continued)

Solvents	Evaporation Rate (Butyl Acetate = 100)	Boiling Point at 760 mmHg (°C)	Vapor Pressure mmHg	°C
Glycols				
Ethylene Glycol	< 1	197.6	0.06	20
Diethylene Glycol	< 0.1	245.8	1	91.8
Propylene glycol	< 1	187.3	0.07	20
Dipropylene Glycol	< 0.1	232.8	0.01	20
Hexylene Glycol	< 1	197	0.05	20
Halogens				
Perchloroethylene	–	121.2	1,847	25
Methylene Chloride	990	40.4	400	21
Trichloroethylene	450	86.7	57.8	20
Trichloro 1,1,1 Ethane	530	74	100	25
Aromatic Hydrocarbons				
Toluene	190	110.5	36.7	30
Xylene	60	140	6.72	21
O-Xylene	54	144	4.9	20
P-Xylene	72	138	6.5	20
Ethyl Benzene	84	136	7.1	20
1,3,5 Trimethyl Benzene	22	163	1.8	20
Alihatic Hydrocarbons				
Pentane	1046	36	422	20
Hexane	620	68.7	150	25
Heptane	290	98.4	40	22
Cyclohexane	440	80.7	10	61

(1) Monopropylene Glycol Monomethyl Ether
(2) Dipropylene Glycol Monomethyl Ether
(3) Tripropylene Glycol Monomethyl Ether

7.3. Other Criteria for Choosing a Solvent

More important properties that are expected in a solvent are: the ability to produce concentrated solvents (solvent power); stable, fluid solutions; and an evaporation rate that adapts to a specific problem. However, beyond economic considerations, the formulator must also consider, in function of the destined application, a number of othe factors that cannot be forgotten for technical, economical or safety reasons.

Technical Factors

- Retention of solvents and hygroscopicity

Safety Factors

- Toxicity, flammability and explosivity

7.3.1. Technical Factors

7.3.1.1. Solvent Retention

The retention of solvents corresponds to the final phase of their evaporation when they are still retained in the film or skin in formation. It can generally be observed that the emanation of solvent vapors is much quicker during the initial phase than during the final phase, i.e., the retention phase.

Knowledge of theoretical evaporation speeds of each solvent allows a good forecast of the phenomena, even if the presence of azeotropes modifies observations.

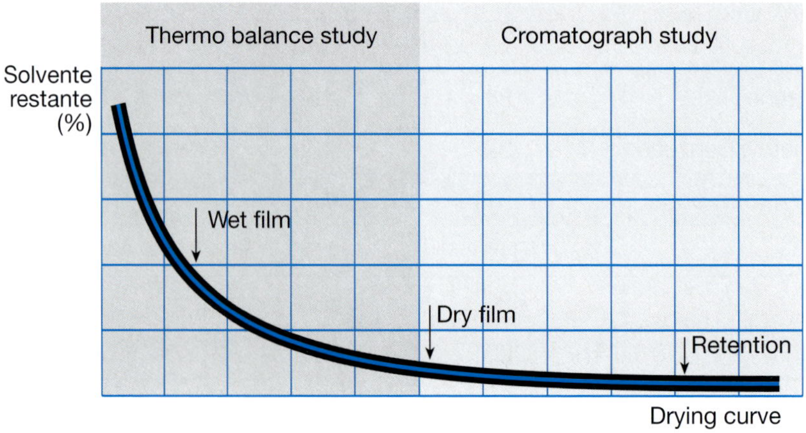

Figure 7.7. Drying Curve of Film[3].

Evaporation can be accompanied by the help of a thermobalance, but when there is a weak percentage of solvents, it is necessary to use another method. The cromatograph in the gaseous phase gives better results because it allows not only knowledge of solvent quantities globally remaining, but also knowledge of percentages of each of the solvents used.

Under certain drying conditions, the higher the speed of diffusion making up the film's elements, lower is the the retention. This speed depends on several factors, among the most important ones are[3]:

- The molecular geometry of solvents;
- The physicochemical interactions that can exist among solvents, resins, or macromolecules in the film;
- Film structure itself, once its aspect is connected to its permeability. A more or less tight mesh strainer can be imagined as intermacromolecular bonds.

Knowledge of solvent retention is indespensible to the formulator and it needs to be studied in all its particularities since a simple relation between evaporation rate of a solvent and its diffusion speed, even for a given component does not exist. It can help with the volatility inversions impossible to forecast using general laws.

Such volatility inversions can lead to the precipitation of one or several polymers or resins during drying by enriching the residual mixture of non-solvents and causing serious defects in the skin.[3]

In order to avoid any precipitation during evaporation, the same solubility volume must always be maintained for the different polymers used and studying the drying phenomena with the help of a precise method is needed.

Chromatography in the gaseous phase is the best method for studying retentions because it allows the separation and dosage of small quantities of solvents to be made.

The details of this method are mentioned in Chapter 9 – Methods for Analyzing Solvents.

7.3.1.2. Hygroscopicity

In a solventborne system, the presence of water is bad because it leads to problems in film drying and in resin dissolution. The advantage of solvents with little or no hygroscopy, even with high relative humidity in the atmosphere is that water droplets formed by decreasing temperature, and deposited on the surface of a film that is drying, do not penetrate deeply and are not inconvenient.[3]

Contrary to this, water is absorbed by the solvents and can cause partial precipitation of the resin, leading to an even more accentuated blush when evaporation rate is high and there is less heavy solvent to cause this deffect is taken into account.

On the other hand, certain types of paint or varnish that harden by chemical reaction (polyester or polyether and polyisocyanates based polyurethanes) must be completely absent of water, even in a small amount, due to the risk of causing a change to the films' mechanical properties or the creation of bubbles from carbonic gas formation.[3]

Consequently, creating lacquers requires solvents that meet two conditions: they must have little water and dehydrated by an appropriate additive; and have a weak ability to absorb water.[3]

In Table 7.2, some common solvents are classsified by increasing hygroscopicity order. Its absorption of water was measured using cromatography in the gaseous stage in identical humidity conditions, after 1, 3, and 10 days. The right column shows the solubility-limit of water in these same solvents.

In general, solvent hygroscopicity seems to be connected directly to the water solubility limit in these same solvents. The highest absorptions of humidity can be found in totally water-miscible solvents.

Table 7.2 Water Absorption of Solvents (% in mass)[3]

Solvent	Quantity of Initial Water	Absorption of Water Afterwards			Maximum Solubility at 20 °C
		1 Day	3 Days	10 Days	
Isobutyl Acetate	0.02	1.05	1.14	1.18	1.64
Butyl Acetate	0.01	1.03	1.12	1.18	1.86
Methylisobutylketone	0.01	1.02	1.28	1.46	2
Isopropyl Acetate	0.07	1.25	1.62	1.66	1.8
Ethyl Acetate	0.01	1.58	2.29	2.8	3.3
Ethyl Glycol Acetate	0.86	1.86	2.81	4.6	6.5
Cycloexanone	0.01	1.64	3.45	4.89	8
Methyl Ethyl Ketone	0.01	1.98	3.60	6.4	10
Butanol	0.03	2.47	4.48	8.72	20.1
Isobutanol	0.02	2.49	4.54	8.87	16.9 (25 °C)
Isopropanol	0.03	2.97	7.12	18.19	Total
Methyl Glycol	0.10	2.95	6.40	18.78	Total
Ethanol	0.03	5.48	10.75	25.8	Total
Acetone	0.01	2.84	7.22	41	Total

Hygroscopicity of solvents

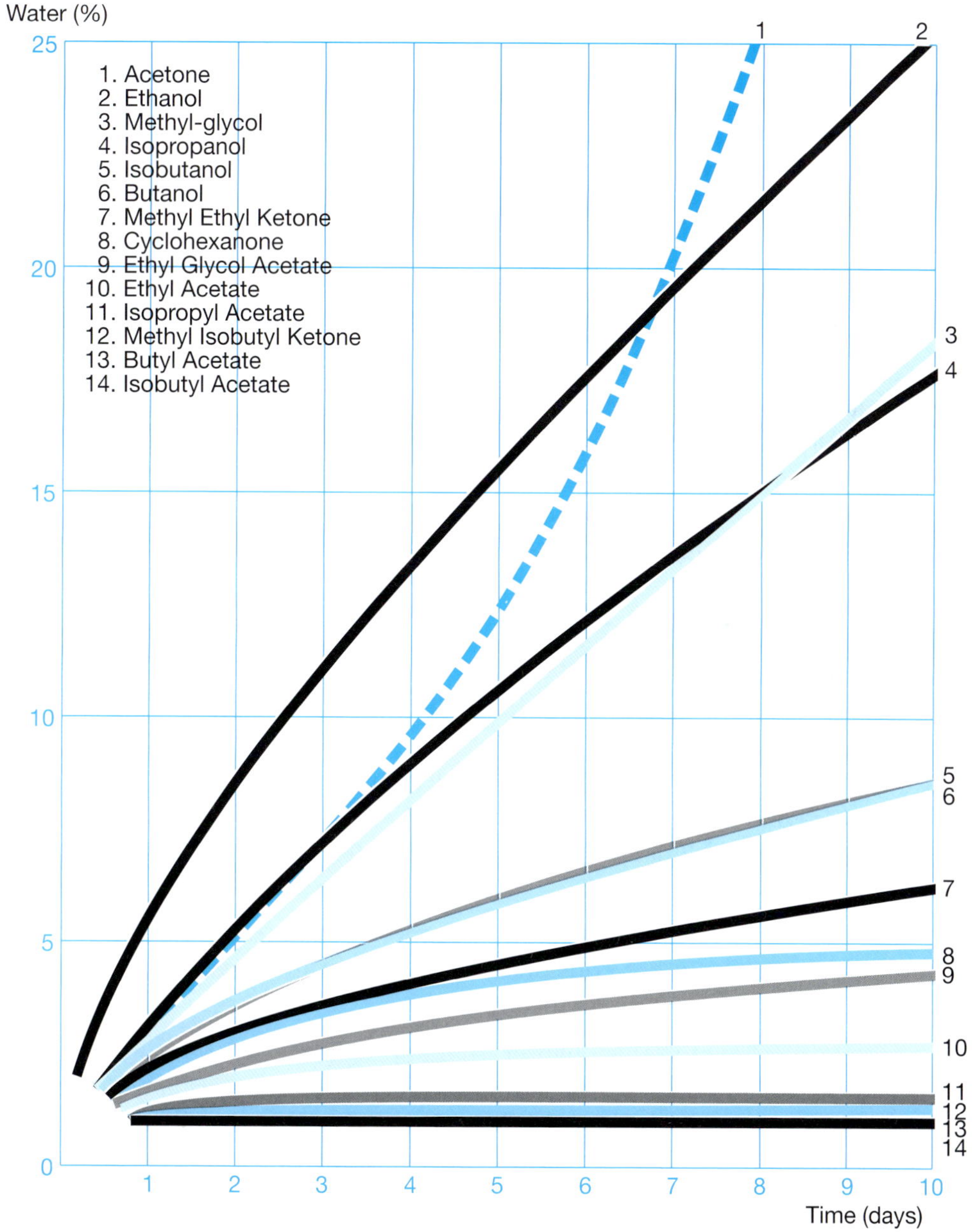

Figure 7.9 Hygroscopicity of Solvents[3].

7.4. Safety, Health, and Environment

The solvent market has been submitted to evermore restricted environmental regulations; therefore it is important criteria that must be considered in choosing a solvent or solvent system. The topic of VOCs is gaining importance and is discussed in Chapter 4 – Green Solvents.

Subjects related to product safety also are important to solvent or solvent system assessment, dealt with in Chapter 10 – Hygene, Safety, and the Environment.

7.5. Example of Solvent Selection in the Separation Process

The separation process involves removing one or more parts of a mixture component. If this substance is solubilized in a lesser amount of a concentration, the lesser amount is denominated a solute, while the greater amount of concentration is denominated the solvent. Generally, the first options for separation techniques are those that use different physicochemical properties, such as:

- Destillation
- Decantation
- Centrifugation

However, if none of these techniques are feasible from a technical and/or economical point of view, a commonly used alternative by the industry is extraction using a solvent. Among the techniques to be noted are liqud-liquid extraction and liquid-solid extraction.

In these cases, solvent selection is a fundamental importance for the appropriate separation, englobing other considerations like operational cost, separation efficiency, and environmental impact.

As previously shown in this chapter, there are several different properties that must be considered in choosing an appropriate solvent for a determined application. Among the properties that define solvent performance, the following are noted.

- Solubility Parameter
- Viscosity
- Density
- Selectivity
- Azeotropy
- Henry Law Constant
- Surface Tension;
- Boiling Point
- Vapor Pressure
- Solvent Loss
- Mixture Viscosity
- Evaporation Rate

Currently, safety, occupational health, and environmental impact have the same importance as factors defining performance. The following are included.

- Toxicity in human beings (e.g.: carcinogenicity, acute toxicity, skin absorption, and inhalation)
- Biological Persistence
- Photochemical Reactivity
- Biodegradability
- Flammability
- Henry Constant in Water
- Solubility in Water
- Flash Point
- Oxygen Chemical Demand (OCD)
- LD50 (Lethal Dose)
- Ozone Layer Destruction Potential

Below is a list of separation techniques using solvents that are often used in industries.

- Liquid-Liquid Extraction
- Liquid-Solid Extraction
- Extractive Destillation
- Azeotropic Distillation
- Stripping
- Absorption
- Lixiviation

Bibliographical References

1. Wesley, L. Archer. *Industrial solventes handbook*. New York: software Included, 1996. 315 p.
2. George, Kakabadse. *Solvent problems in industry*. Londres: Elsevier, 1983. 251 p.
3. Verneret, Hubert. *Solventes industriais*: propriedades e aplicações. São Paulo: Toledo, 1984. 145 p.
4. Fazenda, J. M. R. coordenador. *Tintas e vernizes*. São Paulo: Blucher, 2005. 1043 p.
5. Paul, S. *Surface coating*: science & technology. 2. ed. New York: John Willey, 1997.
6. Payne, H. F. *Organic coating technology*: oil, resins, varnishes and polymers. New York: John Willey. v. 1.

7. Kakabadse, G. *Solvent problems in industry*. New York: Elsevier, 1984.
8. Mellan, I. *Industrial solvents handbook*. 2. ed. New Jersey: Noyes Data, 1977.
9. Durkee, J. B. *How about solvent cleaning*? products finishing, 58-65, 1994.
10. *Formulating fundamentals for coating and cleaners*. American Solvents Council, 2005.
11. Jones, M.; Concepcion, J. G.; Powell, L. *A modern approach to solvent selection*: although chemists and engineers intuition is still important, powerful tools are becoming available to reduce the effort needed to select the right solvent. Chemical Engineering, 2006.
12. Gani, R. Harper; P. M., Hostrup; M. *Solvent based separation*: solvent selection. Disponível em www.cape.kt.dtu.dk/documents/courses/master/notes-lecture4.pdf
13. Rhodia Solventes. São Paulo: Rhodia – Indústria Química e Têxtil S.A.

8 Solvents and Their Applications

This chapter deals with the how solvents work in their principal applications, mainly: coating and varnish, printing ink, and adhesive industries.

Denílson José Vicentim
Edson Leme Rodrigues
Sérgio Martins

8.1. Coatings and Varnishes

Solvents are components in several coating systems and play an important role in film formulation and properties. They impact coating application and properties, including:

- resin solubility and miscibility
- dispersion stability
- application viscosity
- drying time
- film leveling

Many appearance defects associated with the film, like craters and orange peels, are generally results of incorrect solvent equilibrium in coating formulation. Since many properties are affected by the solvent system, any improvement in the formulation has to be done carefully to avoid undesired effects in the other properties.

For example, changing a solvent system to improve resin solubility can decrease coating viscosity or even increase or decrease drying time, causing defects in the film at the end of the process.

The intention of this section is to give information to help formulators in defining solvent systems for coatings according to their composition.

8.1.1. Formulation Principles for Coating Solvent Systems

8.1.1.1. Resin Solubility

Solubility parameters have shown to be one of the most useful tools for coating solvent system formulations with the following advantages:

- They work very well for a larger majority of the systems.
- They are easy to use and visualize, including solvent system formulations for polymer dissolutions or polymer mixtures.
- Experimental data does not need to be used for phase equilibrium.
- Solubility surface is easy to determine experimentally.

Solubility of the majority of polymers needs to be experimentally determined even though for some simple systems, phase equilibrium predictions can be obtained using solution thermodynamic models. The objective here is to quickly review some of the phase equilibrium concepts that in conjunction with Hansen parameter concepts presented in Chapter 6, give a good theoretical basis for understanding solubility and miscibility concepts.

8.1.1.2. Phase Behavior of Polymer Solutions

Polymer solution phase equilibrium behavior is described by the mixture's Gibbs free energy. For complete miscibility and solubility, a system's free energy needs to be negative as presented in Chapter 6

$$\Delta G_m = \Delta H_m - T\Delta S_m < 0$$

where ΔH_m is mixture enthalpy and ΔS_m is mixture entropy.

Through basic thermodynamics, the entropy of a mixture has a positive value, so the term $(-T\Delta S_m)$ contributes to the negative value of Gibbs energy, favoring the solubilization of the polymer in a solvent. Mixture enthalpy is generally a positive value and in this case, enthalpy and entropy need to be correctly balanced to obtain a negative free energy value and total solubility of the polymer in a solvent.

In solutions containing a polymer, mixture entropy value is generally very low because of polymer molecule size being much larger than solvent molecules. Thus, polymer solubility is determined by mixture enthalpy, which needs to have the least value possible or negative values using the correct solvent choice.

Several possible Gibbs energy results in relation to polymer concentration are shown in Figure 8.1 below.

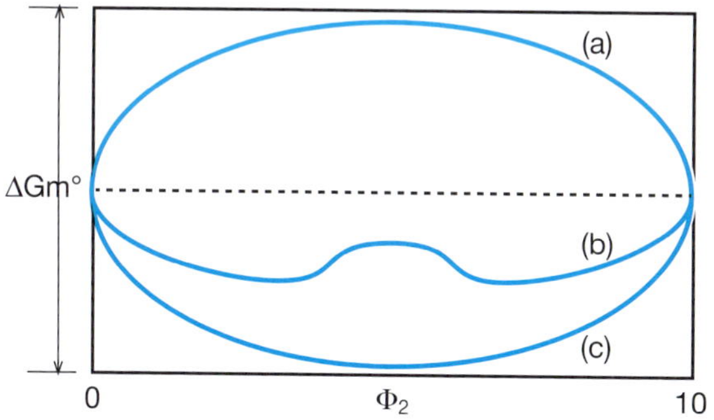

Figure 8.1. Free energy of a mixture and phase stability: (a) completely immiscible, (b) completely miscible, and (c) partially miscible[141].

Based on the figure above, it can be seen that:

- On curve (a) the ΔG_m° value is positive in any of the polymer concentrations (Φ_2) that were evaluated. In evaluating the thermodynamic equation governing solubilization, there are the following considerations:
  - due to solubilizing long polymer chains in very small solvent molecules, the entropy of this solubilization generally gives a very small positive value. Thus, the term $-T\Delta S_m^\circ$ shows a negative value, contributing to the solubilization.

- Since the ΔG°_m value is positive, it means the ΔH°_m value is also positive, and the value is greater than $-T\Delta S^\circ_m$, i.e.,

  $$\Delta H^\circ_m > -T\Delta S^\circ_m$$

- The ΔH°_m value can be obtained using the Hansen parameters.

- On curve (b), the case is exactly opposite that of curve (a), and ΔH°_m as well as ΔS°_m contribute to the solubilization of the polymer in the solvent being considered.
- Curve (c) shows that in certain compositions, ΔH°_m as well as ΔS°_m favor solubilization of the polymer, but between approximately 45% and 75% of the concentration of the polymer in the solvent there is a discontinuity of the curve with the increase of ΔG°_m indicating the existence of other influences in the solubilization. From a practical point of view, the assessment of ΔH°_m shows sufficiently satisfactory results in the evaluation of polymer solubility in solvent systems.

8.1.1.3. Film Formation Mechanism[141]

The conversion of a material in solution into an adherent durable coating is closely connected to the film formation process, which basically includes three main stages: application, adhesion, and drying

The solvent's function plays an essential role at all stages. The solvent or solvent mixture gives the resin solution a viscosity that determines the type of application. During adhesion, which is the coating's stabilizing stage on a surface, the solvent's role is to guarantee a good resin adherence to the surface and a uniform layer. Good adherence depends on, among other factors, the solvent's evaporation rate. Finally, independent of the physical or chemical type, the solvent must, in addition to completely evaporating from the applied layer, leave the system after having solubilized the resin in such a way that the polymer chains intertwine forming a homogenous, uniform and durable layer in the drying process.

The physical process of resin film formation in a solvent base occurs with the evaporation of an organic solvent and can be explained in three stages:

- First, there is quick solvent evaporation from the surface resulting in an increase of polymer concentration and subsequently droplet formation in evaporation areas. At this stage, the evaporation rate is controlled by surface phenomena on the coating's film surface. Air humidity, latent heat in evaporation, and solution surface tension are some important factors at this point.
- Next, more solvent will be evaporated via diffusion through the concentrated polymer layers. Thus, polymer concentration continues to increase and results in the immobilization of macromolecules present. There are seve-

ral ways to explain solvent evaporation at this stage. One of these is that film formation can depend on diffusion processes and solubility parameters, whereas others give importance to solvent activity and solvent concentration.

- Finally, the last traces of solvent in the film are lost in the diffusion and a uniform homogeneous polymer film is formed.

However, a certain amount of solvent is always trapped in the film and this can change mechanical, chemical, and thermal polymer properties, among others.

Solvent or solvent mixture choice for coating and printing press ink use is based on viscosity, solubility, toxicity, flash point, and cost characteristics; reflecting in desirable properties like performance and application. This choice must be made carefully since changes can interfere in the final product appearance and integrity.

Usually, choice is made based on the ability to dissolve the resin, and then on its evaporation rate. There are computational programs that estimate which solvent or solvent mixture is best for a coating formulation. This program includes non-linear properties of solvent mixtures, the method substituting the trial-and-error method in the selection of a solvent or solvent mixture. There are also other techniques for studying solvent evaporation and drying stage like gravimetry, gas chromatography, and infrared.

The Hildebrand solubility parameter is also a valuable tool in choosing a solvent or mixture for a resin since it correlates and predicts resin solubility and compatibility.

As an example of a resin-solvent system, which is made up of a light, medium, and heavy solvent mixture, generally, the light component will evaporate first, so the polymer should be soluble in medium and heavy solvents to avoid precipitation of polymer material occurring. On the other hand, solubility cannot be so great that the polymer is unable to spread on the surface, but should solubilize to the point that polymer molecules are mobile enough to intertwine. The same is true for post-evaporation of a medium solvent in which the heavy solvent is left, giving enough time for the polymer to form a cohesive film, after the evaporation of the three solvent types of the mixture.

Therefore, it is clear that the film formation process is especially guided by the solvent evaporation process, making the importance of the solvent in polymer film preparation clear.

8.2. Solvent Evaporation

Due to non-ideal behavior of solvent mixtures, solvent composition is difficult to predict during evaporation, especially when many solvents are present in the mixture. Evaporation rate is also strongly affected by available surface area for evaporation, as well as mass and heat transfer. Generally, computer models are necessary for precisely solving this type of equation to correctly predict solvent system composition changes during evaporation.

This precision adjustment is associated with the fact that evaporation rate controls the amount of solvent leaving the coating after its application, and consequently determines solvent system composition remaining in the coating during evaporation, thus controlling its viscosity and surface tension. These two properties decide coating flow characteristics, being responsible for film formation with desired characteristics.

For example, quick solvent evaporation is necessary for a quick increase in coating viscosity after its application. However, quick evaporation can result in trapping air bubbles and causing coating film defects.

Further, solvent system composition affects polymer solubility and polymer mixture miscibility that, in turn, are responsible for film integrity and mechanical properties

Coating drying generally occurs in two stages. In the first stage, the film's solvent loss is a function of solvent partial pressure. Resistance to evaporation in this phase is due to solvent molecular diffusion through a thin layer of air above the coating surface

During the second stage of evaporation, the rate is controlled by the solvent molecular diffusion rate through the coating film until it reaches its surface. Generally, this stage is much slower than the first and is only reached when 20-30% of the solvent is evaporated.

The transition from the first stage to the second is shown in Figure 8.3, where the evaporation of methylcyclohexane is shown as a pure solvent and in a mixture with an alkyd resin.

In the pure solvent, evaporation is a function of solvent vapor pressure. However, when a resin is present, a clear transition to the second stage can be seen. Coefficient prediction for diffusion at this stage is very difficult. Because of this, evaporation models generally neglect this stage in an initial approach.

To predict solvent evaporation rate during application, it is necessary to know available surface area for evaporation, which depends on application method and is very difficult to estimate. One way of characterizing solvent evaporation is by using an evaporator model for a thin film. The equipment for this uses a filter paper in which the solvent is placed and paper mass loss is measured over time. Parameters, like temperature, humidity,

The change in other properties during evaporation is a fundamental importance for coating performance. Figure 8.4 shows a solvent system that initially is in a polymer resin's solubility surface.

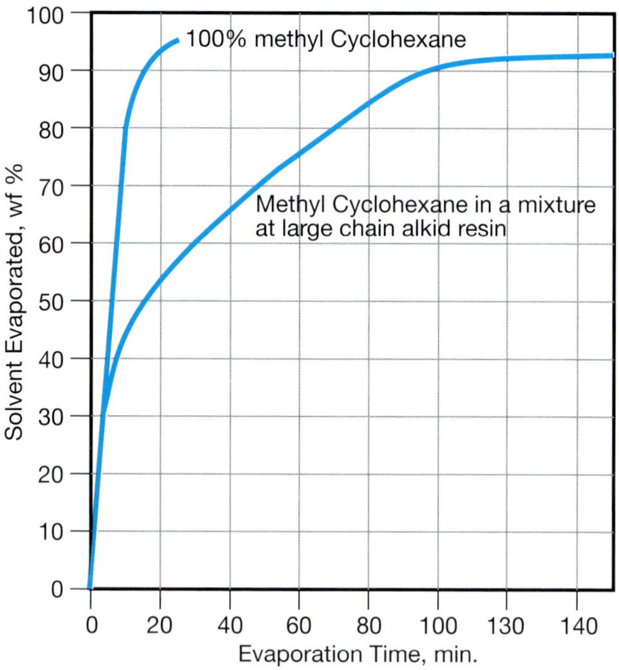

Figure 8.3. Evaporation Transition Stage[141].

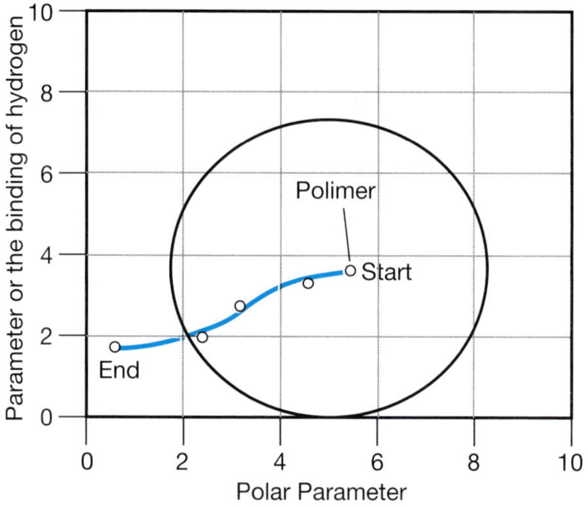

Figure 8.4. Solubility parameter changes during evaporation.

During evaporation, with the solvent system composition change, there is also a change in system solubility parameters that maintain contact with the polymer resin, and in the case shown above, this change takes the solvent system outside the solubility region. Depending on the evaporation speed, there can be phase separation leading to serious defects in the film. That is why during evaporation, the solvent system formulation must include the system staying in the solubility surface throughout the complete process.

Humidity is a factor that needs to be considered in system solvent evaporation, mainly for systems that contain water in its composition or that contain hygroscopic solvents. If humidity is high enough, some film properties can drastically change with retention. On the other hand, there are several organic solvents that create azeotropes with water and then humidity can accelerate evaporation rates.

8.2.1. Formulation Methodology

As previously shown, it is not generally possible to predict coating properties due to the complexity of several interactions between components. However, good results can be obtained through solvent formulation calculations which give a good simplification of the coating formulation problem, given that the solvent system composition affects many of its properties.

Generally, property changes can be satisfactorily estimated with solvent composition change. Following is a methodology to help understand the formulation of coating solvent systems.

8.2.1.1. Establishing a Formulation's Polymer and Miscibility Solubility

First, a polymer or polymer system solubility parameters and surfaces need to be decided when several of them are present in the formulation.

When several are present, their miscibility region needs to be at an intersection of individual polymer solubility surfaces.

Experimental verification of miscibility is necessary because of some complex interactions between some polymers which depend on molecular weight composition and distribution.

Usually, immiscible polymers can be used in coatings during the compatibility effect of solvent systems formulated in the intersection region. In this case, the possibility of phase separation during the solvent system's evaporation must be assessed, especially in the region where there is an increase in viscosity, to insure good film formation. This can be guaranteed through the solvent system formulation remaining in the intersection region of solubility surfaces throughout the evaporation process.

8.2.1.2. Specifying the Evaporation Profile and Other Solvent System Properties

This is to determine the solvent system evaporation profile due to impacts explained in the previous section. Other properties, like viscosity and surface tension, must also be decided.

Usually, viscosity needs to be low enough to facilitate handling the coating so as to avoid running during application. At the same time, surface tension needs to be balanced to allow coating film leveling without causing defects.

8.2.1.3. Formulating a Solvent System with Good Polymer Solubility

There are many methods for getting information related to the previous section, such as using empirical methods in which a large number of solvent systems are analyzed. The disadvantage of this is time spent at getting a result that may not necessarily represent the best system.

Using computational calculations in which solvent evaporation models and their respective solubility parameters are taken into simultaneous account is a significant advantage for saving time and having greater assurance in getting an optimal system in terms of performance and cost. Rhodia uses its SOLSYS® system, which is capable of carrying out this kind of function.

8.2.1.4. Confirming Predicted Results Using Test Formulations

Once the solvent system has been chosen, results need to be experimentally confirmed.

8.3. Solvent System Formulations for Specific Coating Market Segments

8.3.1. Determination of Solubility Parameters and Normalized Distance Concept Definition

8.3.1.1. Measures for Obtaining the Solubility Parameter

The technique for Solubility Parameters consists of testing solubility or miscibility of the substance using a series of pure solvents representing different chemical groups. Examples include Hydrogen Carbons, Ketones, Esters, Alcohols, and Glycols.

Figure 8.5 is an example of a test for determining Solubility Parameters in a resin. It shows three possibilities for evaluating a resin in a solvent system: soluble, partially soluble, or insoluble; and is used to determine resin Solubility Parameters.

Usually, more than one solvent is used for solubilization of a resin or of a resin system mainly because of performance and cost. The most common way to decide on a solvent system is through empirical lab trials.

Using the Hansen Solubility Parameter Theory allows the most appropriate solvent or solvent system for a resin to be chosen using thermodynamic calculations. This theory is fundamentally based on three forces:

- δD: London dispersion force
- δP: attraction/repulsion forces from molecular polarity
- δH: hydrogen bonding force

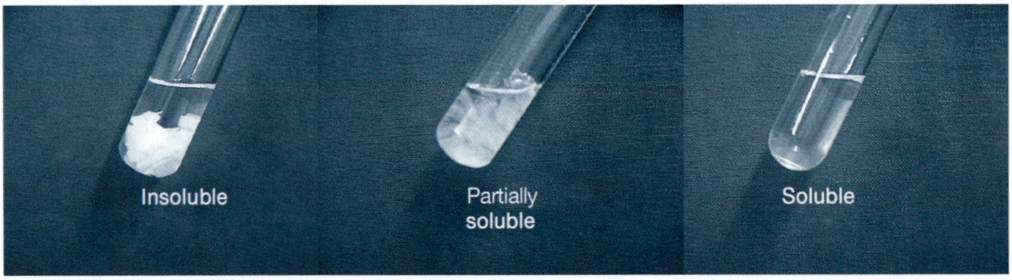

Figure 8.5. Example of a test for determining Solubility Parameters in a resin in a determined solvent. It can be soluble, partially soluble, or insoluble.

Rhodia uses the SOLSYS® simulator which allows resin or resin system Solubility Parameters to be determined as well as determining the best solvent system.

8.3.1.2. Normalized Distance Concept Definition

Assessments will be made through the radius of solubility volume using the normalized distance concept – distance to the center.

Solubility volume is graphically shown in a 3-dimensional figure.

Data is applied to normalization of distances

- the value 0.0 is attributed to the center of the 3-dimensional figure.
- the value 1.0 is attributed to the figure's borders.

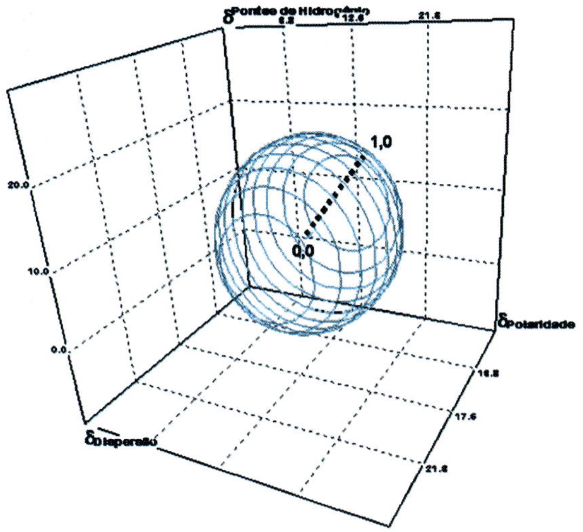

In a simulation for a given polymer, a solvent system results in a normalized distance value:

- values between 0.0 and 1.0: solvent system is efficient for the solubilization of the polymer;
- values above 1.0: low solubility of the polymer in the system.

8.3.2. OEM – Original Equipment Manufacturer

Principal characteristics used in coatings in the market must include:

- Appearance: image shine and distinction
- Durability: color and shine retention
- Mechanical properties and stone chip resistance
- Adhesion
- Good performance against corrosion and humidity
- Resistance to oil and solvents
- Chemical resistance of acids
- Hardness and resistance to sea air
- Properties that allow its repair

The application of automotive refinishing involves the cleaning and treatment of the metallic substrate surface before coating application. Layers making up original automobile refinishing are shown in Figure 8.6 below.

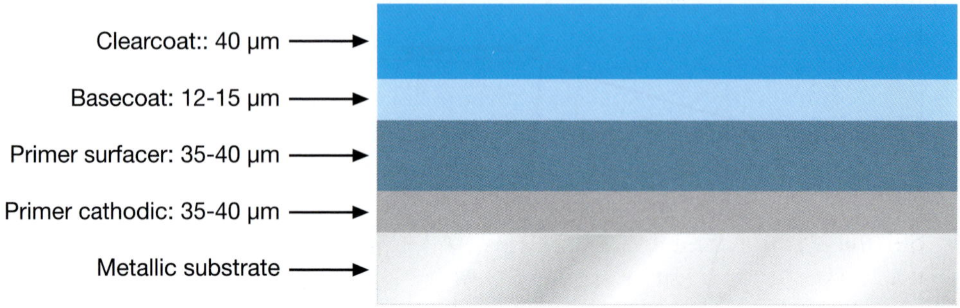

Figure 8.6. Layers of an original coating with their respective thicknesses.

a) Phosphate Layer

b) Anti-corrosive layer or cathodic primers

After initial pre-treatment with phosphate agents, a primer mainly for stopping corrosion is used. This primer contains anti-corrosive pigments in its resin system in order to improve mechanical and anti-corrosive properties. Its application consists of the car chassi immersed in a waterborne coating formulated with an anodic or cathodic resin system.

Due to adhesion reduction problems in a phosphate layer and low resistance to saponification, anodic primers were substituted with cathodic primers. These primers are mostly based on aminoepoxy systems stabilized in water by neutralizing them with several acids. Their structure contains cross-linking elements which are shown to be excellent properties in the film that is formed, with their being cured in an oven at a temperature between 165-180°C.

c) *Primer Surfacer*

The layer applied after the cathodic coating is the primer surfacer used for leveling and preparing the surface for the next layers from a mechanical and chemical point of view, increasing stone chipping resistance, and acting as a filter or barrier to degradation caused by ultraviolet light in the cathodic coating.

Essentially, primers used worldwide by the automobile industry use hydroxy polymer systems, preferentially saturated polyesters, cured with melamine/formaldehyde resins.

d) *Basecoat*

The basecoat system was developed to maximize the appearance of new metallic colors and improve pigmented film protection. In metallic basecoats containing aluminium flakes, the particles need to be appropriately oriented in a lamellar structure parallel to the substrate and to the surface for there to be shine reflectance and light mirrored in the coating layer. For this to work, the following processes need to occur:

- Appropriate aluminium particle flow promoted by low film viscosity in its initial stage of formation due to low linking solid content that forms the film.
- Setting aluminium lamellar particles after being appropriately oriented by the wax component action in the basecoat formulation.
- Final setting of lamellar particles in the film matrix through rapid increase in viscosity by solvent evaporation.
- Shrinking in the coating's volume of film which pressures the particles into its final parallel orientation.
- The ability of the basecoat film to resist an attack of solvents in the clearcoat ensures appropriately keeping aluminium flakes in their position during the subsequent application of the clearcoat.

Resins used in basecoats are generally polyesters or a modified acrylic resin with cellulose acetate butyrate (CAB) to promote faster drying. These formulations present solid content lower than 20% with a film thickness of about 15 µm. These resins are generally cured with melamine-based resins. CAB is used to promote quick drying while the presence of polyethylene waxes help in correcting the orientation of metallic flakes.

e) *Topcoat* or *Clearcoat*

The clearcoat can be applied along with the basecoat (wet/wet) or by using a drying process just before the basecoat called "flash off." The main advantages for using a clearcoat are:

a) Improving appearance – its use allows excellent shine properties and produces a leveling of the film surface of the coating.

b) Improving durability – it gives additional protection to the basecoat film, improving shine retention and reducing chemical damage to the aluminium flakes.

Clearcoats are based on reticulated acrylic resins with melamine resins. The development of acrylic/urethane 2K (two-component) systems for use as clearcoats in original coatings are also important, mainly for applications where there is a need to use low oven temperatures. Acrylic resins with functional hydroxyls are used with polyisocyanate as cross-linking agents. Desirable properties for polyurethane systems in original coatings are summarized below:

- Rapid curing at room temperature or at low oven temperatures.
- High levels of exterior durability, especially with color and shine retention.
- Excellent hardness, flexibility and resistance to abrasion.
- Good adhesion to the basecoat and a variety of other substrates.
- Good transparency.
- Capacity for formulations with high solids[141].

They follow some formulation examples found on the original paint markets for some resin systems. Considering an acrylic-melamine resin system, a solvent system with good application performance is shown in the following table along with the respective solubility parameters and normalized distances in relation to the system considered.

As shown in previous data, the solvent system has an excellent performance in terms of acrylic resin solubilization and in particular, the analyzed melamine resin. In relation to normalized distances to the center, values approaching 0.0 indicate the best possible solubility of the resin in the system. With melamine resin, this value is practically at the center at any given moment of evaporation. This is not necessarily always the best situation, considering that these distances indicate an excellent interaction of solvent system with the resin, which can represent a difficulty at the moment of its evaporation.

Analyzing solubility surfaces of the two evaluated resins shown in Figure 8.7, it can be verified that the surface in reference to the acrylic resin is contained in the melamine resin surface.

Another important factor to be considered is that the radii from the two surfaces, with the melamine resin being the largest. This means the degrees of freedom in the formulation of a solvent system are greater in relation to the acrylic resin.

By the positioning of the acrylic resin surface in relation to the melamine, any solvent system common to both needs to be positioned in the interior of the acrylic resin, as shown below. Even during solvent system evaporation, this remains common to both surfaces.

Table 8.1. Solvent System Composition, Solubility Parameters, and Normalized Distances in a Melamine-Acrylic Resin System

Solvent System Composition		Values in % m/m
Solvents	Butyl Acetate	60.0
	Ethyl Acetate	13.0
	Xylene	27.0
	Total	100.0
Solubility Parameters	δD	16.36
	δP	3.15
	δH	5.68
	δG	17.6
Standard Acrylic	Initial Normalized Distance	0.72
	50% Evaporated Mass	0.70
	90% Evaporated Mass	0.65
Standard Melamine	Initial Normalized Distance	0.07
	50% Evaporated Mass	0.06
	90% Evaporated Mass	0.04

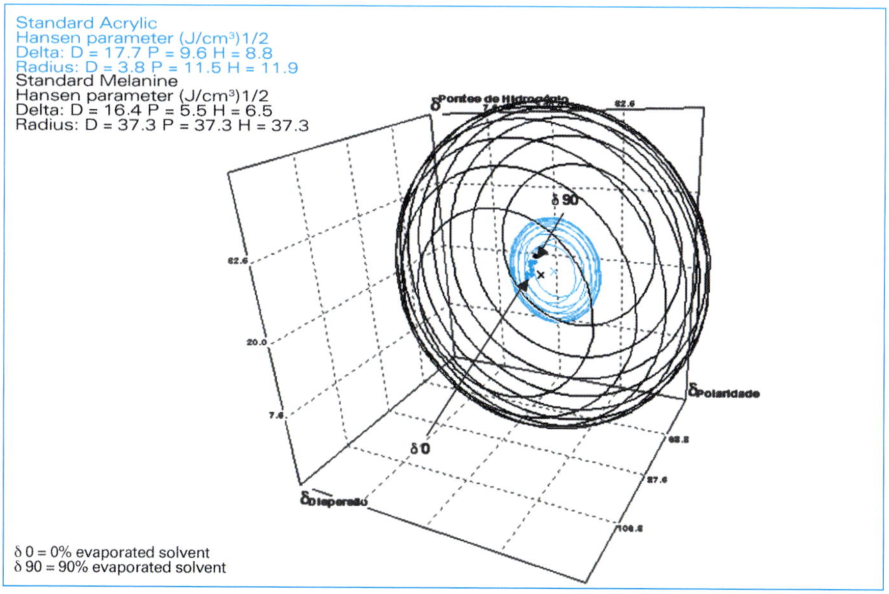

Figure 8.7. Solubility Surfaces of Melamine and Acrylic Resins.

This figure shows that even with 90% of the solvent mass evaporated, the system remains common to both surfaces. The solvent system composition that remains in the film at each moment of evaporation is what determines the solubility parameters and normalized distances.

To demonstrate the variation of solvent system composition, Figure 8.8 shows the progression of the system make-up which remains in the film in relation to the quantity of evaporated solvent.

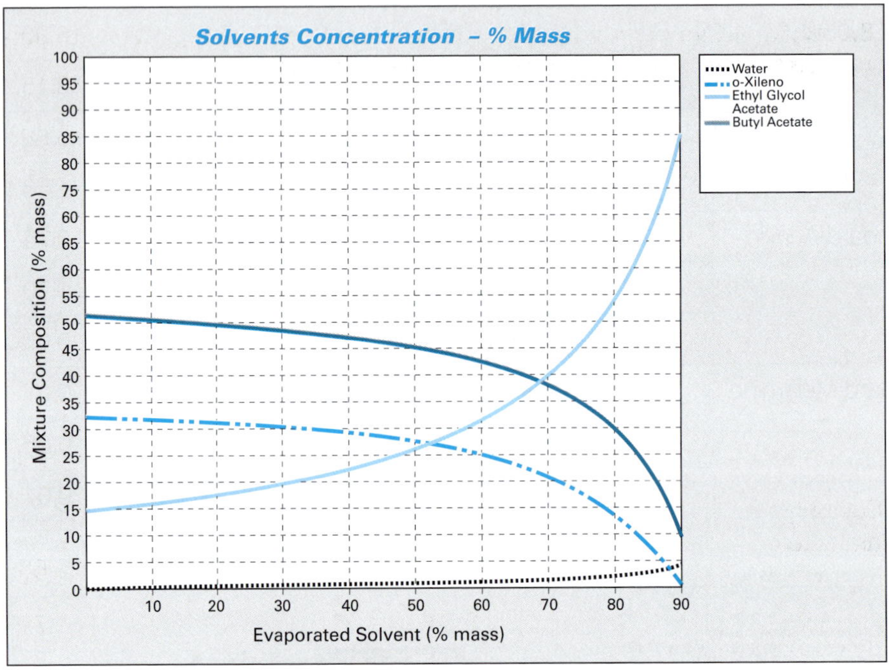

Figure 8.8. Variation of solvent system composition in relation to the evaporated solvent mass.

This simulation also takes into account the absorption of water present in the atmosphere on the part of the solvent system remaining in the film during evaporation. In this case, it shows the quantity is low and this can be credited to the fact of its low affinity to this solvent system.

The following system shows the same resin system, but the solvent system was balanced to show a presence of light compounds without affecting the solubility parameters and normalized distance. The main advantage of this system in relation to the previous one is the better drying speed. On the other hand, if the system is not well balanced, the presence of light compounds in the formulation can show some defects from its quick drying such as orange peel and blistering.

In the presence of these light components, spray gun application that starts system evaporation, even before the coating reaches the substrate, needs to be considered since it causes problems of resin system spreading, among others.

Due to all these possibilities, it is extremely important that the solvent system be balanced to accelerate drying speed without affecting the perfect film formation on the other. Therefore, it is important that the presence of light solvents always be conditioned to the presence of those with medium and low evaporation rates.

Table 8.2. Solvent System Composition, Solubility Parameters, and Normalized Distances in a Melamine-Acrylic Resin System

Solvent System Composition		Values in % m/m
Solvents	Methyl Ethyl Ketone	6.0
	Ethyl Acetate	5.0
	Butyl Acetate	52.0
	Propylene Glycol Monomethyl Ether Acetate	13.0
	Xylene	24.0
	Total	100.0
Solubility Parameters	δD	16.31
	δP	3.63
	δH	5.75
	δG	17.67
Standard Acrylic	Initial Normalized Distance	0.69
	50% Evaporated Mass	0.70
	50% Evaporated Mass	0.65
Standard Melamine	Initial Normalized Distance	0.05
	50% Evaporated Mass	0.06
	50% Evaporated Mass	0.04

Evaporation Conditions:
25 g/m^2/Temperature – 25 °C/Relative Air Humidity – 70%/Air Speed – 0.0667 — Substrate – Metallic

Based on the table on the previous page, it can be verified that:

- Solvents with a high evaporation rate were added to the system. These were ethyl acetate and MEK, and they led to the reduction of the butyl acetate and xylene concentration, solvents having intermediate evaporation rates.
- Even with the change in composition, the system was balanced to show the same solubility parameters in relation to the system that did not have light ones in its make-up.
- Observing the normalized distances and Figure 8.9, it can be seen that the solvent system is positioned internally in the solubility surface of the acrylic resin at all moments of evaporation, and consequently solubilizing the melamine resin. This result indicates the presence of compounds with a faster evaporation rate do not affect resin system solubility during solvent system evaporation mainly because of the correct balance with low and medium evaporation rate solvents like butyl acetate and PMA, respectively.
- The main solvents with medium evaporation rates used for this application are butyl acetate and MIBK. And in relation to low evaporation rates, PMA stands out.

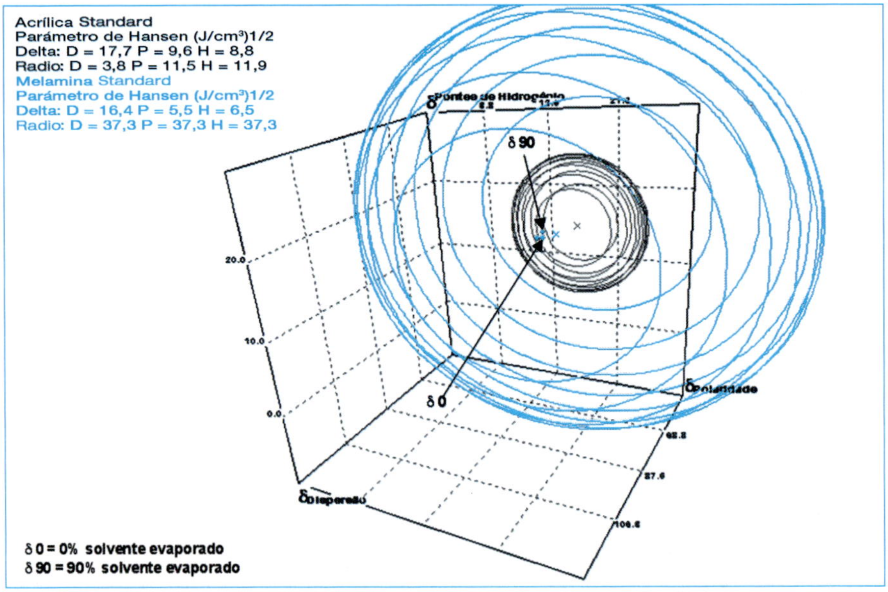

Figure 8.9. Solubility Surfaces of Melamine and Acrylic Resins.

In analyzing Figure 8.10, the change in solvent system composition during evaporation, light solvents quickly leave the film surface transferring the work from the resin solubilization to the solvents with low and medium evaporation rates. For them to show appropriate performance, good resin system solubility in these solvents is fundamental.

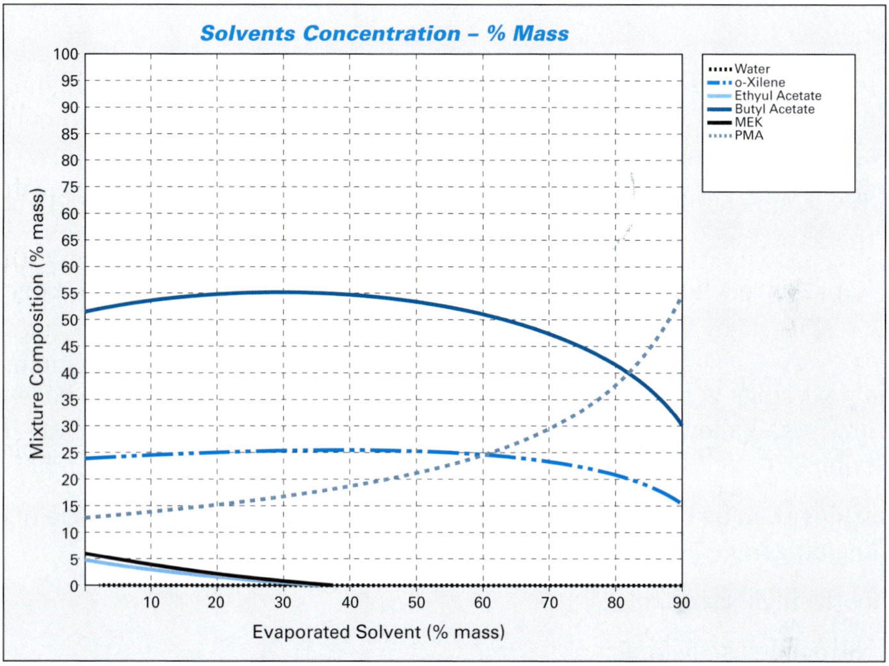

Figure 8.10. Variation of the solvent system composition in relation to the evaporated solvent mass.

8.3.3. Automotive Refinishing

The objective of automotive refinishing is to reproduce the appearance and the durability of the original finish as close as possible. To meet the needs of this market, coating formulations must provide easy application, fast drying with a good surface spread at room temperature, and have similar properties to that of the original refinishing.

A resin system currently used on the market is a compound system with nitrocellulose modified with alkyd resins. One of the big advantages of using this system is the production of fast drying coatings, with satisfactory performance, even under unfavorable application conditions. Due to good polishing properties in the film, defects can be removed using a good finishing obtained in terms of final appearance.

However, one of the main deficiencies with this system is low UV resistance of the film with the recoated areas having the tendency of discoloring faster than the original. The use of UV absorbers minimizes the problem but does not eliminate the effect. Despite these problems, this system is still used because of its easiness to use even in unfavorable application conditions.

Polyurethane (PU) system formulations were introduced in Germany at the start of the 1980's. Its use became worldwide when coating companies exported their products to other countries, mainly when German assemblers, like BMW and Mercedes, guaranteed the use of two-component (2K) PU systems in repairing its automobiles.

In terms of refinishing, layers have a composition similar to the original paint functions. When a metal sheet is unprotected, it needs to be leveled and corrected before applying leveling primers. Existing masses on the market are:

- Two-component, high-quality polyester mass that substitutes rapid and synthetic masses.
- Rapid nitrocellulose mass used a lot due to its easy application and drying.

The next stage is the use of a primer, which can be a nitrocellulose base, for filling in grooves. It allows fast drying or a PU-base that presents greater filling-in with rapid drying.

Finishings can be classified as smooth, metallic or pearl effect, with the qualities depending on garage installations and the year of the car to be recoated.

a) Smooth single layer colors:

- Nitrocellulose lacquer
- Polyester
- Polyurethane polish

b) Two-layer metallic, pearl or smooth colors

- Acrylic lacquer
- Double- or triple- layer polyester metallic base
- Acrylic varnish
- 2K PU varnish

PU systems for automotive refinishing are mostly based on a polyol, generally polyester or acrylic with a polyisocyonate. They have the following characteristics when compared to other technology used for recoating:

- The possibility of working with a greater number of solids, restricting VOC emissions.
- Application properties: the disadvantage of 2K PU systems is the need to mix them with polyisocyanate along with the polyol before use. These systems contain three components because there is also a need to mix them with a thinner to adjust the application viscosity. When the application is made in areas with higher humidity, it is also necessary to add an accelerator.

- Drying: 2K PU systems cure by reacting with hydroxyl groups from polyols and isocyanate groups from polyisocyanate which have mechanical and durability properties greater than the film formed. N/C and TPA cure using solvent evaporation.
- Greater dry film thickness, as a result of the greater amount of solids in the application.
- 2K PU systems show greater color and shine retention than the other two, and generally are equivalent to the system used in the original paint in terms of performance.
- 2K PU systems also show an excellent chemical resistance, differentiating it from the other two systems.[142]

Table 8.3 presents an example of solvent formation for a polyester-based coating created using the SOLSYS® system developed by RHODIA. Also represented in this table are initial solubility parameters of the mixture, in addition to normalized distance during solvent system evaporation.

The main points that can be highlight in Table 8.3 are:

- The solvent system is made up of an ethanol composition, which is a light solvent; butyl acetate and xylene which are solvents with an intermediate evaporation rate; and propylene glycol monomethyl ether acetate (PMA), which is the heaviest solvent in the system.
- Solvent system solubility parameters result from the sum of the individual solvent parameters.
- Normalized distance results in the polyester resin are shown for the original system, or in other words, without evaporation of the solvent system; with 50% of the mass of the evaporated solvent system; and finally with 90% of the mass of the evaporated solvent system. It can be shown that the original system value is 0.56, thus lower than 1.00 which means the resin has good solubility initially in the solvent system. With the evaporation of the solvent system there is a change in the solubility to 0.76 at 50% of evaporation and 0.72 at 90% evaporation. This can be shown graphically as in Figure 8.11. Atmospheric conditions considered in the simulation include relative humidity of 70% at 25 °C.
- This figure shows that throughout evaporation, the solvent system remains in the solubility surface of the polyester resin. This fact determines that the resin has good solubility during the complete evaporation stage of the solvent system. This is a fundamental condition for good accommodation of polymer chains, resulting in good film formation.

- If at any time during the evaporation the solvent system normalized distance is greater than 1.0, it may indicate the possibility of resin precipitation at that moment, which would negatively affect film formation.
- This good polymer resin solubility can be understood in relation to selective evaporation of solvents that make up the formulation.

Table 8.3. Composition, initial solubility parameters, and solubility development during solvent system evaporation in a polyester-based system

Solvent System Composition		Values in % m/m
Solvents	Butyl Acetate	18.0
	Propylene Glycol Monomethyl Ether Acetate	13.0
	Ethanol	25.0
	Xylene	51.0
	Total	100.0
Solubility Parameters	δD	16.83
	δP	3.68
	δH	8.06
	δG	19.02
Standard Polyester	Initial Normalized Distance	0.56
	50% Evaporated Mass	0.76
	90% Evaporated Mass	0.72

Evaporation Conditions:
25 g/m^2/Temperature – 25 °C/Relative Air Humidity – 70%/Air Speed – 0.0667 — Substrate – Metallic

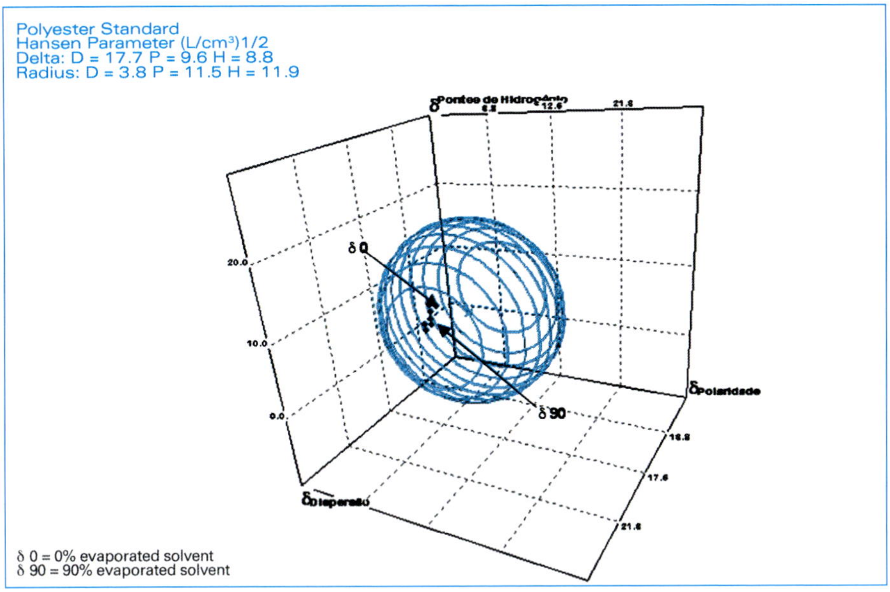

Figure 8.11. Development of Solvent System Solubility Parameters on the polyester resin solubility surface.

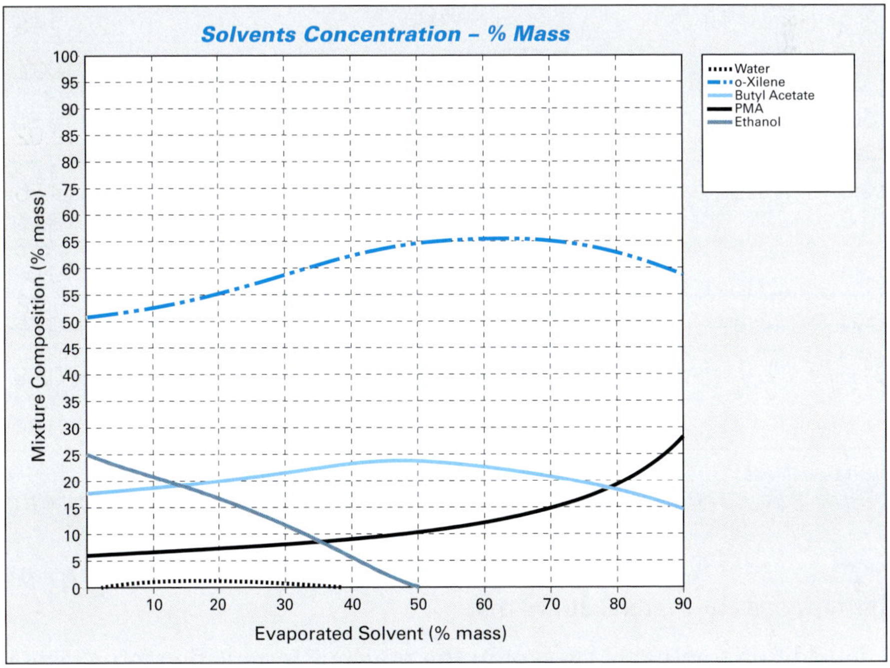

Figure 8.12. Individual solvent evaporation present in the polyester resin formulation.

- As seen in Figure 8.12, ethanol completely evaporates when 50% of the solvent system is evaporated. It is also possible to verify a small amount of water absorption from the humidity in the atmosphere, but which is practically nothing in the solvent system at any time of the evaporation. Therefore, it does not compromise resin solubility during film formation.

Another solvent system could be proposed for the same polyester resin, using different solvents in relation to what is shown in Table 8.4. Again, the formulations were analyzed using the SOLSYS® system.

Table 8.4. Composition, initial solubility parameters, and solubility development during solvent system evaporation in a polyester-based system

Solvent System Composition		Values in % m/m
Solvents	Ethyl Acetate	17.0
	Butyl Acetate	12.0
	Ethyl Glycol Acetate	7.0
	Ethanol	22.0
	Xylene	42.0
	Total	100.0
Solubility Parameters	δD	16.65
	δP	4.03
	δH	8.29
	δG	19.04
Standard Polyester	Initial Normalized Distance	0.56
	50% Evaporated Mass	0.69
	90% Evaporated Mass	0.64

Evaporation Conditions:
25 g/m^2/Temperature – 25 °C/Relative Air Humidity – 70%/Air Speed – 0.0667 — Substrate – Metallic

This new solvent system shows that:

- In addition to ethanol present in the previous formulation, ethyl acetate was added as another light component. This addition, along with the use of 7% ethyl glycol acetate, contrary to PMA, resulted in an expressed reduction of butyl acetate and xylene concentrations.

- In relation to the previous formulation, solubility parameters as well as normalized distances change very little. What may be shown is just a little reduction in distance when 50% and 90% of the solvent system was evaporated. Since the values are extremely close and below 1.0, i.e., within the solubility surface, the two systems can be considered the same in terms of performance. This development can be graphically confirmed in Figure 8.13.

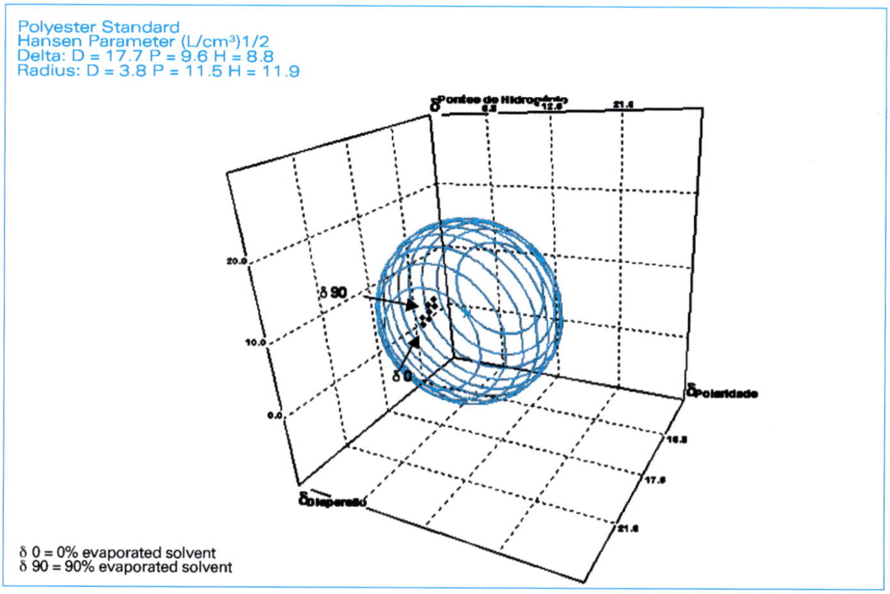

Figure 8.13. Development of solvent system solubility parameters on the polyester resin solubility surface.

- It can be shown again that at any time during solvent system evaporation, it is within the polyester resin solubility surface.
- Solvent system composition during evaporation is shown in Figure 8.14.
- When 90% of the solvent system was evaporated, the formulation composition is approximately 12% butyl acetate, 40% xylene, and 48% ethyl glycol acetate, which is a good solubility of polyester resin at this stage, guaranteeing a good film formation.

A third formulation alternative can be proposed, mainly modifying the heavy solvent. In this case, the solvent system, Rhodiasolv TV101 was used to substitute the PMA and the ethyl glycol acetate. Table 8.5 shows a summary of the properties obtained for the polyester resin.

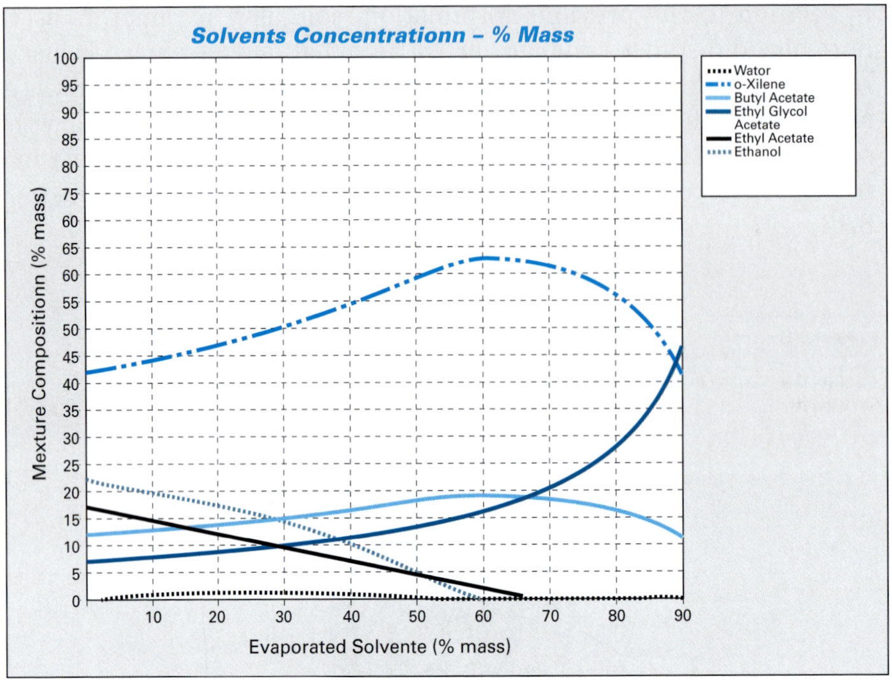

Figure 8.14. Individual solvent evaporation present in the polyester resin formulation.

This table, compared to the other two, shows the versatility of this type of formulation simulation. Therefore, even with significant changes in terms of composition, solubility parameters are adjusted to remain practically the same throughout all stages of evaporation, not compromising system solubility and thus, a good film formation.

Figure 8.15 shows solubility parameter progression in relation to the solvent system evaporation.

The same conclusion can be reached in relation to solubility parameter maintenance of the solvent system in the solubility surface during the full evaporation stage, guaranteeing a good performance of film formation.

These three cases show it is possible to maintain solvent system performance even with significant changes from the qualitative and quantitative point of view of their components, using the correct balance of formulations when considering individual component characteristics in each.

Another important system in automotive refinishing is the PU system, mainly used as a topcoat. This system characteristically shows a cure by chemical reaction as the solvent evaporates.

Table 8.5. Composition, initial solubility parameters, and solubility development during solvent system evaporation in a polyester-based system		
Solvent System Composition		Values in % m/m
Solvents	Ethyl Acetate	12.0
	Butyl Acetate	13.0
	Rhodiasolv TV 101	8.0
	Ethanol	18.0
	Xylene	49.0
	Total	100.0
Solubility Parameters	δD	16.77
	δP	3.63
	δH	7.03
	δG	18.54
Standard Polyester	Initial Normalized Distance	0.59
	50% Evaporated Mass	0.78
	90% Evaporated Mass	0.79

Evaporation Conditions:
25 g/m²/Temperature – 25 °C/Relative Air Humidity – 70%/Air Speed – 0.0667 — Substrate – Metallic

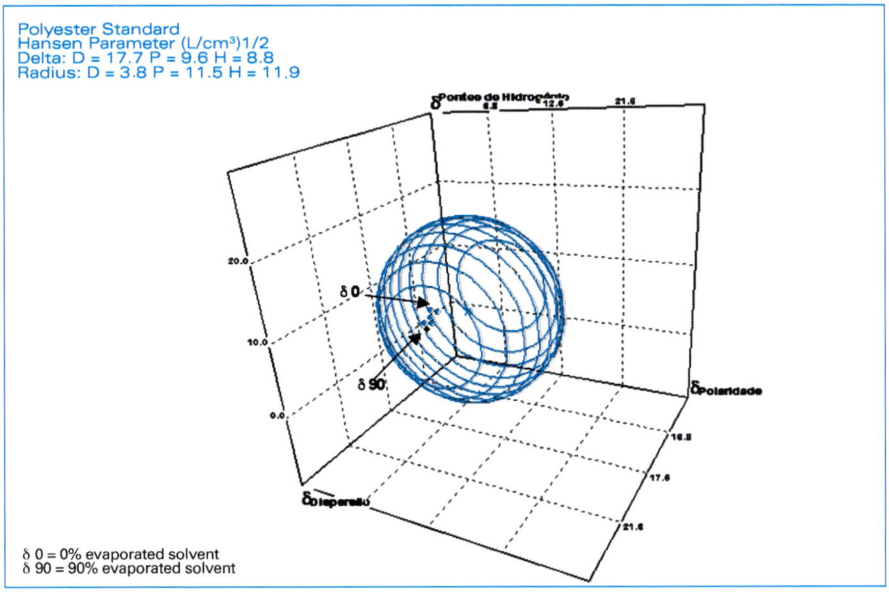

Figure 8.15. Development of Solvent System Solubility Parameters on the polyester resin solubility surface.

Next, examples of solvent system formulations for a 2K PU system will be presented using:

- Hydroxyl polyester polyol;
- Polyisocyanate Tolonate HDB 75 MX.

The solvent systems presented correspond to the solvent mixture composition in the polyol and polyisocyanate formulations before application, with this mixture being responsible for all film formation stages from the initial moment until the whole solvent system is evaporated. Just as for the film formation, there is also a chemical reaction even more important for the maintenance of good solubility of the two components during system evaporation. In addition to good polymer chain organization, it also gives mobility which increases the probability of polyol and polyisocyanate colliding into each other and consequently increasing the probability of a chemical reaction between them.

Table 8.6 shows composition and performance data from the three formulations used for a 2K PU system with a varnish application on the recoating market.

The main observations from this comparative table are:

- Despite formulation composition differences, when solubility parameters are analyzed there is no significant differences among them.
- When normalized distances are evaluated, it is important to consider that they are related to the two individual polymers; that is, to polyol and polyisocyanate. Figure 8.16 shows spatial location of formulation 3 in relation to these 2 surfaces.
- The polyol solubility surface is located inside the polyisocyanate surface. Solvent system solubility parameters are different from the two polymers, i.e., normalized distance data needs to have values showing the solvent system is positioned such that it is common to the 2 solubility surfaces.
- Normalized distance data indicate that all solvent systems act at the limit of the polyol surface solubility with values very close to 1.0, while with the polyisocyanate they are found in its surface with average values of 0.6. Thus, all solvent systems are common to the two surfaces.

Table 8.6. Composition, initial solubility parameters, and solubility development during solvent system evaporation for 2K PU systems

Solvent System Composition		Formulation 1 (% m/m)	Formulation 2 (% m/m)	Formulation 3 (% m/m)
Solvents	Ethyl Acetate		14.5	17.0
	Butyl Acetate	25.0	32.0	30.0
	Propylene Glycol Monomethyl Ether Acetate	7.0		
	Ethyl Glycol Acetate		8.0	
	Rhodiasolv TV 101			8.0
	Methyl Isobutyl Ketone	33.0		
	Toluene			25.0
	Xylene	35.0	45.5	20.0
	Total	100.0	100.0	100.0
Solubility Parameters	δD	16.35	16.73	16.74
	δP	3.65	2.78	3.0
	δH	4.59	5.32	4.57
	δG	17.36	17.77	17.61
Tolonate HDB 75 MX	Initial Normalized Distance	0.65	0.66	0.68
	50% Evaporated Mass	0.64	0.67	0.66
	90% Evaporated Mass	0.61	0.52	0.64
Standard Polyester	Initial Normalized Distance	1.01	1.01	1.01
	50% Evaporated Mass	1.01	1.02	1.01
	90% Evaporated Mass	1.0	0.93	1.01

Evaporation Conditions:
25 g/m^2/Temperature – 25 °C/Relative Air Humidity – 70%/Air Speed – 0.0667 — Substrate – Metallic

- Another important consideration is that during evaporation, they continue to maintain the same distances in relation to the two surfaces, guaranteeing good resin solubility during evaporation, and favoring all factors for a good film formation.
- Figures 8.17, 8.18, and 8.19 show selective evaporation of the formulation for the three solvent systems analyzed.

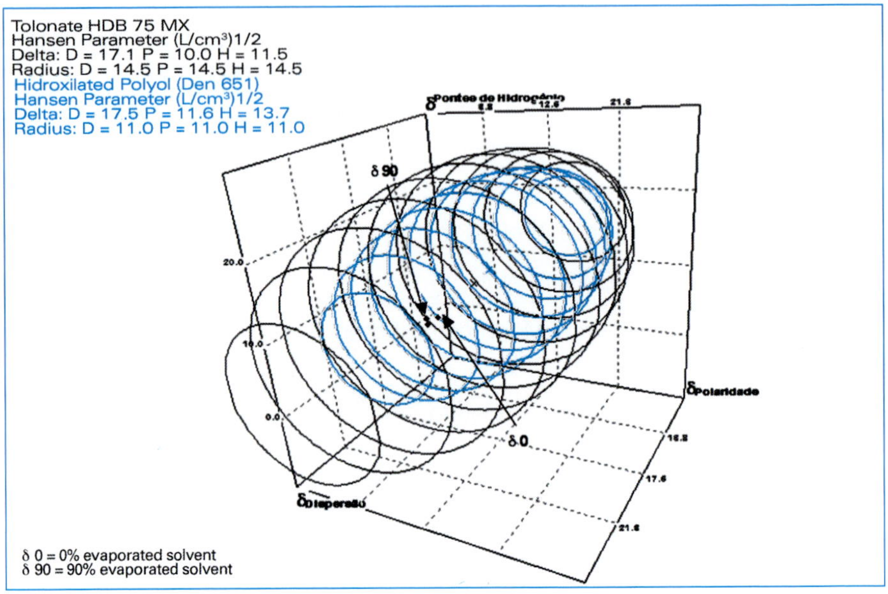

Figure 8.16. Development of Solvent System Solubility Parameters on the polyester resin solubility surface.

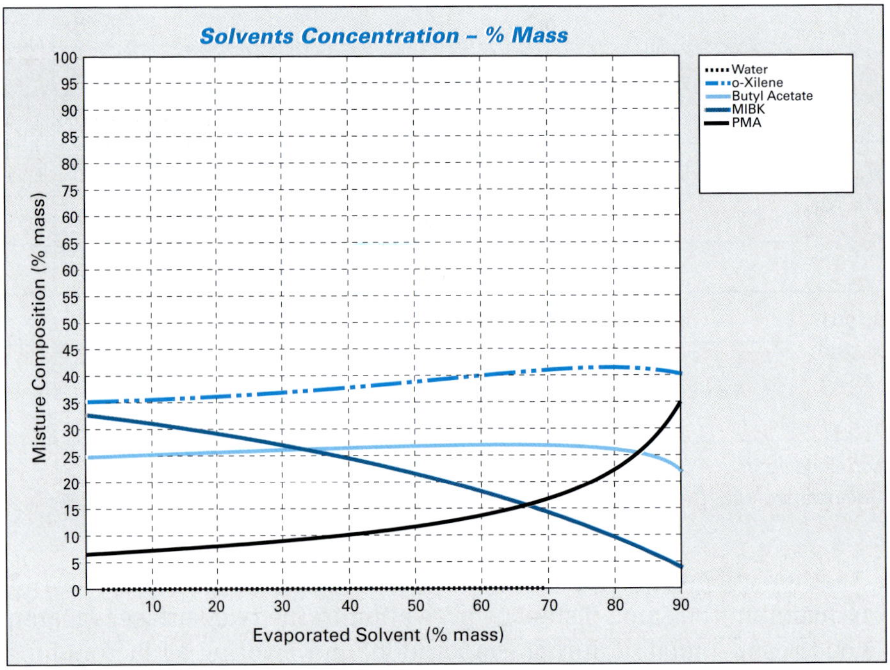

Figure 8.17. Individual solvent evaporation present in formulation 1.

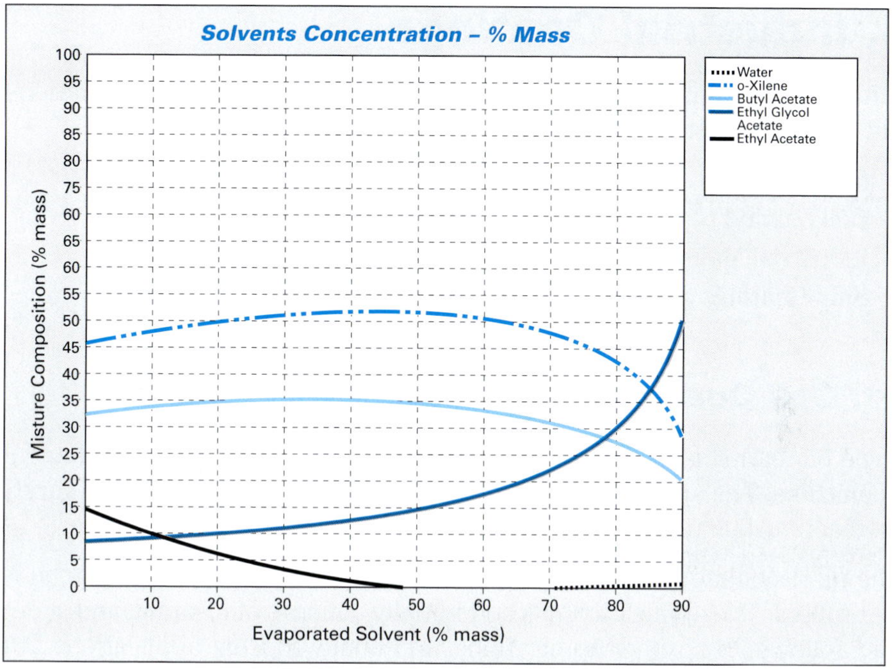

Figure 8.18. Individual solvent evaporation present in formulation 2.

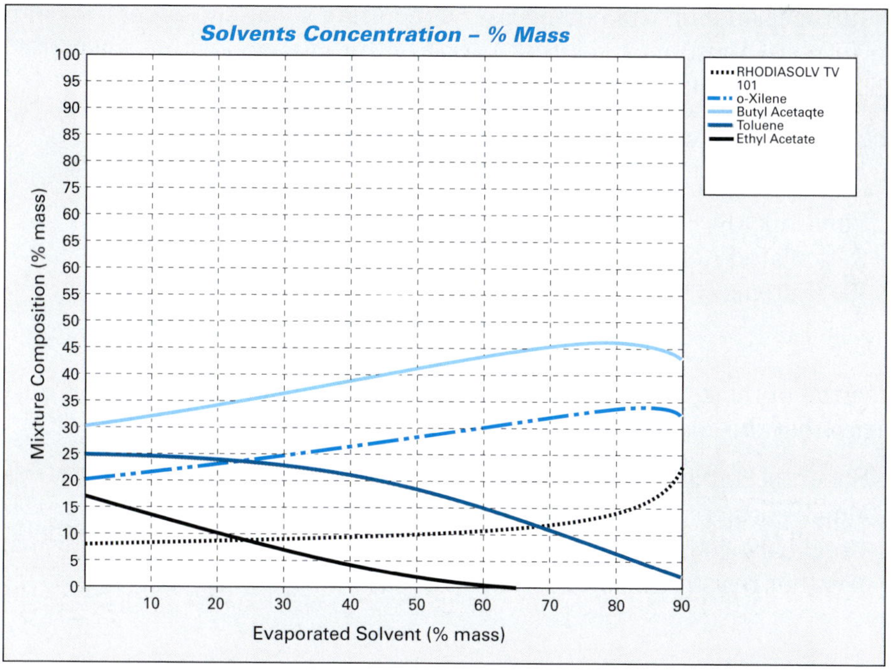

Figure 8.19. Individual solvent evaporation present in formulation 3.

8.4. Industrial Coatings

In addition to automotive coating and recoating, other important applications involving the use of industrial coatings are:

- Can coating
- Coil coating
- Industrial Maintenance
- Ship Coatings

8.4.1. *Coil Coating*

This type of coating is used for metal sheets that will later be re-rolled and shaped for specific use. The application process in general is a continuous high-speed application, reaching 200 meters per minute.

The most common method for applying a liquid coating to sheet metal is using transfer rollers. The application method initially consists of cleaning and sheet metal treating, followed by primer applications, and finally topcoat applications. After application, the coating needs to be quickly cured, producing a dry, hard film between varying times of 10 to 60 seconds, and then cooled.

This sheet metal is subsequently used in assembling for specific ends, or in other words, further handling will be necessary, requiring some film properties of coating formed to resist these changes that vary from producing coating for metal beams used in roof constructions to automotive oil filters.

The most common resins for coating formulation are:

- Amino-Alkyds;
- Vinyl-Alkyds;
- Reticulated Acrylics;
- PU Systems

In virtue of these application characteristics for sheet metal, there is a series of properties that the coating needs to have including:

- Shearing stability;
- Quick curing;
- Good balance between flexibility and hardness;
- Weather resistance.[141]

Table 8.7 analyzes the normalized distance between the solvent system and the melamine resin and shows interaction between them. The simulation in the example represents solvent system evaporation in ambient conditions. If this were so, total solvent evaporation would be extremely slow for two reasons: solvents show a slow evaporation rate and there is an affinity between them and the resin, which makes its complete elimination difficult.

Table 8.7. Composition, initial solubility parameters, and solubility development during solvent system evaporation for a melamine resin

Solvent System Composition		Values in m/m %
Solvents	Butyl Glycol	60.0
	AB9	40.0
	Total	100.0
Solubility Parameters	δD	15.97
	δP	4.52
	δH	7.65
	δG	18.28
Standard Melamine	Initial Normalized Distance	0.04
	50% Evaporated Mass	0.18
	90% Evaporated Mass	0.19

Evaporation Conditions:

25 g/m²/Temperature – 25 °C/Relative Air Humidity – 70%/Air Speed – 0.0667 — Substrate – Metallic

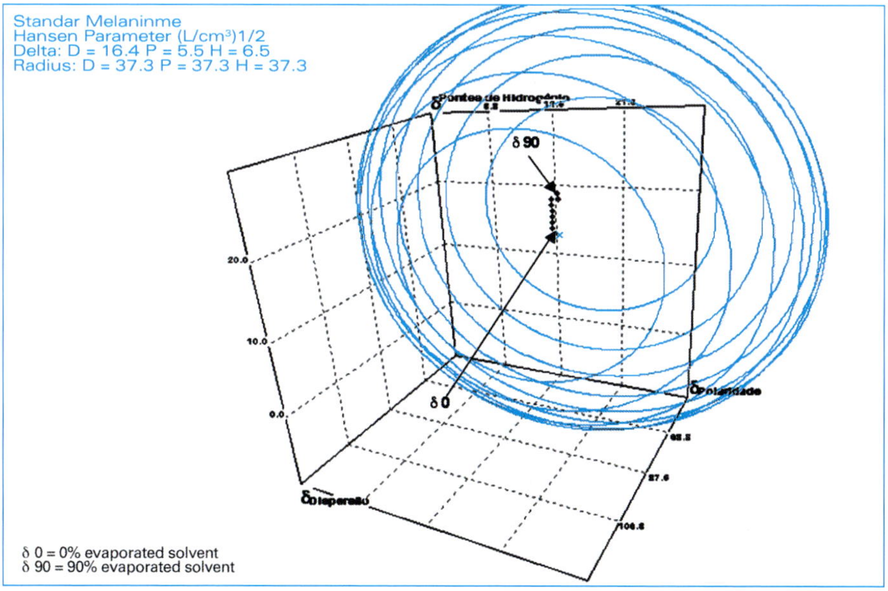

Figure 8.20. Development of Solvent System Solubility Parameters on the melamine resin solubility surface.

In general, as was shown, these systems are cured in ovens for quick solvent elimination on the film surface.

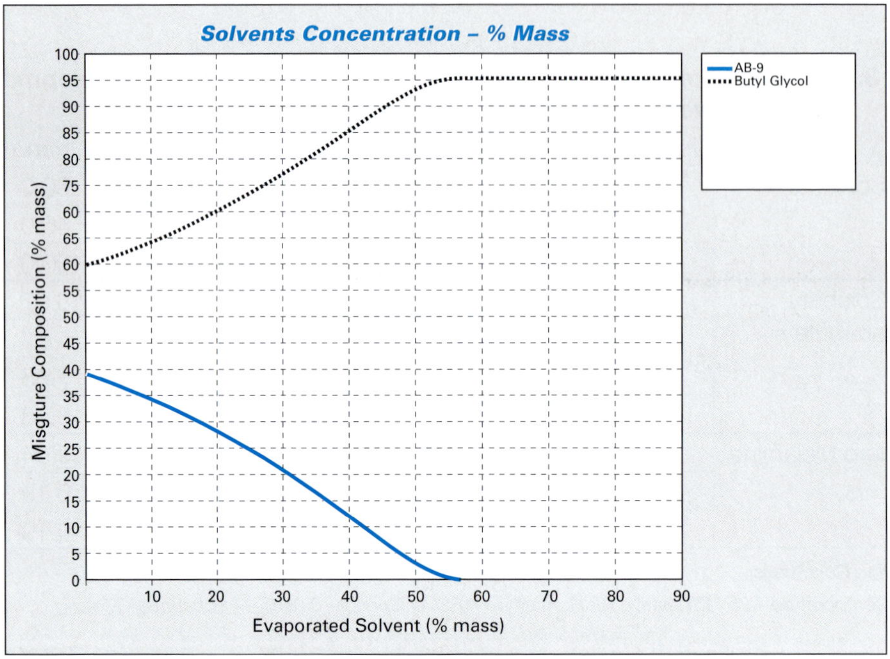

Figure 8.21. Individual solvent evaporation present in the formulation of the melamine resin.

8.4.2. *Can Coating*

This coating is used for several types of drums and containers for different ends, including food storage. What determines the type of coating necessary for packaging is:

- the nature of the product that will be stored
- the type of container construction
- the manufacturing method and container filling.

For example, for storing food products, what should mainly be considered is the possibility of absorption of coating substances on the part of the food product that can impact the quality. So for this specific application there are coatings approved by the FDA (Food and Drug Administration) resistant to chemicals and acids and can be sterilized.

The main resins used in these markets are phenyls, epoxies, vinyls, and alkyds. Phenyl resins are used in can coating and they are excellent for resisting chemicals in general, mainly solvents. Vinyl resins show excellent chemical resistance and flexibility, but have low resistance to heat and abrasion. The use of epoxy-based resins are excellent for the majority of properties, compared to the systems used, and for this they stand out in this market, having an application that includes food product storage.[141]

Table 8.8 gives an example of the solvent system formulation used in this market.

Table 8.8. Composition, initial solubility parameters, and solubility development during solvent system evaporation for an epoxy resin

Solvent System Composition		Values in m/m %
Solvents	Ethyl Glycol	20.0
	Butanol	25.0
	Xylene	40.0
	AB9	15.0
	Total	100.0
Solubility Parameters	δD	16.77
	δP	4.21
	δH	8.15
	δG	19.12
Standard Epoxy	Initial Normalized Distance	0.78
	50% Evaporated Mass	0.71
	90% Evaporated Mass	0.86

Evaporation Conditions:
25 g/m^2/Temperature – 25 °C/Relative Air Humidity – 70%/Air Speed – 0.0667 — Substrate – Metallic

Figures 8.22 and 8.23 show the solvent system solubility parameter progression and the system composition in the polymer film during evaporation.

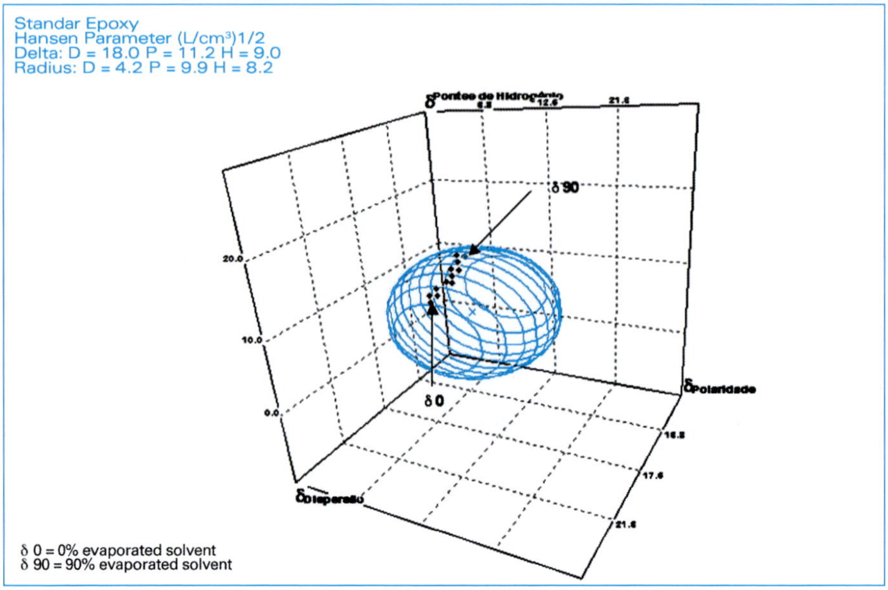

Figure 8.22. Development of Solvent System Solubility Parameters on the epoxy resin solubility surface.

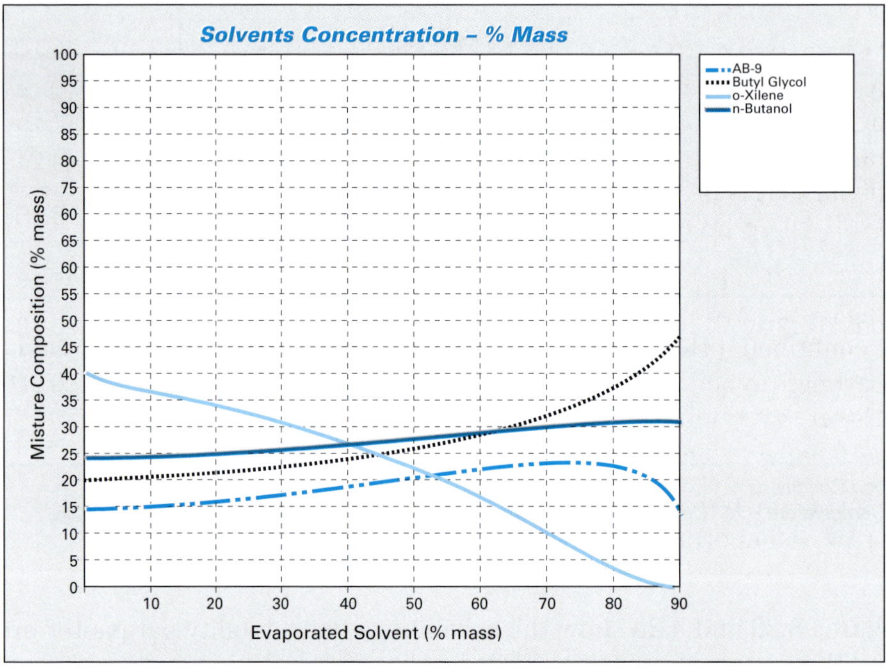

Figure 8.23. Individual solvent evaporation present in the formulation of the epoxy resin.

8.4.3. Industrial Maintenance

Industrial coating is a process for applying a coating film on metal, concrete, or masonry surfaces for protection, safety, hygiene, productivity, marketing, economy, and quality.

For metal surfaces, the coating film is to protect the substrate, isolating it from an aggressive medium through its adhesion, waterproofing, and flexibility properties. Primers, in addition to the necessity of having high quality waterproofing, also need to include anti-corrosive pigments in their formulation.

For concrete, the film must be resistant to the alkaline properties in concrete when it has pH properties of 12 to 13. With reinforced concrete, the film is for waterproofing the pores so as to avoid, for example, water absorption in its interior by capillarity. There are situations, as in the case of food and pharmaceutical industries and also hospitals in which there is the need to guarantee environments free of humidity and that are easy-to-clean, allowing a decontamination efficient, in which the coating has the basic role of guaranteeing these properties. For example, there are coatings that have anti-mold properties, stopping the development of microorganisms on its surface.

Table 8.9 shows two examples of resin systems used in industrial coatings. Each system shows an important difference in performance and cost for the resin and solvent system characteristics.

The same table also verifies that alkyd resins, with their main characteristic of long carbon chains, have low polarity and therefore the most appropriate solvents are aromatics and aliphatics which have similar characteristics. Actually, the most important solubility parameter that governs this system is dispersion (δD). Thus, the ethanol that is present is exercising the function of a dilutent since its impact on the solvent system performance is little considering this resin.

Using the same evaluation for the 2K PU system, the presence of groups with polar characteristics and with a tendency to form hydrogen bonds in the chains can be confirmed, principally those of isocyanate and polyurethane formed by the polyisocyanate reaction with the polyol. This chemical composition directly impacts the constitution of the solvent system which contains esters with low, medium, and high evaporation rates.

In this resin system, in addition to dispersion, polarity and hydrogen bonding forces play an important part in the definition of the solvent system.

Table 8.9. Composition, initial solubility parameters, and solubility development during solvent system evaporation for a 2K PU system and an alkyd system

Solvent System Composition		Alkyd System	2K PU System
Solvents	Ethyl Acetate		23.0
	Butyl Acetate		40.0
	Ethyl Glycol Acetate	7.0	7.0
	Ethanol	25.0	
	Tolueno	38.0	17.0
	Xylene	30.0	13.0
	Total	100.0	100.0
Solubility Parameters	δD	17.25	16.45
	δP	3.36	3.40
	δH	7.28	7.28
	δG	19.02	17.72
		Short Alkyd in Soybean Oil	Tolonate HDB
	Initial Normalized Distance	0.27	0.61
	50% Evaporated Mass	0.32	0.61
	90% Evaporated Mass	0.29	0.46
			Polyol
	Initial Normalized Distance		1.05
	50% Evaporated Mass		1.05
	90% Evaporated Mass		0.85

Evaporation Conditions:
25 g/m^2/Temperature – 25 °C/Relative Air Humidity – 70%/Air Speed – 0.0667 — Substrate – Metallic

Another point to highlight is that in the case of the alkyd resin, as a single-component system, there is a degree of greater freedom for the solvent system composition, as shown in Figures 8.24 and 8.25 in relation to the 2K PU system, which the solvent system needs to solubilize both resins during all evaporation stages.

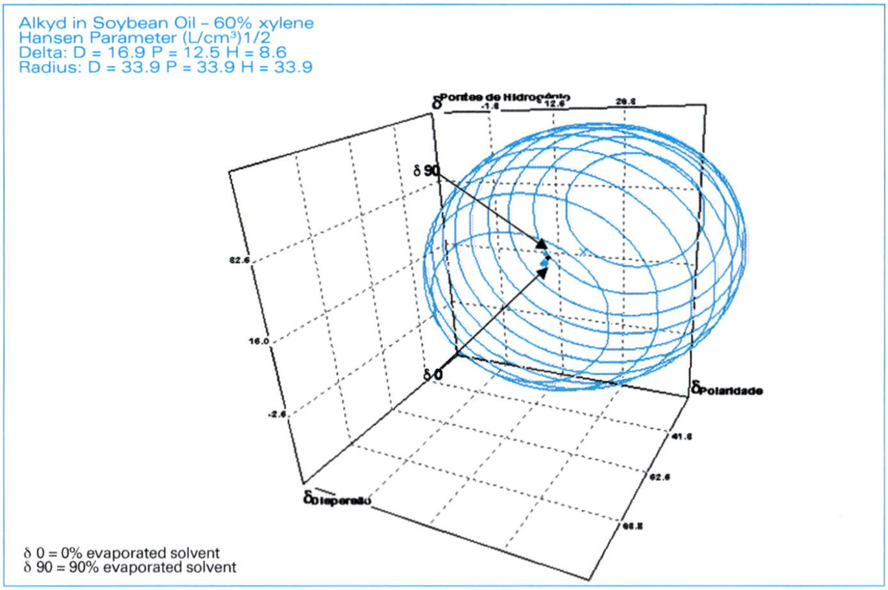

Figure 8.24. Development of Solvent System Solubility Parameters on the alkyd resin solubility surface.

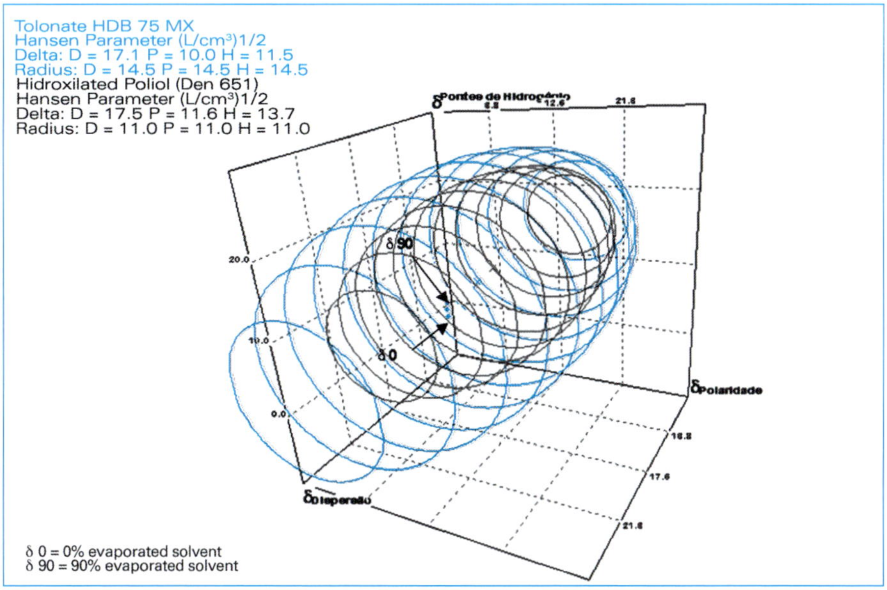

Figure 8.25. Development of Solvent System Solubility Parameters on the 2K PU solubility surface.

When the progression of the solvents systems for the two coatings are compared, during evaporation in the alkyd resin there is an increase in the xylene concentration, making it is a true solvent for it. However, in the 2K PU system, the concentration increases with the butyl acetate which shows greater solvency power for the polyisocyanate and for the polyester.

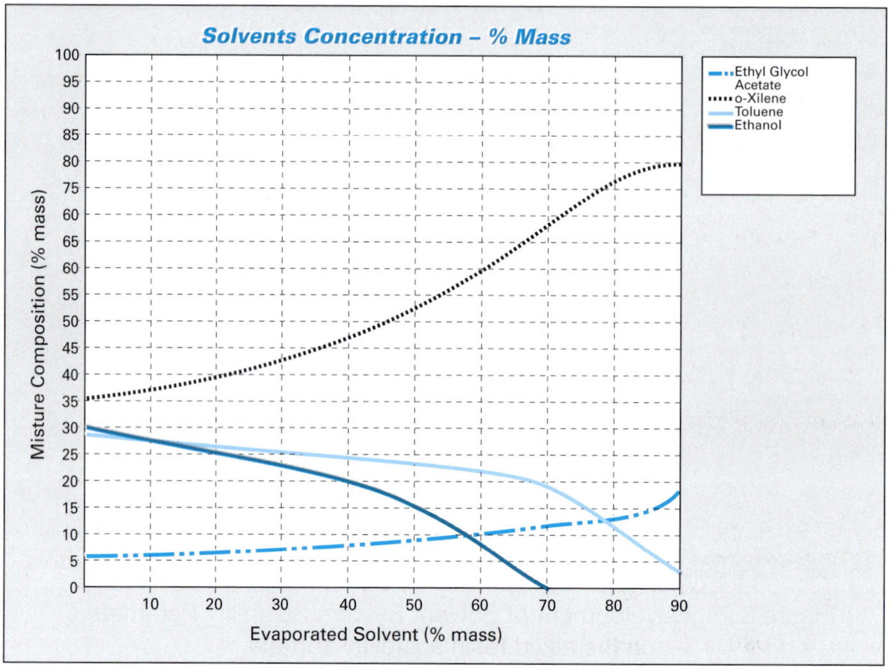

Figure 8.26. Individual solvent evaporation present in the alkyd resin formulation.

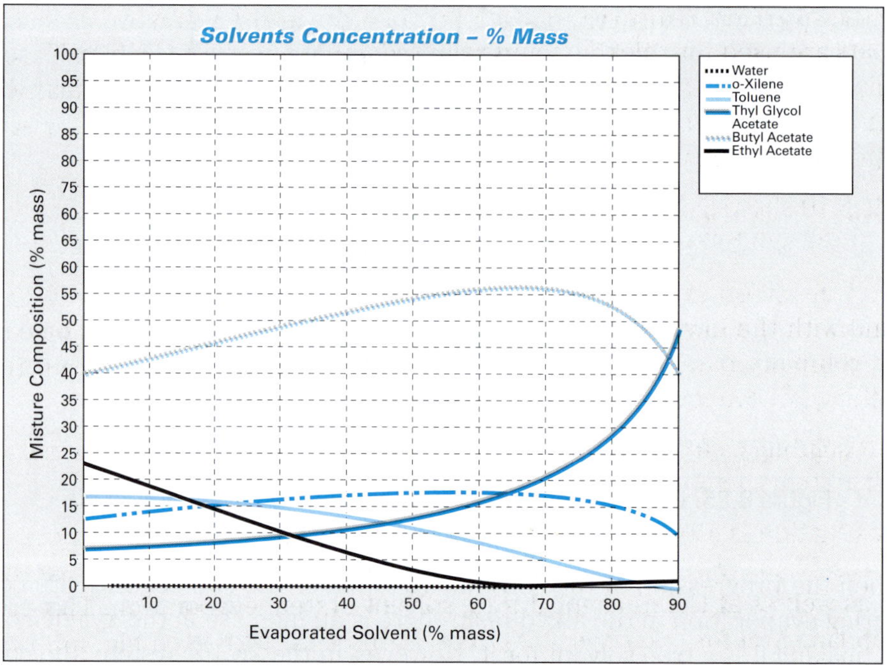

Figure 8.27. Individual solvent evaporation present in the 2K PU system formulation.

8.4.4. Wood Varnish

Varnishes on wood surfaces are for improving appearance and resisting effects from weathering. Main properties for wood finishings are:

- Fast drying: the substrate quickly allows handling and storage;
- Good wetting penetration: for restoring wood color;
- Low cost;
- Good flow: The varnish penetrates the wood and does not clog pores;
- Appearance: soft appearance even at low levels of shine;
- Low levels of shine: satin finishes are frequently used in this market.

Finishes with nitrocellulose meet these criteria well. PU systems are out in terms of cost and, depending on the formulation, are very slow drying, have a poor flow, and wetting of substrates can be even worse.

Weather exposure can lead to photo degradation.

A wood coating system includes the primer, undercoat, and topcoat. Primers are mainly for appropriately joining the substrate and the subsequent finishing layer. In addition, they must also work as a wood sealant and offer additional resistance to weathering during the period before the final finishing coat is added. Undercoats are used for thickness and required opaqueness for the film. Topcoats are applied for their properties of desired surfaces, like shine and durability. For these purposes there are waterborne and solventborne systems, with the use of each depending on finishing needs.

Table 8.10 shows some typical formulations used on the market, along with an analysis of the solubility parameters during solvent system evaporation.

The composition of both solvent systems is very similar, changing only in the compound with the low evaporation rate, also known as retarder. It can be confirmed that solubility parameters are practically the same as the two systems in any given instance of evaporation.

Resin solubility surfaces are shown in Figures 8.28 and 8.29, demonstrating that both systems act in the same region.

This situation is a typical example of the intersection of response surfaces in which the solvent system needs to be located at their intersection at the initial moment as well as at the moment during solvent system evaporation. This condition is fundamental for good miscibility between resins, with good film formation, guaranteeing its mechanical properties and in terms of appearance. Considering the two systems, it can be seen that they act at the solubility surface limit of the nitrocellulose resin, and approaches the maleic resin center.

Table 8.10. Composition, initial solubility parameters, and solubility development during solvent system evaporation for a nitro-maleic system

Solvent System Composition		Formulation 1	Formulation 2
Solvents	Ethanol		13.0
	Ethyl Acetate	15.0	16.0
	Methyl-isobutyl-ketone	17.0	8.0
	Toluene	10.0	33.0
	Xylene	30.0	25.0
	Butyl Glycol	23.0	
	Diacetone Alcohol		5.0
	Total	100.0	100.0
Solubility Parameters	δD	16.88	16.99
	δP	3.74	3.60
	δH	6.47	5.98
	δG	18.46	18.36
Nitrocellulose BT 1464½" AN ETA	Initial Normalized Distance	0.97	0.61
	50% Evaporated Mass	1.07	0.61
	90% Evaporated Mass	1.01	0.46
60% Maleic in xylol	Initial Normalized Distance	0.27	0.28
	50% Evaporated Mass	0.36	0.4
	90% Evaporated Mass	0.35	0.23

Evaporation Conditions:
25 g/m^2/Temperature – 25 °C/Relative Air Humidity – 70%/Air Speed – 0.0667 — Substrate – Metallic

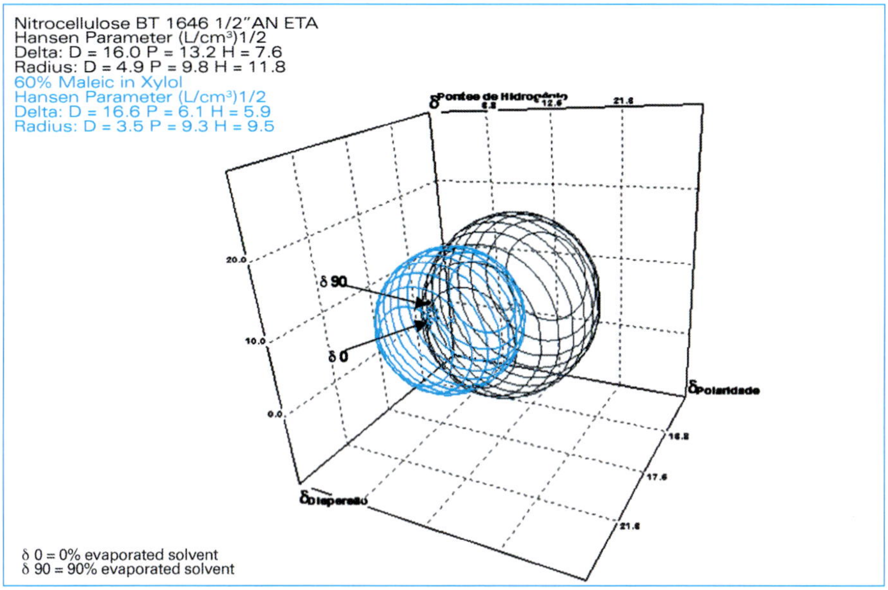

Figure 8.28. Development of Solvent System Solubility Parameters for a Nitro/Maleic resin system – Formulation 1.

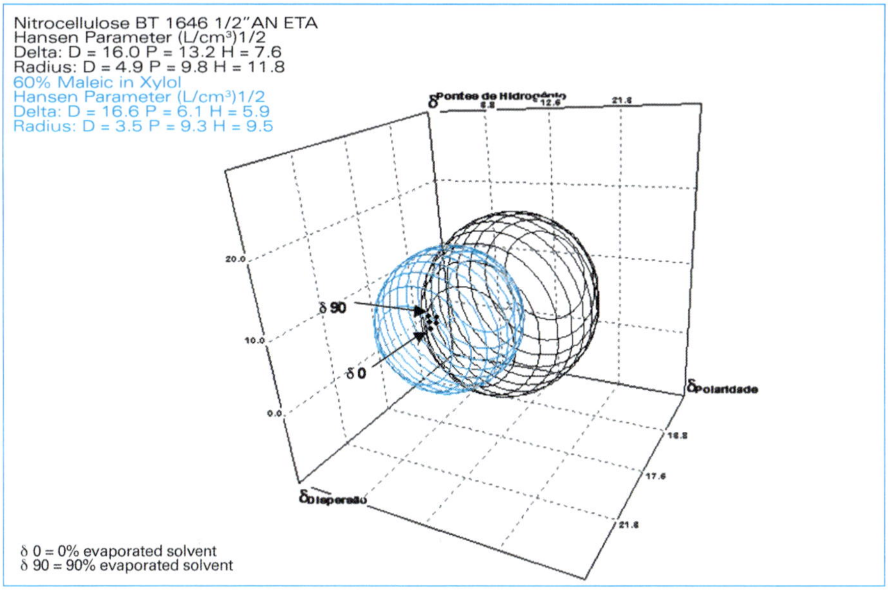

Figure 8.29. Development of Solvent System Solubility Parameters for a Nitro/Maleic resin system – Formulation 2.

Solvent system composition during evaporation is shown in Figures 8.30 and 8.31.

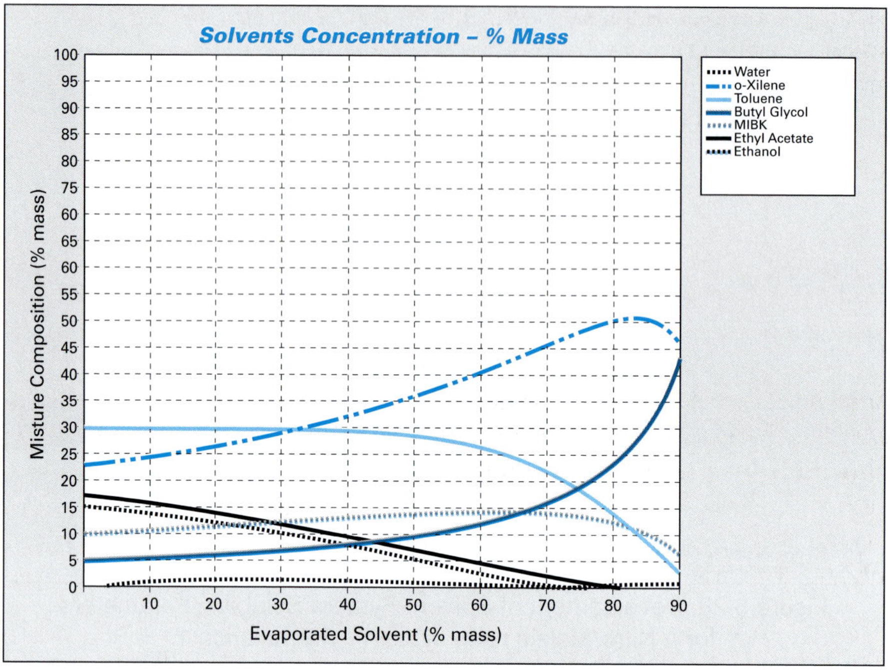

Figure 8.30. Individual solvent evaporation present in the nitro-maleic resin – Formulation 1.

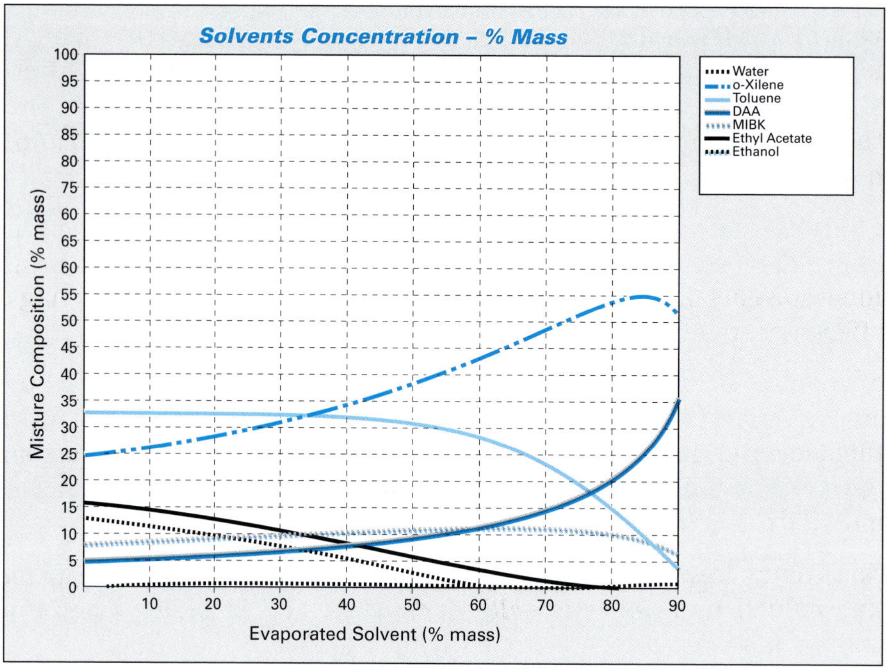

Figure 8.31. Individual solvent evaporation present in the nitro-maleic resin – Formulation 2.

Looking at the graphs, both systems show practically the same quantitative film composition in the coating, with 90% of the evaporated solvent mass, only changing the retarder, that in this case is butyl glycol and in the other, diacetone alcohol.

Since in the two cases the normalized distances are the same, system performance of the two, and in the last instance, of the two retarders, is the same.

8.4.5. Printing Inks

8.4.5.1. Introduction

The graphics area fulfills an important role in the perpetuation and recording the development of humanity. Within this context, the packaging market becomes more demanding each year. Expansion of the global market led to the surge of a large variety of ink processes capable of meeting new needs, making it is necessary for the ink and coating processes to accompany development and new market tendencies. Thus, solvents have also had to accompany these changes.

Solvents continue being used in large quantities in the main ink printing markets, like lithography, flexography, rotogravure, offset and silkscreen. Comparing the solvent with other formulation, pigment, resin, synthetic, additive, and filler constituents, organic solvents have a higher volume and generally a lower cost.

Solvent power and volatility are extremely important properties that influence coating characteristics used in different types of printing. The solvent used in coatings for lithography, offset, flexography, and rotogravure is discussed in relation to the solvent properties that influence coating stability, easiness in printing use, and drying on the substrate.

The formulation used for flexography and rotogravure coatings to print food packages requires a careful choice of solvents in order to avoid odor problems and residual solvents on the printed material. Solvent properties and drying equipment efficiency are equally important.

Two factors influence technological progress for coatings: the need to print on different types of substrates and the demand that coatings print with the same efficiency and quality as they do with high-speed machines. Components used in greater volume and have a lower cost are solvents that have an important role in meeting these demands.

Market trends lean toward complete elimination of petrochemical materials, decreasing influence on human health and safety, and including legislation on environmental pollution.

8.4.5.2. Solvent Function in Printing Inks

Solvents are for dissolving resins used in coating preparation, forming polymer films, facilitating the orientation of the elements and the pigmented solutions so they produce coatings. Solvents also help in wetting and pigment dispersion.

Solvent system evaporation is the main mechanism for drying in flexography, rotogravure, lithography, and offset printing processes.

Solvent or solvent system absorption in paper and film pores is normal for all printing processes, but the real contribution for drying depends on the substrate and the printing process. For example, the combination of absorption and oxidation is fundamental for quick drying in lithography printing processes on coated paper.

The most important property in the solvent is its solvency power, followed by the evaporation rate, which controls drying, and then stability in the printing machine. Solvent effects on the printed substrate are also important. Residual solvent on the printed package after final drying can cause undesirable odors in this market, mainly in the food sector. In this application, the use of solvents with low levels of odor is desirable.

8.4.5.3. Flexography

The U.S. Flexographic Technical Association defines flexography as being the method of direct rotary printing that uses resilient relief image plates made of rubber or photopolymer material which can be affixed to cylinders of various repeat lengths.

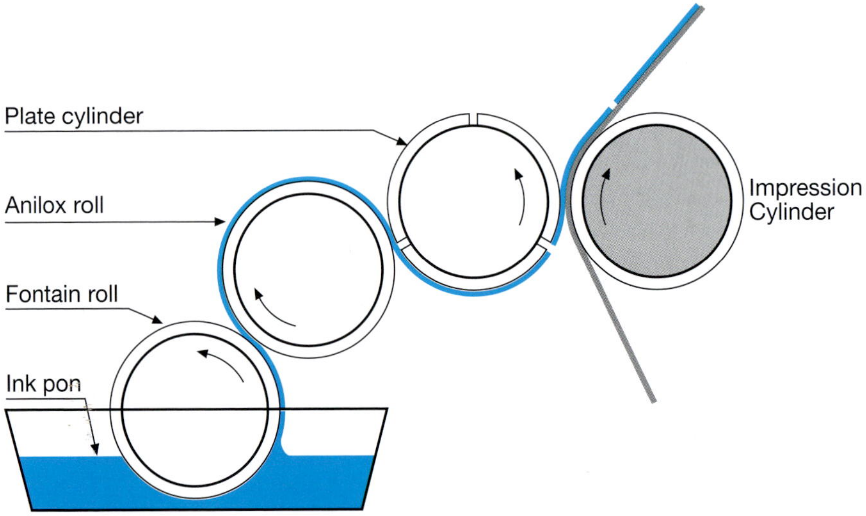

Figure 8.32. Simplified diagram of flexographic printing[133-134].

Flexography started in the United States in the 1920's and is considered a new printing method in rapid transformation. At that time it was called aniline printing. In 1952 the process became known as flexographic printing[131-132].

The ink is applied with an anilox roll and the excess is removed using doctor blades, capable of transporting fluid or paste coatings to virtually any substrate.

Due to liquid coating characteristics and raised lettering used, the flexographic printing system is one of the most versatile, being able to print on different substrates like paper, plastic, shopping bags, corrugated cardboard, ceramic, etc. (using machines especially manufactured for each kind of use).

Printing on waterproof substrates can be achieved with liquid coating and extremely fast drying (using heaters and fans installed in the substrate process after printing).

Cleaning must be done using the same solvent used in coating to help with coating removal on the plate surface and guarantee solvent compatibility with the plate. The coating will never completely dry on the cliche. When the machines are stopped, they should be cleaned using a soft-bristle brush.

Even from one day to the next, the plate must be completely cleaned, removing the coating that penetrates in the negative text space ("empty") and in small letters. Excess coating on the cliche is one of the most common enemies in quality flexographic printing.[135]

Solvents routinely used in flexographic coatings include alcohol (ethyl, isopropyl, n-butyl, and isobutyl), esters (ethyl acetate, propyl acetate, and isopropyl acetate), ethyl glycol, propylene glycol monomethyl ether, and diacetone alcohol.

The perfect transfer from the anilox to the photopolymer depends mostly on the solvents. This process must be balanced, i.e., drying must be adjusted so that the image is printed on the substrate with the desired quality and ideal drying for appropriate residual levels of solvents for the end market.

Table 8.11 shows some typical market formulations, analyzed according to the Hansen solubility parameter concept using the SOLSYS® – Design de Sistemas Solventes – Rhodia tool. Resin solubility parameters were determined using the description in part 8.3.1.1. The resin system considered for this analysis was fumaric/nitrocellulose.

The solvent system used in the formulation found on Table 8.11 shows the normalized distance, 90% of evaporated mass, approaching the solubility volume limit for the fumaric and nitrocellulose resins. The solvent system's initial solubility performs well with resins.

Table 8.11. Solvent System Composition, Solubility Parameters, and Normalized Distances in a Fumaric/Nitrocellulose Resin System

Solvent System Composition		Values in m/m %
Solvents	Ethanol	52.0
	Ethyl Acetate	18.0
	Isopropanol	20.0
	Butanol	6.0
	Propylene Glycol Monomethyl Ether	4.0
	Total	100.0
Solubility Parameters	δD	15.79
	δP	7.41
	δH	16.17
	δG	23.79
Standard Nitrocellulose	Initial Normalized Distance	0.76
	50% Evaporated Mass	0.85
	90% Evaporated Mass	0.83
Standard Fumaric	Initial Normalized Distance	0.75
	50% Evaporated Mass	1.01
	90% Evaporated Mass	0.90

In the ink coating market, formulations are purposely formulated at the solubility surface limit, making easy elimination of solvents, thus decreasing the retention possibility in the final printing. Here, quick system evaporation favors the absence of defects in the film.

This graph also indicates δD, δP and δH orce values during solvent system evaporation. Therefore, it is possible to analyze possible defects in the film's final formation.

Using 2 or more resins in preparing the coating, solubility surfaces of the polymers involved should be considered for choosing the best solvent system, thus meeting solubility and evaporation needs for the polymers in question.

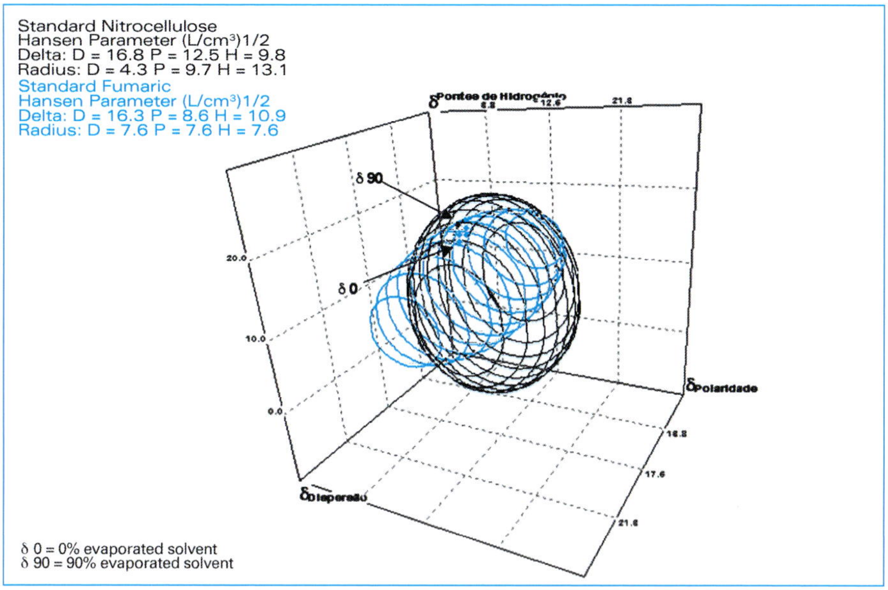

Figure 8.33. Positioning of the solvent system on the solubility system surface of the Fumaric and Nitrocellulose Resins.

Figure 8.34 shows the individual solvent evaporation during drying and film formation, indicating the concentration at each evaporation stage. In this graph it is possible to analyze the influence of each solvent on film performance.

It should be noted in this case that the appearance of water in the graph is from relative humidity considered in the simulation's atmospheric conditions. This phenomenon occurs in the other solvent system evaporation graphs presented.

In Table 8.12, another solvent system is evaluated using the same resin system.

As discussed in the previous example, the solvent system used here also shows the normalized distance approaching the solubility volume limit for the fumaric resin. This is particular to the ink coating market.

The graph in Figure 8.35 also indicates δD, δP and δH force values during solvent system evaporation.

The greater the unideal behavior of a solvent mixture solution is, the greater the system's evaporation rate will be. This property is fundamental for the flexography and rotography printing market.

Figure 8.36 shows individual solvent evaporation during drying and film formation. Thus, the solvent that will be present in this phase can be analyzed. Ethyl acetate evaporates almost completely at 60% of the solvent system evaporated.

In Table 8.13, we work with another solvent system using the same pair of resins.

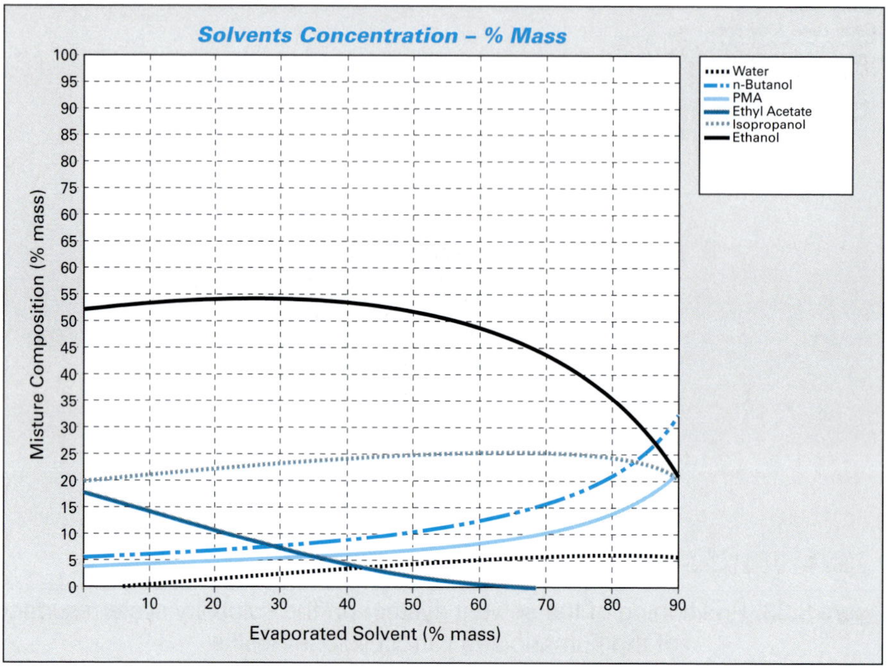

Figure 8.34. Solvent System Composition related to evaporated solvent mass.

Table 8.12. Solvent System Composition, Solubility Parameters, and Normalized Distances in a Fumaric/Nitrocellulose Resin System

Solvent System Composition		Values in m/m %
Solvents	Ethanol	55.0
	Ethyl Acetate	21.0
	Isopropanol	18.0
	Ethyl Glycol	6.0
	Total	100.0
Solubility Parameters	δD	15.82
	δP	7.6
	δH	15.99
	δG	23.75
Standard Nitrocellulose	Initial Normalized Distance	0.74
	50% Evaporated Mass	0.83
	90% Evaporated Mass	0.77
Standard Fumaric	Initial Normalized Distance	0.72
	50% Evaporated Mass	1.02
	90% Evaporated Mass	0.96

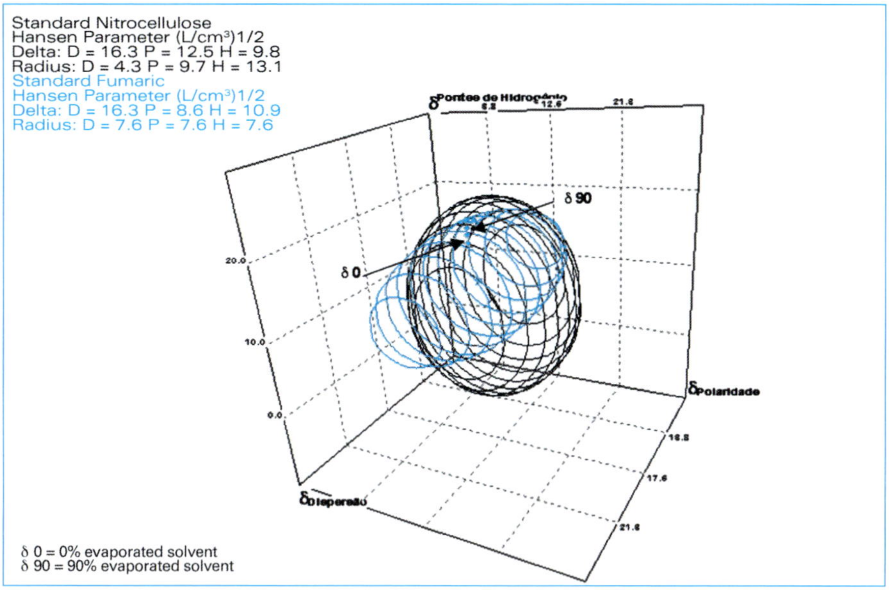

Figure 8.35. Positioning of the solvent system on the solubility system surface of the Fumaric and Nitrocellulose Resins.

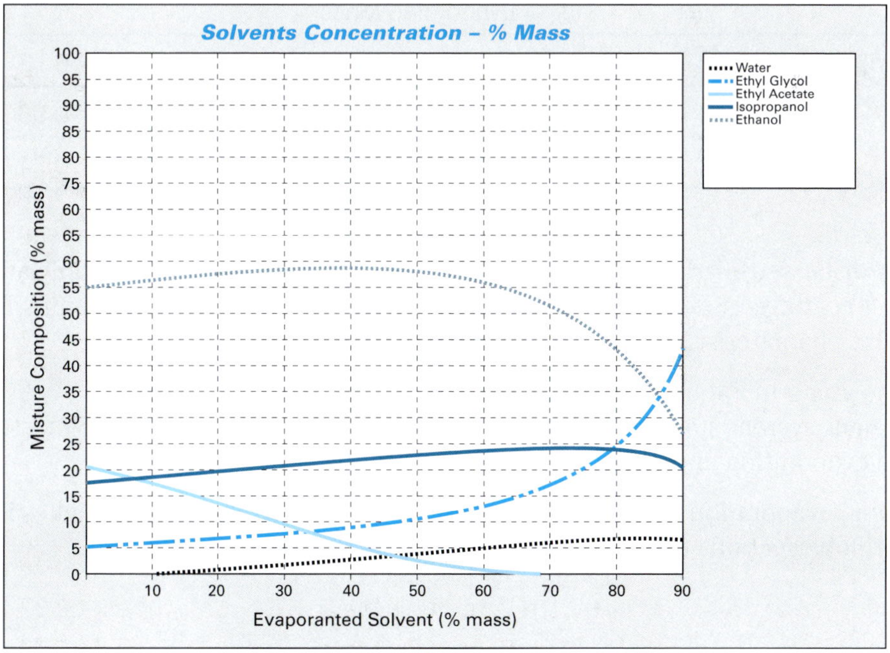

Figure 8.36. Solvent System Composition related to evaporated solvent mass.

Table 8.13. Solvent System Composition, Solubility Parameters, and Normalized Distances in a Fumaric/Nitrocellulose Resin System

Solvent System Composition		Values in m/m %
Solvents	Ethanol	52.0
	Ethyl Acetate	18.0
	Isopropanol	25.0
	Diacetone Alcohol	5.0
	Total	100.0
Solubility Parameters	δD	15.8
	δP	7.47
	δH	16.02
	δG	23.71
Standard Nitrocellulose	Initial Normalized Distance	0.75
	50% Evaporated Mass	0.83
	90% Evaporated Mass	0.71
Standard Fumaric	Initial Normalized Distance	0.73
	50% Evaporated Mass	0.98
	90% Evaporated Mass	0.74

As discussed in the previous examples, the solvent system for application in printing coatings is defined at the polymer system solubility surface limit. In this case, this characteristic can also be observed.

The graph in Figure 8.37 shows that despite variations in the δ*D*, δ*P*, δ*H* values, the solvent system stays in the solubility surface throughout the complete solvent system evaporation and film formation.

Rapid evaporation of the solvent system favors film formation and makes it difficult for imperfections to occur.

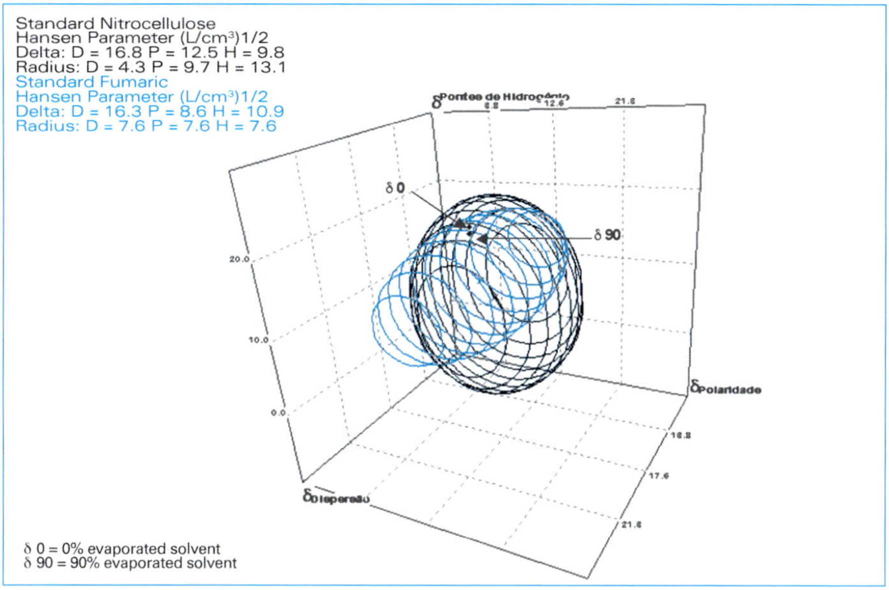

Figure 8.37. Development of Solvent System Solubility Parameters on the nitrocellulose and fumaric resins solubility surface.

Figure 8.38 of the solvent system evaporation graph is important because solvents present at various stages of film formation can be identified and associate them with δD, δP, δH force changes. This analysis allows the understanding of possible film defects.

The next 3 examples were made using the Nitrocellulose/Polyamide resin system. This resin blend is often used in the ink coating segment.

The solvent system studied in this case shows the normalized distance is slightly outside the solubility surface for the Polyamide resin at 50% evaporated mass. At this stage of film formation there is the possibility of ink precipitation. Rapid evaporation of the solvent system and film formation do not favor this problem occurring.

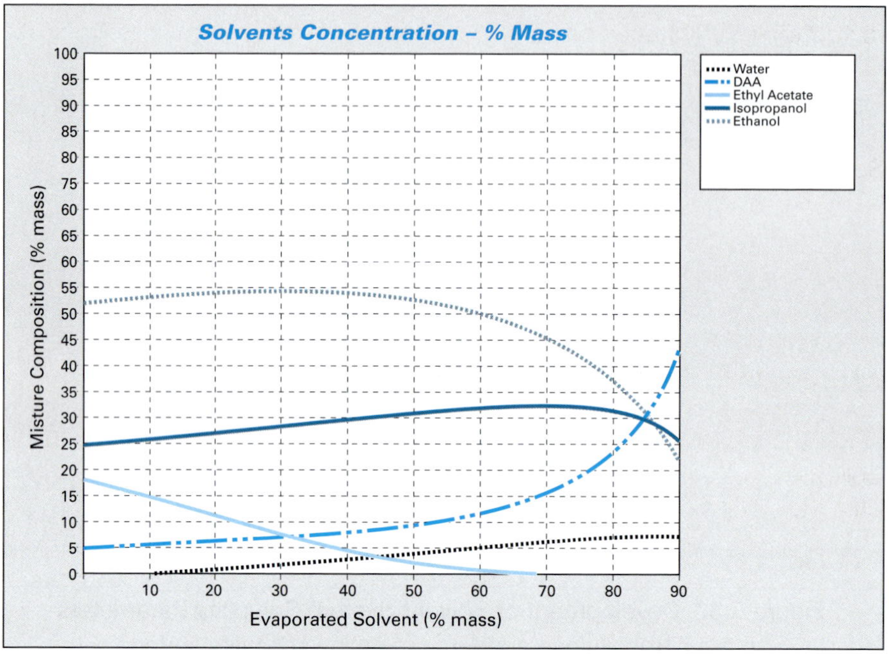

Figure 8.38. Solvent system evaporation.

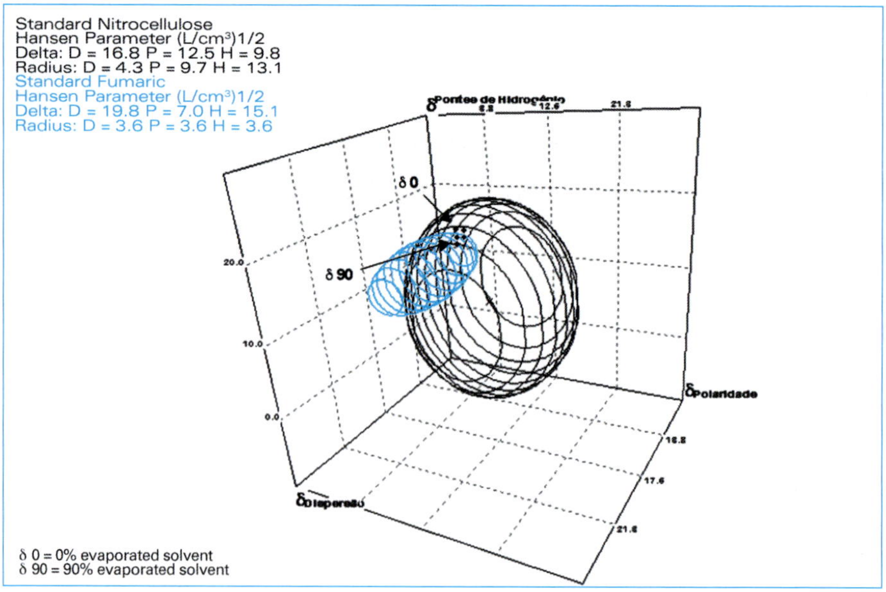

Figure 8.39. Positioning of the solvent system on the solubility system surface of the polyamide and nitrocellulose resin solubility.

The graph in Figure 8.39 shows δD, δP, δH force values during solvent system evaporation.

Table 8.14. Solvent System Composition, Solubility Parameters, and Normalized Distances in a Nitrocellulose/Polyamide Resin System

Solvent System Composition		Values in m/m %
Solvents	Ethanol	52.0
	Ethyl Acetate	12.0
	Isopropanol	14.0
	Butanol	15.0
	Propylene Glycol Monomethyl Ether	7.0
	Total	100.0
Solubility Parameters	δD	15.8
	δP	7.48
	δH	16.59
	δG	24.1
Standard Nitrocellulose	Initial Normalized Distance	0.77
	50% Evaporated Mass	0.84
	90% Evaporated Mass	0.81
Standard Polyamide	Initial Normalized Distance	0.99
	50% Evaporated Mass	1.22
	90% Evaporated Mass	0.92

Figure 8.40 shows individual solvent evaporation during drying and film formation. This allows the detection of which solvent is influencing the different stages of film formation.

At 90% of evaporated mass, butanol is the most prevalent solvent, followed by propylene glycol monomethyl ether. They are the main ones responsible for guaranteeing the final solubility and the film formation.

Again, it should be remembered that the water registered in the figure is from atmospheric conditions used in the data simulation.

Table 8.15 shows another solvent system for the same nitrocellulose/polyamide resin blend.

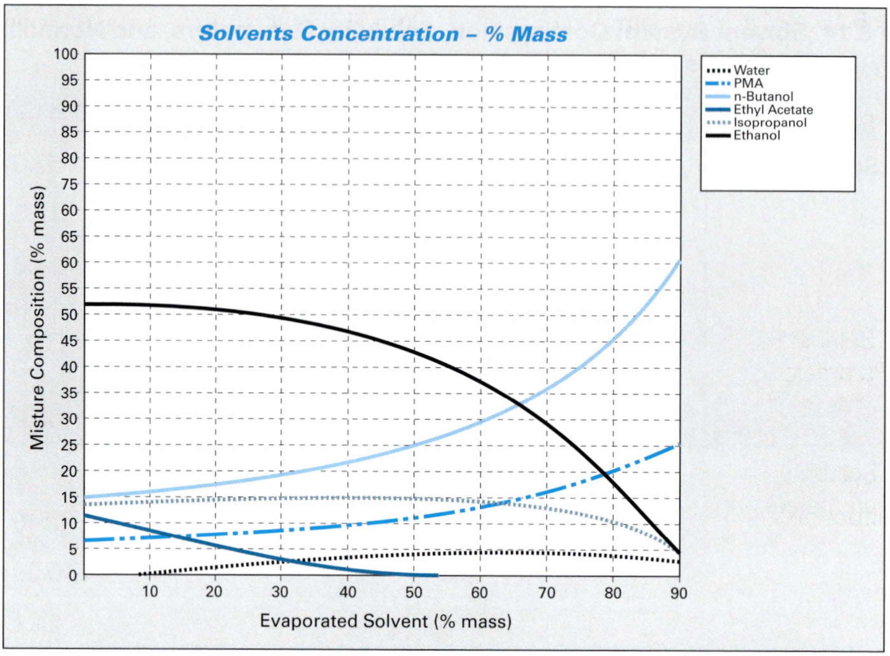

Figure 8.40. Solvent system evaporation during film drying.

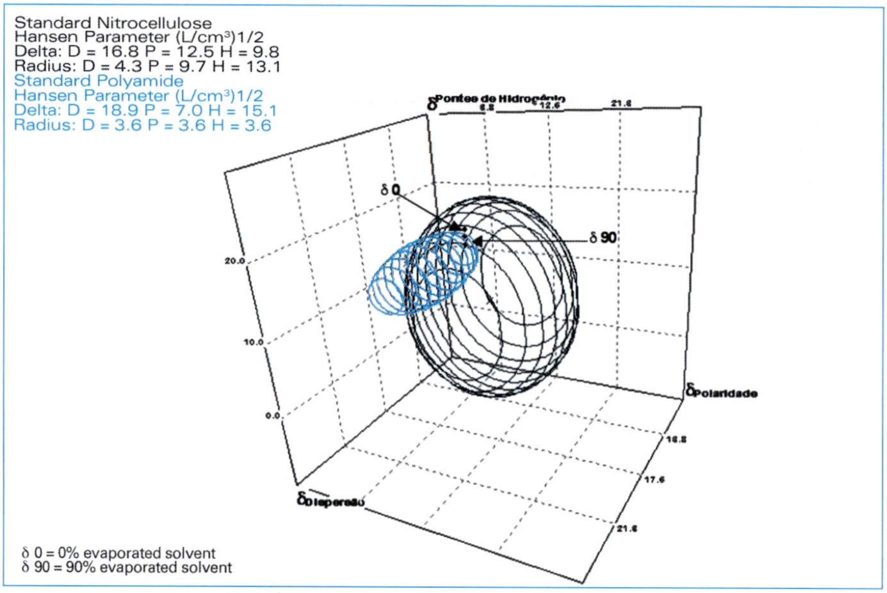

Figure 8.41. Positioning of the solvent system on the solubility system surface of the nitrocellulose and polyamide resin solubility.

Table 8.15. Solvent System Composition, Solubility Parameters, and Normalized Distances in a Nitrocellulose/Polyamide Resin System

Solvent System Composition		Values in m/m %
Solvents	Ethanol	55.0
	Ethyl Acetate	17.0
	Isopropanol	19.0
	Butanol	5.0
	Diacetone Alcohol	4.0
	Total	100.0
Solubility Parameters	δD	15.81
	δP	7.51
	δH	16.23
	δG	23.87
Standard Nitrocellulose	Initial Normalized Distance	0.75
	50% Evaporated Mass	0.83
	90% Evaporated Mass	0.73
Standard Polyamide	Initial Normalized Distance	0.95
	50% Evaporated Mass	1.28
	90% Evaporated Mass	0.92

This graph maintains the same tendency as the prior ones, showing that the solvent system is slightly outside the solubility surface for the polyamide resin at 50% evaporated mass.

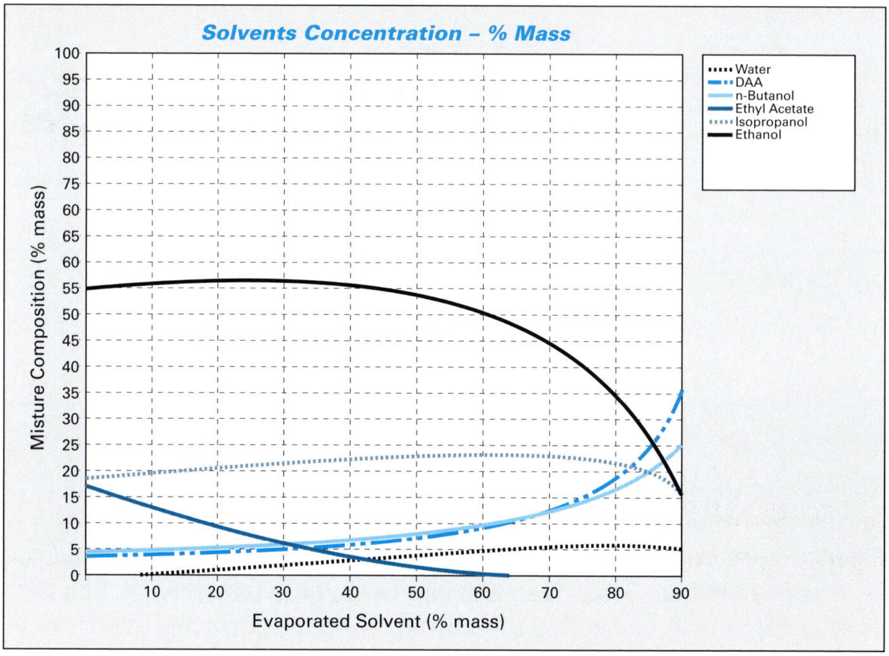

Figure 8.42. Solvent System Evaporation.

Solvents used in the formulation of the solvent system are shown in Figure 8.42 with the graph indicating, as previously shown, the concentration of each solvent during film drying. Ethyl acetate evaporates completely at 60% of the solvent system evaporated.

Table 8.16 shows the solvent system with normalized distance within the solubility volume for the nitrocellulose resin, indicating that in this case, the polymer has good solubility with this resin.

Table 8.16. Solvent System Composition, Solubility Parameters, and Normalized Distances in a Nitrocellulose/Polyamide Resin System

Solvent System Composition		Values in % m/m
Solvents	Ethanol	58.0
	Ethyl Acetate	17.0
	Isopropanol	19.0
	Ethyl Glycol	6.0
	Total	100.0
Solubility Parameters	δD	15.82
	δP	7.72
	δH	16.45
	δG	24.09
Standard Nitrocellulose	Initial Normalized Distance	0.75
	50% Evaporated Mass	0.84
	90% Evaporated Mass	0.77
Standard Polyamide	Initial Normalized Distance	0.98
	50% Evaporated Mass	1.37
	90% Evaporated Mass	1.27

In the case of the polyamide resin, the initial solubility is at the solubility volume limit. During the evaporation of the solvent system, it can be seen that the normalized distance goes beyond the solubility volume indicating a low solubility of the polymer as seen in Figure 8.43. Again it must be pointed out that the film formation is favored by rapid evaporation of the solvent system, however, in this case resin precipitation can occur.

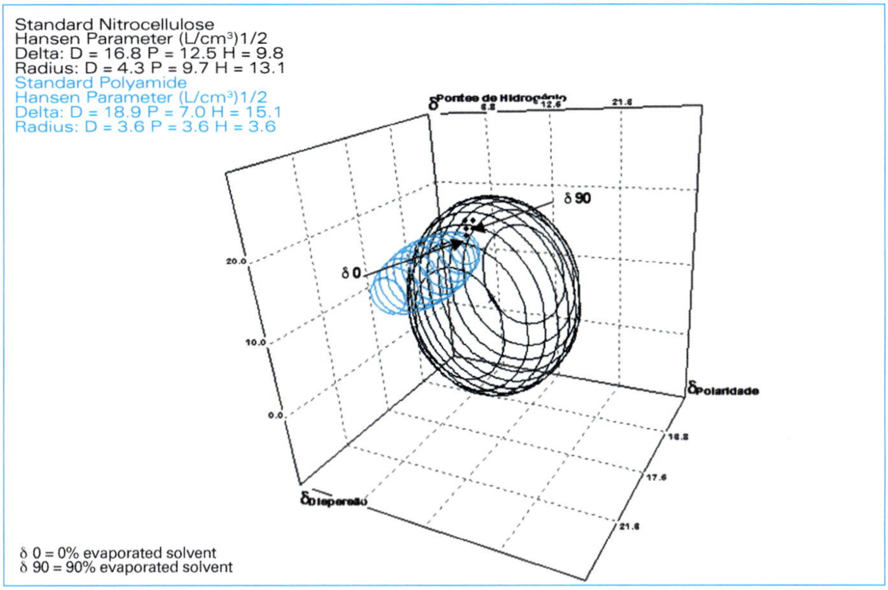

Figure 8.43. Positioning of the solvent system on the solubility system surface of the resins.

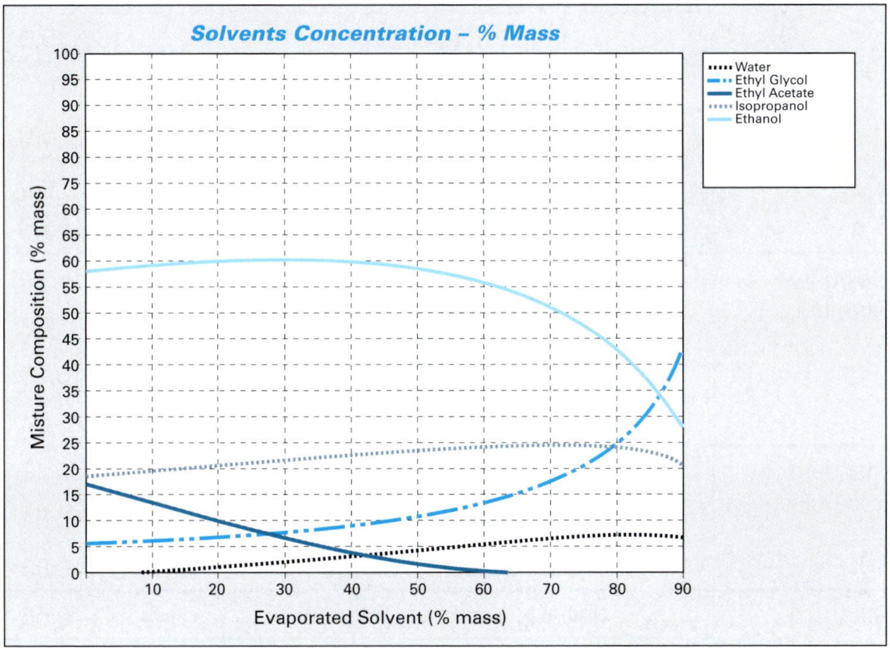

Figure 8.44. Solvent system evaporation.

As already mentioned, the ink coating market works with the formulations purposely formulated at the solubility surface limit, making easy elimination of solvents, and thus decreasing the possibility of solvent retention on the final printing.

This graph also indicates δ*D*, δ*P* and δ*H* force values during solvent system evaporation. Thus, it is possible to analyze possible defects in the film's final formation.

Figure 8.44 shows individual solvent evaporation during drying and film formation. Therefore, it is possible to analyze which solvent is present during film formation.

Analyzing solubility surface and solvent system evaporation graphs, the ethyl glycol and ethanol are the principle solvents at 90% evaporated mass, concluding that they are the ones that most influence final solubility of the resin system.

The next 3 examples were made using the Nitrocellulose/Polyurethane resin system; another which is used in this market.

Table 8.17. Solvent System Composition, Solubility Parameters, and Normalized Distances in a Nitrocellulose/Polyurethane Resin System

Solvent System Composition		Values in % m/m
Solvents	Ethanol	47.0
	Ethyl Acetate	15.0
	Isopropanol	28.0
	Butanol	6.0
	Propylene Glycol Monomethyl Ether	4.0
	Total	100.0
Solubility Parameters	δD	15.79
	δP	7.3
	δH	16.29
	δG	23.84
Standard Nitrocellulose	Initial Normalized Distance	0.77
	50% Evaporated Mass	0.84
	90% Evaporated Mass	0.83
Standard Polyurethane Resin	Initial Normalized Distance	0.58
	50% Evaporated Mass	0.73
	90% Evaporated Mass	0.65

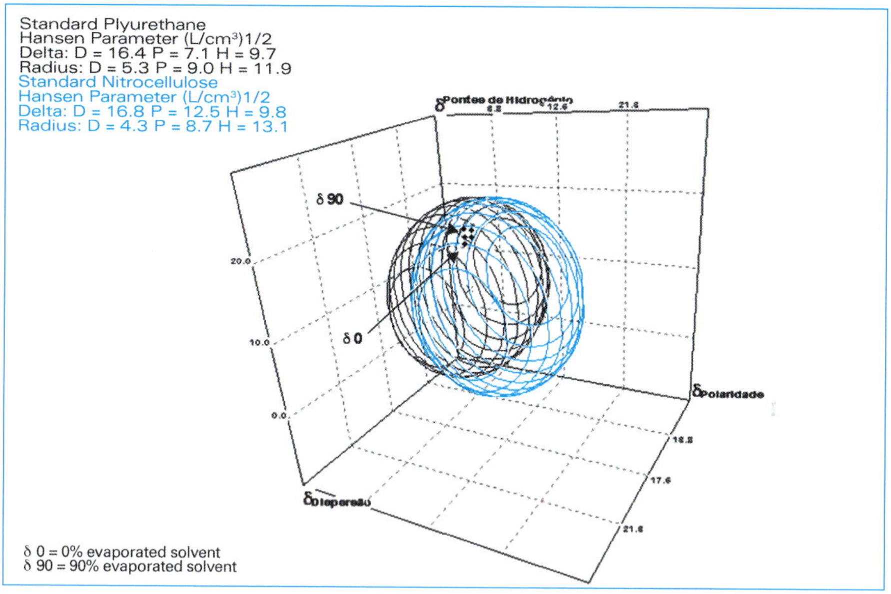

Figure 8.45. Positioning of the solvent system on the solubility system surface of the Nitrocellulose and Polyurethane Resins.

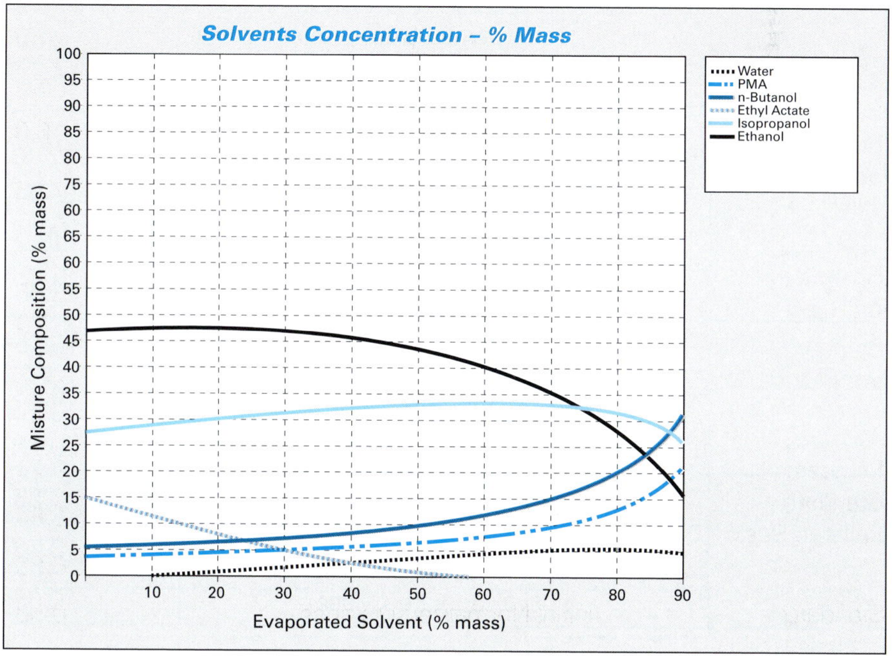

Figure 8.46. Solvent system evaporation.

Figure 8.45 maintains the same tendency as the others, showing that the solvent system approaches the solubility volume limit for the nitrocellulose resin. With the polyurethane resin, the normalized distance at 90% evaporated mass is 0.65, indicating that the resin has an excellent interaction with the solvent system, however, this effect increases the tendency of solvent retention in the final packaging.

Figure 8.46, as with previous ones, shows pure solvents evaporating during drying and film formation. At 90% evaporated mass, isopropanol, ethanol, butanol, and propylene glycol monomethyl ether solvents are still present. Ethyl acetate was practically eliminated at 60% evaporated mass.

In this solvent system a formulation containing diacetone alcohol as the only heavy solvent in the solvents used is analyzed.

Table 8.18. Solvent System Composition, Solubility Parameters, and Normalized Distances in a Nitrocellulose/Polyurethane Resin System

Solvent System Composition		Values in % m/m
Solvents	Ethanol	56.0
	Ethyl Acetate	18.0
	Isopropanol	22.0
	Diacetone Alcohol	4.0
	Total	100.0
Solubility Parameters	δD	15.8
	δP	7.55
	δH	16.2
	δG	23.86
Standard Nitrocellulose	Initial Normalized Distance	0.75
	50% Evaporated Mass	0.83
	90% Evaporated Mass	0.75
Standard Polyurethane Resin	Initial Normalized Distance	0.58
	50% Evaporated Mass	0.76
	90% Evaporated Mass	0.66

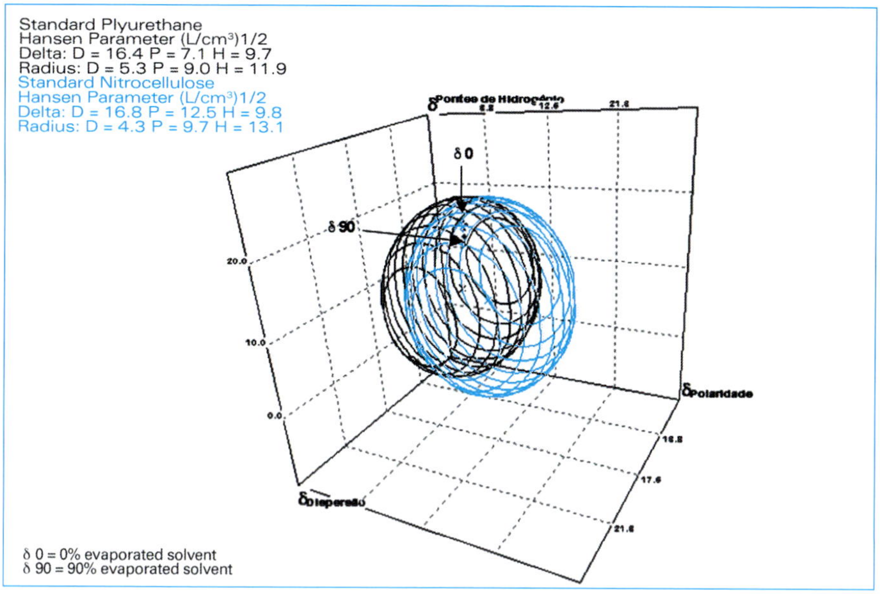

Figure 8.47. Positioning of the solvent system on the solubility system surface for the nitrocellulose/PU resin system.

Figure 8.48 also indicates δ*D*, δ*P* and δ*H* force values during solvent system evaporation. Therefore, it is possible to analyze possible defects in the final film formation.

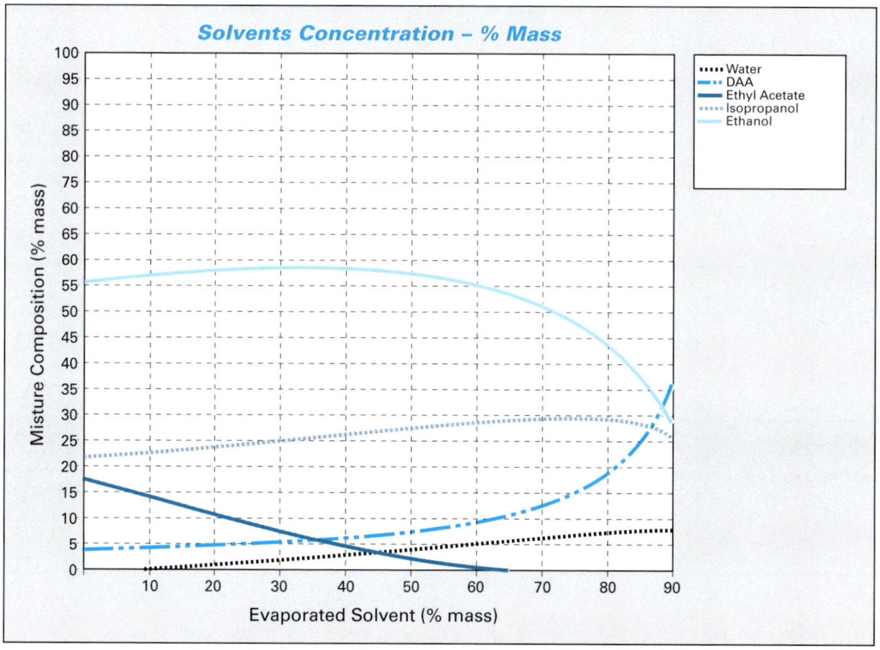

Figure 8.48. Solvent system evaporation.

The PU resin shows better solubility than the nitrocellulose in this solvent system.

Solvent system performance needs to be analyzed at the intersection of the resins used in the formulation.

Figure 8.48 shows individual solvent evaporation during drying. Therefore, it can be analyzed as to which solvent is present during film formation and the possible interferences. Diacetone alcohol, ethanol, and isopropanol are solvents responsible for the final characteristics in the film since they are present at drying.

The system suggested in Table 8.19 shows normalized distance values within the solubility surface of nitrocellulose and polyurethane resins.

Table 8.19. Solvent System Composition, Solubility Parameters, and Normalized Distances in a Nitrocellulose/Polyurethane Resin System

Solvent System Composition		Values in % m/m
Solvents	Ethanol	54.0
	Ethyl Acetate	19.0
	Isopropanol	20.0
	Ethyl Glycol	7.0
	Total	100.0
Solubility Parameters	δD	15.83
	δP	7.62
	δH	16.13
	δG	23.85
Standard Nitrocellulose	Initial Normalized Distance	0.74
	50% Evaporated Mass	0.83
	90% Evaporated Mass	0.74
Standard Polyurethane Resin	Initial Normalized Distance	0.57
	50% Evaporated Mass	0.76
	90% Evaporated Mass	0.71

Figure 8.49 shows δD, δP and δH force values during film drying that always show the solvents in the solubility surface.

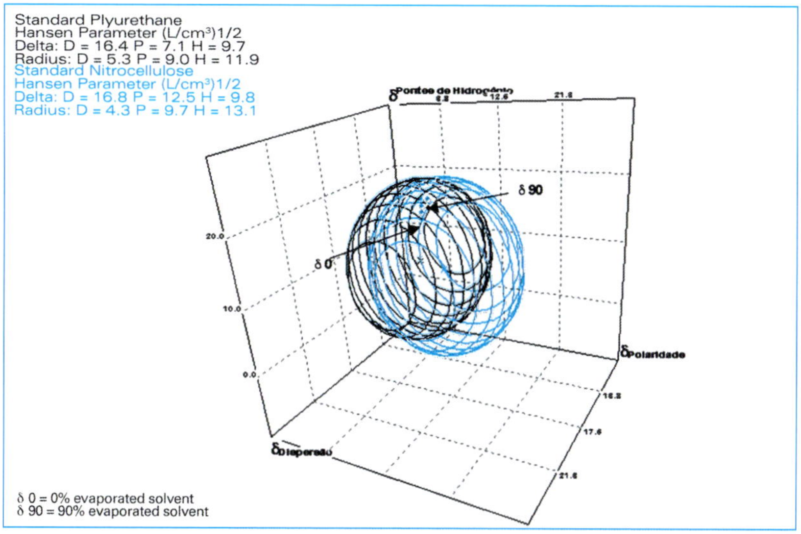

Figure 8.49. Positioning of the solvent system on the solubility system surface of the Nitrocellulose/Polyurethane resin system.

Solvents used in the formulation of the solvent system are shown in Figure 8.50 indicating, as previously shown, the concentration of each solvent during film drying.

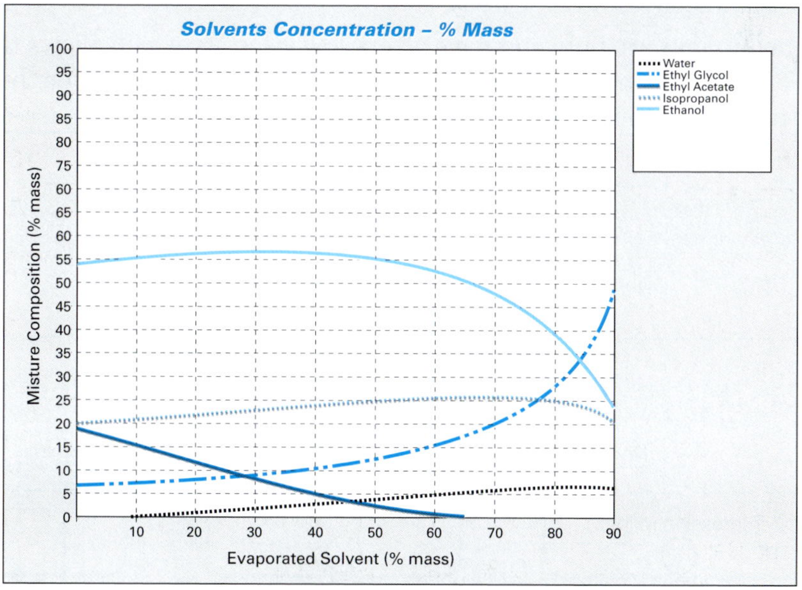

Figure 8.50. Solvent system evaporation.

The main printing substrates are: flexible packages cellophane, polyethylene, polypropylene, nylon, polyester, aluminium, paper, corrugated cardboard, wrapping paper, paper towels, banners, among others.

8.4.5.4. Rotogravure

This is a direct printing process in which its name comes from the cylinder form and from the rotary principle of the press used. It is different from the other methods in that the complete original has to pass through a reticulation process, including the text. The rotary printing method is used on several types of surfaces.

The first project with a machine to use this printing form, demonstrating engraved graphism was patented in 1784 by Thomas Bell. However, Karl Klic, considered the father of Rotogravure, started using it in 1860.[136]

During the process, the gravure printing cylinder applies quantities of ink to the surface substrate. This is possible because of cells in the gravure cylinder made of copper and chrome.[136]

Engraving can be done chemically or electromechanically. Chemical engraving is the older method, but is still used in some companies. Electromechanical engraving, which uses a small diamond stylus on the rotating copper cylinder surface, is more precise and is more commonly used. Recently, laser engraving is making a large advance in the sector.

Image tone is determined by the cell depths: the deeper cells hold more ink and therefore print in darker tones; shallow cells (less ink) create lighter tones. After engraving is done on the copper cylinder, it is given a chrome plating, allowing the cylinder to last longer.

Rotogravure inks are fluid and have a very low viscosity which allows them to be extracted from the cylinder cells and transferred to the substrate. For the purpose of drying the ink and eliminating residual solvents, the substrate is passed through an oven.

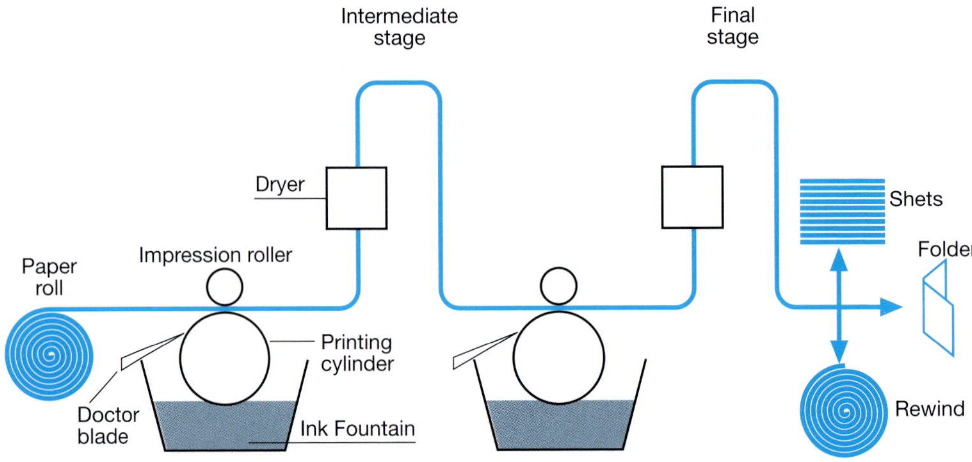

Figure 8.51. Simplified diagram of rotogravure printing[136].

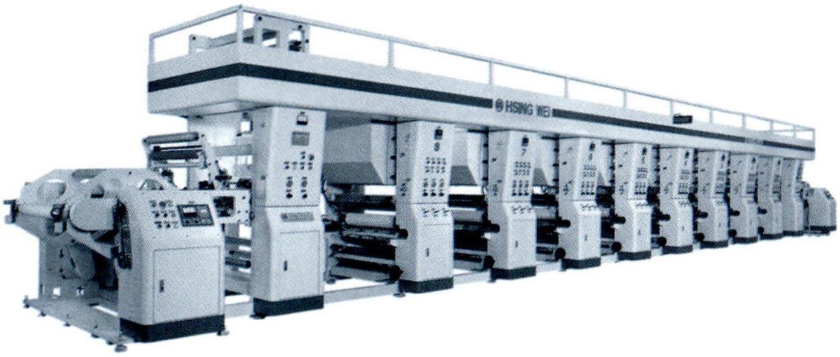

Figure 8.52. Rotogravure[136].

The ink dries before the substrate receives the next layer. This is to allow printing overlay without running or smudging. Consequently, ovens are located after each printing station. Flexography and rotogravure inks are very similar and the components are essentially the same.

Solvents show how resin flow becomes the main characteristic for it to fill and leave the cylinder cells. Another characteristic includes how it slightly penetrates the substrate surface, carrying the resin with it and allowing it to stick to the substrates.

After the printing, the solvent needs to be eliminated immediately, thus allowing film formation. It must be practically absent in the printing product.

Although solvent in the final packaging is practically nonexistent, it plays an extremely important part in all the stages previous to printing.

As explained in Chapter 7, one of the criteria for choosing a solvent is resin solubility used to help with coating adhesion to the substrate. Coating solvent can still include 2 or 3 different solvents with specific characteristics.

Usually, a true resin solvent that readily dissolves is easily able to be retained when the coating film is drying. This phenomenon occurs because of the resin-solvent physicochemical affinity. In these cases, it becomes necessary to add co-solvents or dilutants to weaken the interaction and help eliminate solvents in the final printing.

These cannot solubilize resins on their own. Thus, solvent choice must be considered:

- Resin type;
- Printing speed;
- Restrictions imposed by the final product.

Solvent evaporation has a fundamental importance in rotogravure, but some points need to be observed. If the solvent is eliminated too quickly, the ink dries on the inside of the cells and ink transfer to the substrate does not happen, and could block up cells. If the solvent is eliminated too slowly, there could be problems with tearing, wrinkling, stickiness, and at certain levels, solvent retention in the printing could lead to mold in the end product.

Restrictions on solvent quantity retained are characteristic for each.

Table 8.10 shows some typical market formulations, analyzed according to the Hansen solubility parameter concept using the SOLSYS® – Design de Sistemas Solventes — Rhodia tool. Resin solubility parameters were determined using the description in part 8.3.1.1. The resin system considered for this analysis was fumaric/nitrocellulose.

Table 8.10. Solvent System Composition, Solubility Parameters, and Normalized Distances in a Fumaric/Nitrocellulose Resin System

Solvent System Composition		Values in % m/m
Solvents	Ethanol	20.0
	Ethyl Acetate	80.0
	Total	100.0
Solubility Parameters	δD	15.8
	δP	6.0
	δH	9.64
	δG	19.46
Standard Nitrocellulose	Initial Normalized Distance	0.7
	50% Evaporated Mass	0.68
	90% Evaporated Mass	0.67
Standard Fumaric	Initial Normalized Distance	0.37
	50% Evaporated Mass	0.31
	90% Evaporated Mass	0.45

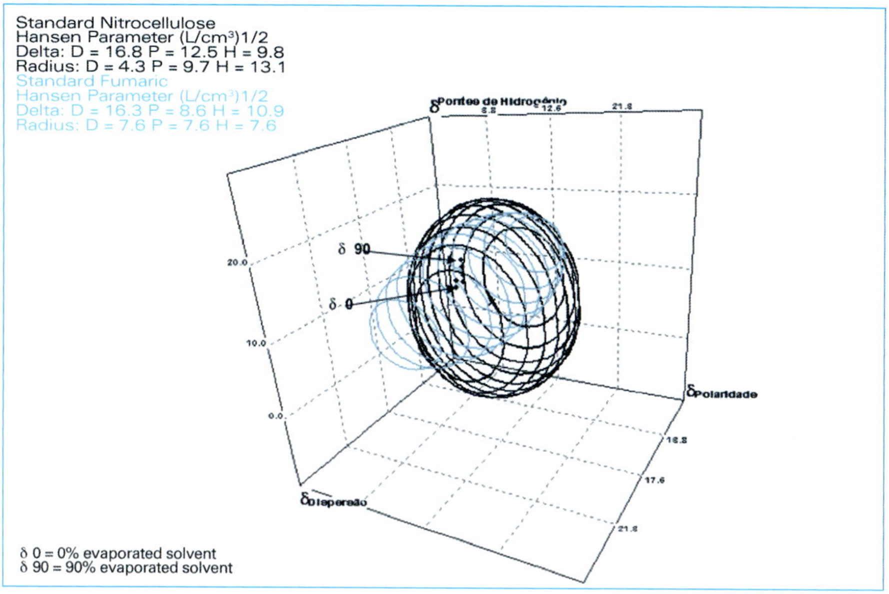

Figure 8.53. Positioning of the solvent system on the resin solubility system surface.

The rotogravure has the same characteristic as the flexography, i.e., the solvent system normalized distance is within the solubility volume for the nitrocellulose and fumaric resins, indicating that the solvent system has good solubility for the resins used.

Solubility is very important because it guarantees the coating has low enough viscosity for it to get in and out of the cells efficiently. Remember, solubility and viscosity are directly proportional.

In the ink coating market, formulations are purposely formulated with a normalized distance at the solubility limit, i.e., nearest 1, making easy elimination of solvents, thus decreasing the probability of retention on the final printing.

With the fumaric resin, normalized distance values are closer to the solubility volume center, indicating an excellent interaction with the solvent system. This can complicate solvents evaporating, increasing the tendency for retention.

With rotogravure, this is less critical because of the drying between printing stages, forcing solvents out.

Graph 8.54 shows the individual solvent evaporation during drying and film formation, indicating the concentration at each evaporation stage. It is possible to analyze the influence of each solvent on film performance.

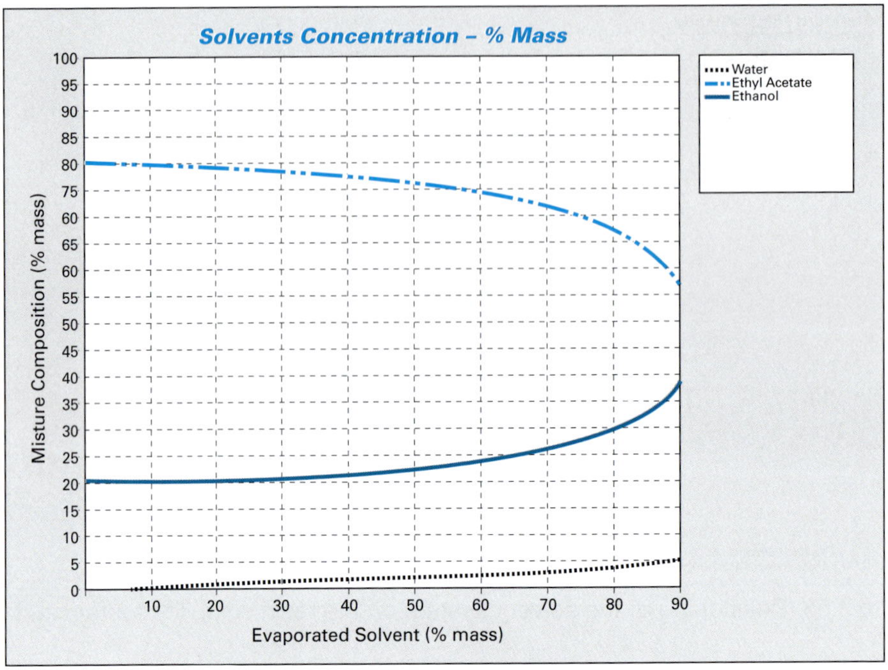

Figure 8.54. Solvent System Evaporation.

It should be noted that the appearance of water in the graph is from relative humidity in the simulation's atmospheric conditions. This phenomenon occurs in the other solvent system evaporation graphs shown.

Figure 8.55 maintains the same tendency as the others, showing the solvent system is within solubility volume limit for the nitrocellulose/polyurethane resin.

Table 8.11 shows that polyurethane resin has excellent solubility in the proposed solvent system and increases the normalized distance with evaporation. This makes solvent retention difficult since 90% of the evaporated mass has a normalized distance of 0.87.

As previously mentioned, being closer to the solubility volume limit, solvent evaporation becomes easier during the final film drying.

It is worth remembering that with rotogravure, ovens help solvents evaporate more easily after each printing stage.

Table 8.11. Solvent System Composition, Solubility Parameters, and Normalized Distances in a Nitrocellulose/Polyurethane Resin System

Solvent System Composition		Values in % m/m
Solvents	Ethanol	30.0
	Ethyl Acetate	60.0
	Isopropanol	10.0
	Total	100.0
Solubility Parameters	δD	15.8
	δP	6.43
	δH	11.78
	δG	20.73
Standard Nitrocellulose	Initial Normalized Distance	0.68
	50% Evaporated Mass	0.7
	90% Evaporated Mass	0.91
Standard Polyurethane Resin	Initial Normalized Distance	0.24
	50% Evaporated Mass	0.41
	90% Evaporated Mass	0.87

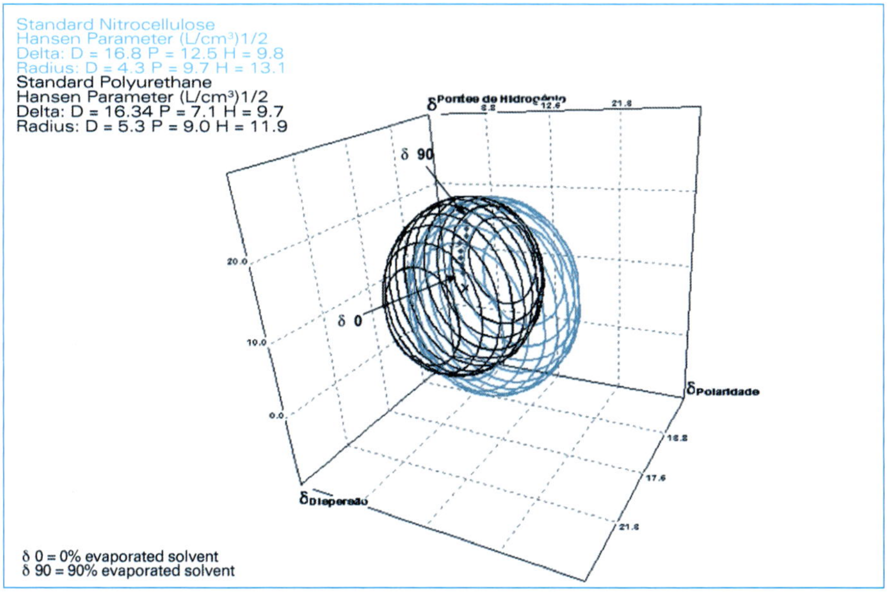

Figure 8.55. Positioning of the solvent system on the solubility system surface.

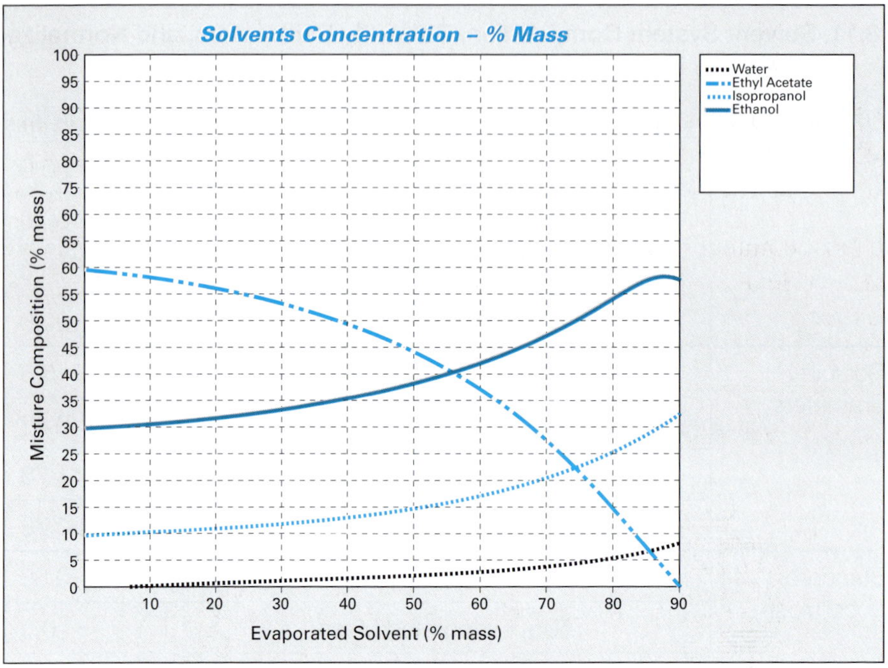

Figure 8.56. Solvent System Evaporation.

Figure 8.56 also indicates δ*D*, δ*P* and δ*H* force values during solvent system evaporation. Thus, it is possible to analyze possible defects in the final film formation.

Solvents habitually used in rotogravure include alcohols (ethyl, isopropyl), ethyl acetate. Ethyl glycol, propylene glycol monomethyl ether, and diacetone alcohol can also be found.

8.4.5.5. Serigraphy

Serigraphy or *silk-screen* is a printing process in which the coating is spread across the substrate using a roller or squeegee, over a screen usually made of silk or nylon. The screen is usually stretched out over a wood or aluminium frame.

The coating is applied to the screen using a fill bar (also known as a floodbar) to fill the mesh openings with ink and a squeegee to pull the coating across the screen starting at the rear with the screen raised up off the support. Then using a slight amount of downward force the fill bar is pulled to the front of the screen. This effectively fills the mesh openings with ink.

The process can be used for printing on several types of material (paper, plastic, rubber, wood, glass, cloth, etc.), different surfaces (cylinder, spheres, irregular surfaces, clear, dark, opaque, shiny, etc.), and different types of thicknesses or shapes with different types of coatings or colors. It can also be done by hand or using machine automation.[137]

In addition to the equipment and appropriate components, the serigraphy process needs specialized, qualified labor to guarantee the printing job. Therefore, it is important to use the right coating for each substrate.

The most commonly used solvents in this market are: isophorone, cyclo-hexanone, toluene, xylene, methyl, isobutyl ketone, aromatic solvents C9, 2-butoxyethanol, and diesters.

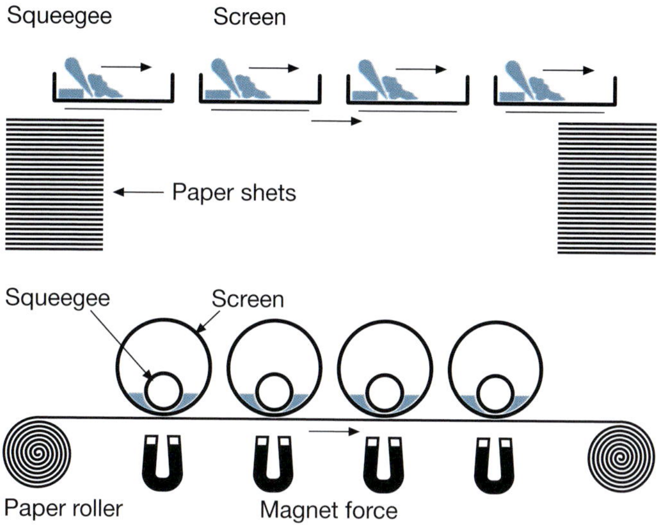

Figure 8.57. Simplified diagram of serigraphy (silk screen) printing[137].

8.4.5.6. Offset

Offset printing is a technique that uses a thin flexible aluminium plate on a printing cylinder which is treated chemically so that the image repels a water solution and accepts the ink. The inverse occurs on the non-image area.

Offset printing uses machines that have three cylinders; the printing plate cylinder where the plate goes; the rubber blanket cylinder to which the image is transferred wrong-side up; and the impression cylinder where the substrate, like paper or plastic, is mounted.[138]

With thermoplastic printing (rigid or flexible) a dry offset process is used with only the coating and the engraving on the plate is raised so that the coating is transferred to the plate, then from the plate to the rubber blanket cylinder, and from there to the support (plastic).[138]

The main characteristics of offset printing are:

- Several formats in black and white or in color can be printed at a relatively low cost;
- It requires more attention than typography or rotogravure to keep image uniformity throughout the printing process;
- Printing plates are low cost and are prepared quickly as compared to other printing methods;
- The plates can be made of aluminium, metal stainless steel or specific resins;
- Excellent printing quality.[139]

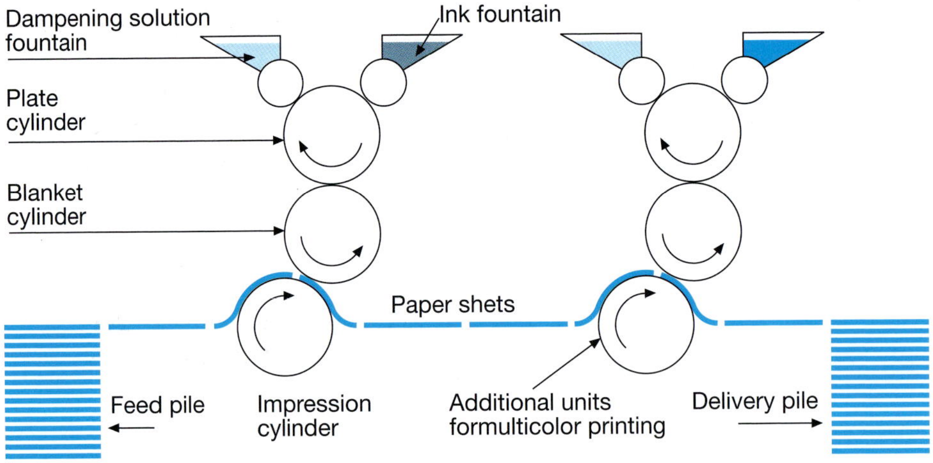

Figure 8.58. Simplified diagram of offset printing[139].

8.5. Adhesives

Adhesives are used in a large range of applications from flexible packages and cloth to structural application as in gluing automobile parts. Sealants are used to prevent gas and/or a liquid seeping between two surfaces; they are used to fill in, harden, and give some flexibility to spaces. In some cases, they are also used because of their adhesive properties. Among the several classes of adhesives, solventborne adhesives that stand out will be discussed later in this chapter.

One of these classes is contact adhesives. This class is characterized by the adhesive being applied on both surfaces to be glued. After being applied, some of the solvent evaporates, the surfaces are put together, and the adhesive joins the two. Contact adhesives are, in large part, use polychloroprene rubber (neoprene). The market for this type of adhesive is furniture making and carpets. Another important market is glue for shoes. Solvent-based and pressure sensitive adhesives are also generally used in natural rubber, styrene block copolymers, rubber, styrene-butadiene, and acrylic polymers.

In the majority of solventborne adhesives, the solvent is added to reduce viscosity and improve application in addition to drying time. Generally, adhesives are made with solvent mixtures that allow optimal drying time.

A huge variety of solvents can be used but the most usual are toluene, heptane, acetone, ethyl acetate, methyl ethyl ketone, chloroform, naphtha, and mineral oil. The variety of solvents used in the adhesive industry is less than the coating industry because with adhesives, appearance is a less important factor. Surfaces coated with a *solvent base* generally contain a mixture of low and high vapor pressure solvents, and according to the evaporation rate, determining its final appearance. Light solvents that evaporate quickly are used with adhesives, making drying time an important parameter. With coatings, xylene (heavy solvent) and toluene (light solvent) are important, but with adhesives and sealants, xylene is hardly used. The same happens with ethyl acetate and MEK (light) and butyl acetate and MIBK (heavy solvent). Additionally, contrary to coatings, it is not necessary to reduce viscosity before application with adhesives. Below are some formulation examples developed for PU-based adhesives. Table 8.22 shows one of the formulations used in the market.

Table 8.22. Composition, initial solubility parameters, and solubility development during solvent system evaporation for a PU-based adhesive

Solvent System Composition		Values in % m/m
Solvents	Acetone	40.0
	Methyl Ethyl Ketone	60.0
	Total	100.0
Solubility Parameters	δD	15.8
	δP	9.56
	δH	5.86
	δG	19.37
PU Adhesive	Initial Normalized Distance	1.0
	50% Evaporated Mass	0.84
	90% Evaporated Mass	0.84

Evaporation Conditions:
25 g/m^2/Temperature – 25 °C/Relative Air Humidity – 70%/Air Speed – 0.0667.

In this case, only two solvents in the developed system were used. Based on the solubility parameters and normalized distances, the solvent system maintains a good resin solubility in any evaporation situation, as can also be seen in Figure 8.59.

Another important point is that the solvent system has an extremely fast evaporation, and the formulation is developed to act close to the PU resin solubility limit. In other words, it has a minimum interaction with the resin to promote its solubilization and a good film formation, but is close enough to the solubility limit indicating low interaction between the solvent and resin, thus decreasing the probability of its retention in the film.

As for an adhesive, mechanical properties are the most important in the film: the less solvent retained, the less solvent there will be in the adhesive polymer chains, and consequently greater the resolve between them, favoring the increase in crystallinity that in turn is responsible for the mechanical properties of the adhesive.

The solubility at the limit at the initial moment also guarantees viscosity adjustment necessary for adhesive application, usually using a brush.

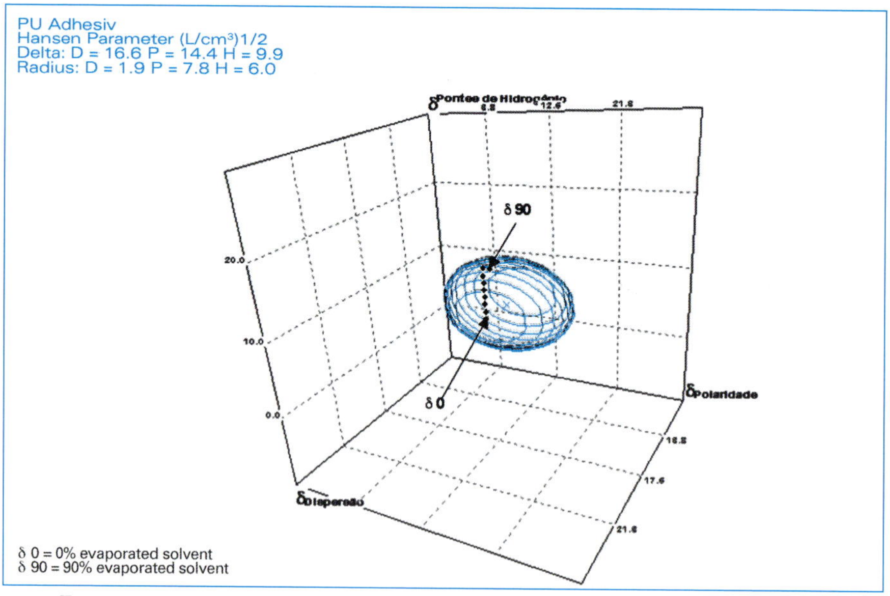

Figure 8.59. Development of Solvent System Solubility Parameters for a PU-based adhesive.

An important point is the absorption of water by the system. Remember, absorption is a function of relative ambient humidity in which the solvent was applied and of the hygroscopicity of the solvents involved. In this particular case, the amount of absorbed water was not so little, and it also contributes to resin solubility in any evaporation situation (Figure 8.60).

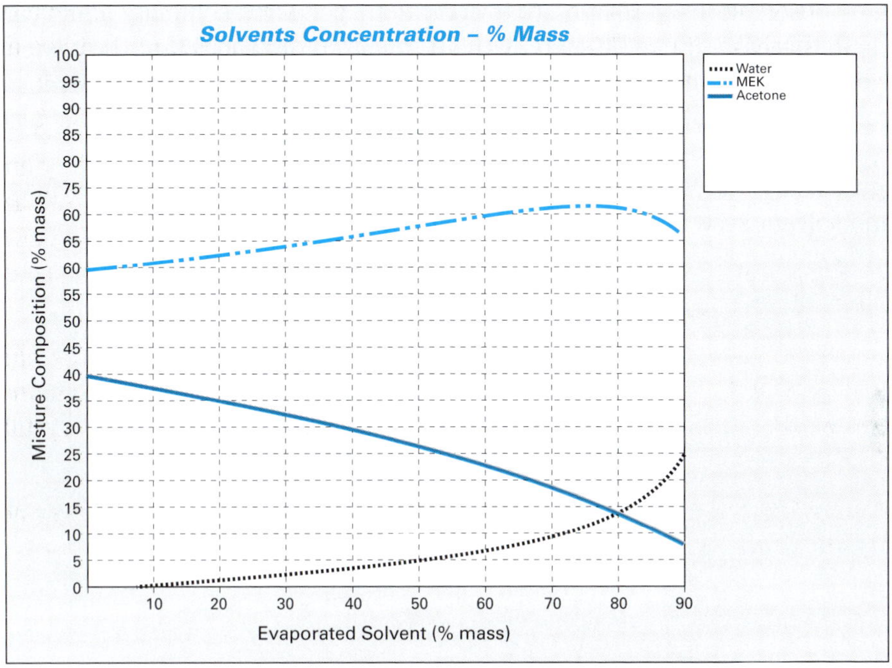

Figure 8.60. Individual solvent evaporation present in the formulation of a PU-based adhesive.

Table 8.23 shows another formulation also used for PU based systems.

Table 8.23. Composition, initial solubility parameters, and solubility development during solvent system evaporation for a PU-based adhesive

Solvent System Composition		Values in % m/m
Solvents	Acetone	60.0
	Ethyl Acetate	40.0
	Total	100.0
Solubility Parameters	δD	15.62
	δP	8.36
	δH	7.08
	δG	19.08
PU adhesive	Initial Normalized Distance	1.0
	50% Evaporated Mass	0.95
	90% Evaporated Mass	1.0

Evaporation Conditions: 25g/m^2/Temperature – 25°C/Relative Air Humidity – 70%/Air Speed – 0.0667.

The same observations in relation to evaporation speed and solubility limits can also be made in this case. It can be seen by the normalized distances that the system acts at the solubility limit, guaranteeing a good resin solubility, and allowing quick solvent evaporation because of its relatively low interaction with the resin. This fact also guarantees an adequate viscosity for adhesive application.

Figures 8.61 and 8.62 show solubility progression on the solubility surface of the resin and solvent system progression during evaporation.

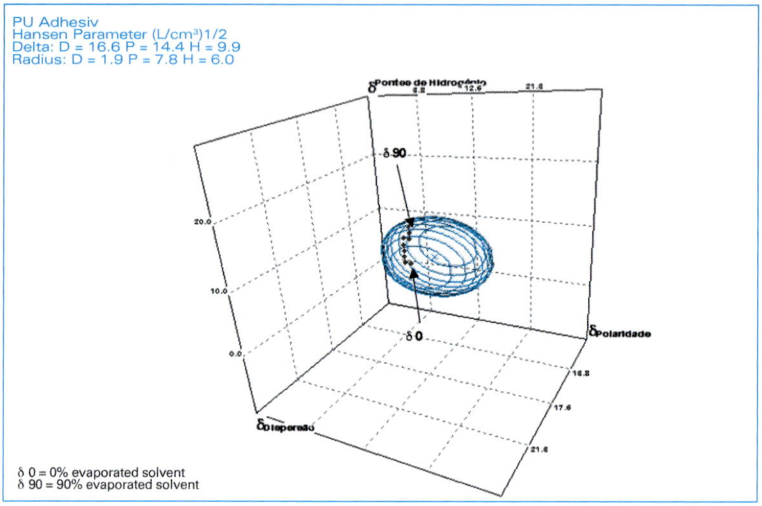

Figure 8.61. Development of Solvent System Solubility Parameters for a PU-based adhesive.

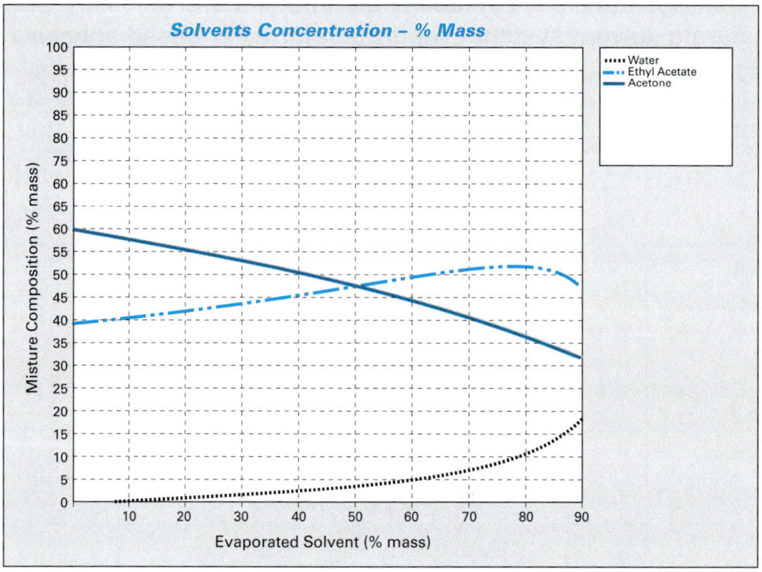

Figure 8.62. Individual solvent evaporation present in the formulation of a PU-based adhesive.

Another example is the polychloroprene-based adhesive. In this case, the evaluation is a little more complex because normally in these systems there is a resin mixture, usually polychloroprene-based with different viscosities, in addition to phenol resins contributing to the mechanical resistance of the adhesive.

Table 8.24 and Figures 8.63 and 8.64 are examples of a polychloroprene resin solubilization.

Table 8.24. Composition, initial solubility parameters, and solubility development during solvent system evaporation for a Polychloroprene-based adhesive

Solvent System Composition		Values in % m/m
Solvents	Rubber Solvent	8.0
	Hexane	7.0
	Acetone	15.0
	Toluene	70.0
	Total	100.0
Solubility Parameters	δD	17.25
	δP	2.54
	δH	2.5
	δG	17.61
Polychloroprene	Initial Normalized Distance	0.57
	50% Evaporated Mass	0.56
	90% Evaporated Mass	0.56

Evaporation Conditions: 25 g/m²/Temperature – 25 °C/Relative Air Humidity – 70%/Air Speed – 0.0667.

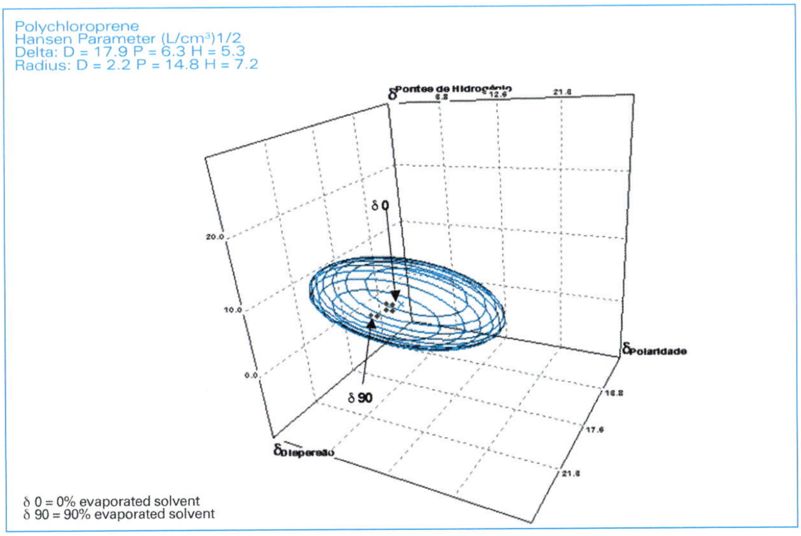

Figure 8.63. Development of Solvent System Solubility Parameters for rubber, polychloroprene-based.

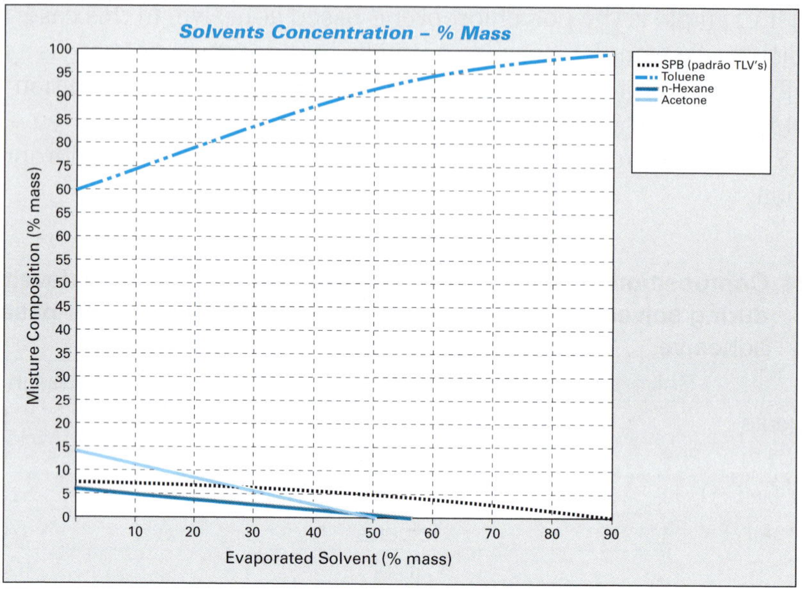

Figure 8.64. Individual solvent evaporation present in formulation for polychloroprene rubber.

Another example is given in Table 8.25 and Figures 8.65 and 8.66.

Table 8.25. Composition, initial solubility parameters, and solubility development during solvent system evaporation for a Polychloroprene-based adhesive

Solvent System Composition		Values in % m/m
Solvenes	Rubber Solvent	16.0
	Cyclohexane	24.0
	Acetone	25.0
	Ethyl Acetate	20.0
	Toluene	15.0
	Total	100.0
Solubility Parameters	δD	16.32
	δP	3.87
	δH	3.64
	δG	17.16
Polychloroprene	Initial Normalized Distance	0.78
	50% Evaporated Mass	0.72
	90% Evaporated Mass	0.62

Evaporation Conditions: 25 g/m^2/Temperature – 25 °C/Relative Air Humidity – 70%/Air Speed – 0.0667.

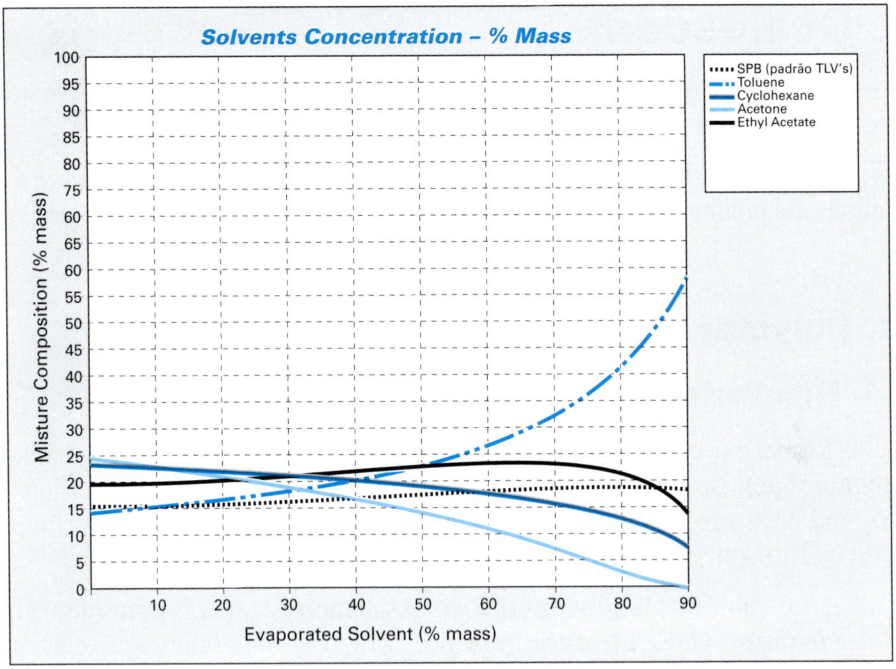

Figure 8.65. Individual solvent evaporation present in the formulation of polychloroprene rubber.

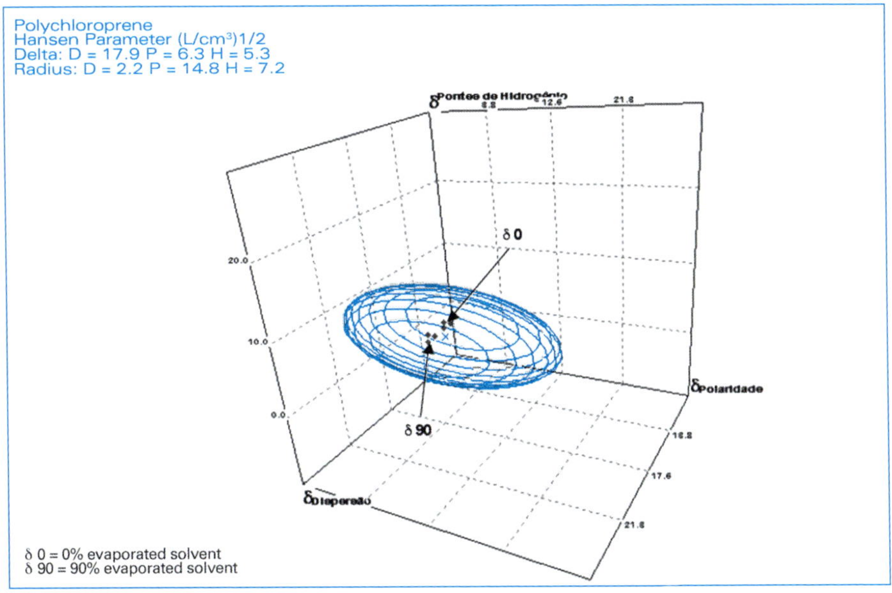

Figure 8.66. Development of Solvent System Solubility Parameters for a polychloroprene rubber.

8.6. Coalescent

Coalescence is the process in which latex particles come into contact with each other, joining together to form a continuous homogenous layer.

Las propiedades físicas y mecánicas de las películas poliméricas se ven afectadas por la naturaleza del polímero y también por las condiciones y métodos de preparación.

8.6.1. Polymer Film Preparation

8.6.1.1. Film Formation

Latex film formation comes from the coalescence of individual latex particles that normally approach each other but held apart because of stabilization forces, (electrostatic and/or steric) resulting from a charged polymer chain. These forces are increased by the continuous evaporation of the aqueous phase.

Transparent film formation without cracks depends on the minimum film formation temperature (MFFT) of the polymer, which is dependent on resistance to particle deformation, and to a lesser extent, polymer viscosity. If the film is formed above the MFFT, coalescence of latex particles can occur. However, below the MFFT, a discontinuous film or compact powder can form.

The tendency is that the MFFT approaches the Tg of a given polymer, but there are several polymers in which the MFFT is above or below the Tg. Both the Tg and MFFT are influenced by molecule characteristics.

As described by Paul A. Steward in his article, *Literature Review of Polymer Latex Film Formation and Particle Coalescence* (see reference), the solvent drying process can be described as a sigmoid curve divided into three stages. The process can be complicated because it is non-uniform (different film areas can have different drying rates).

Stage 1

Water evaporates from the latex surface, concentrating the latex. The evaporation rate is the same as water by itself, or water in a diluted solution of a surfactant plus electrolyte, such as that which constitutes the aqueous phase of a latex. This stage, the longest of the three, lasts until the polymer has reached 60 to 70% of the total volume and the surface area of the liquid-air interface starts to decrease as a result of disjoining or solid film formation.

Initially, particles move with a Brownian motion, but due to the electrical double layer, there is significant interaction once a critical volume of the water has evaporated.[140]

Stage 2

This stage starts from the moment in which particles come into irreversible contact. The evaporation rate per unit area of wet latex remains constant, but the total rate of evaporation decreases greatly during this stage.

Reducing the rate of evaporation can give a better quality film by allowing the particles more time to compact into an ordered structure. In high temperature, particles acquire enough energy to overcome their repulsion of each other and the film forms before the particles are ordered.[140]

Stage 3

This stage starts with the initial formation of the continuous film. The remaining water leaves the film initially through some remaining channels between the particles and by diffusion through the polymer itself. It is during this stage that the latex becomes more homogeneous and acquires its mechanical properties. The rate at which the water is removed can be decreased by film additives that are impermeable or hydrophilic.

Drying will also depend on film thickness and the latex solid index. A film with low solid content will dry faster than one with high solid content despite the lower quantity of water to be removed from the latter.

When the aqueous phase evaporates, there will be three distinct regions: a dry region, a wet region, and an intermediate region of flocculated latex. These regions vary according to the Tg of the polymer and can be observed in a circular film in which they appear as concentric bands. With low Tg polymers, the flocculated and dry regions are continuous, which results in hard polymers having visible fine radial cracks. The flocculated regions have some mechanical strength resulting from the van der Waals forces.[140]

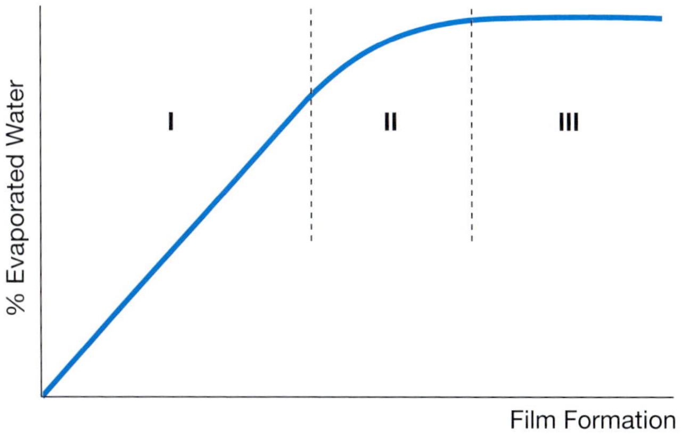

Figure 8.67. Film Formation Stages[140].

The constant rate (r_c) period of drying can be assessed in relation to latex particle size, using an equation to account for the non-uniform drying. If the three regions previously described were expressed in terms of film areas ($A1$ = area of wet latex, $A2$ = area of flocculated region, $A3$ = area of dry film in which the evaporation rate was assumed to be negligible), we have the following equation:

$$r_c = \frac{1}{A1 + A2}\frac{dW}{dt}$$

where W is the gross film weight, and t is time.

The evaporation rate (r_c) increases with particle size. The rate increase with particle size is the result of the differences in water content in the double layer or the particle surface area available for surfactant absorption. Studies showed that films prepared with large particles are of poor quality than those films prepared from smaller particle sizes because larger particles had a lower resistance to corrosion due to larger interparticle voids. This is explained by the fact that larger particles have less coalescence.[140]

Several theories describe the formation of polymer films; among them are dry sintering, wet sintering, and the capillary theory.

- Dry sintering: is driven by the polymer-air surface tension, and the coalescence process is discussed in terms of the viscous flow of the polymer. This viscous flow results from shearing stresses caused by the decrease in the polymer-particle surface area that decreases in polymer surface energy. The forces acting on the interparticle holes (spheres of radius R) are expressed in the Young-Laplace equation:

  $$P_i - P_e = \frac{2\gamma}{R}$$

  (P_i = Internal pressure; P_e = external pressure, γ = polymer-air interfacial tension, R = radius of curvature of the sphere).

  The extent of coalescence is then related to the following equation:

  $$\theta^2 = \frac{3\gamma t}{2{\neq}\eta r}$$

  (θ = angle seen in Figure 8.68; γ = surface tension; t = time; η = polymer viscosity coefficient; r = particle radius).

- Wet sintering: driven by the polymer-water interfacial tension, leading to the deformation of particles during drying. Taking into account the forces acting both for and against the coalescence of the latex particles, there should be a difference in which the capillary force Fc (resulting from the surface tension of the interstitial water, caused by the formation of a small radius of the curvature between the particles and evaporated water) must overcome the forces of resistance to deform, Fg of the latex spheres.

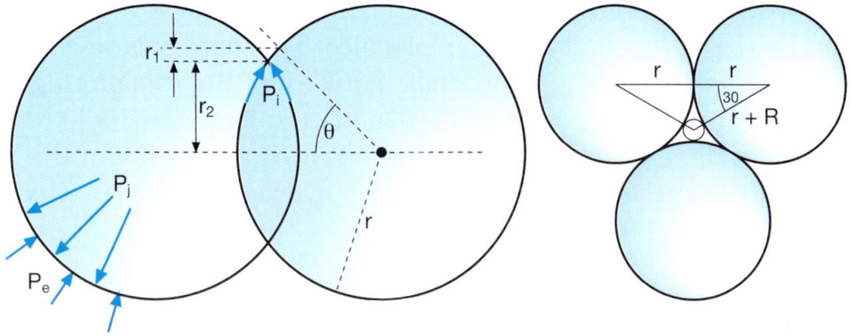

Figure 8.68 Cross section of sintered latex particles.[140]

Coalescence can also be explained in qualitative terms: as water evaporates, the latex becomes more concentrated and flocculation occurs as the repulsive forces of the particles are overcome. Particles at the latex-air interface are then subject to the forces of capillarity and therefore coalescence, which leads to compaction and deformation of the particles under the surface. Water in the film's interior must then diffuse through the upper layers to escape and this generates an additional compressive force on the film's surface. The mechanism is based on wet sintering. The source of the energy for fusion of the particles is the heat from water evaporation.

In the theories above, deformed latex particles stay joined together by physical forces, but studies show these forces alone are insufficient for the physical and mechanical properties of the film. Film formation can be analyzed in the following manner: evaporation of the solvent moves the polymer molecules closer together, making deformation possible due to the capillary forces. Then in the final stage, the film acquires a mechanical strength because solvent molecules migrate between polymer molecules. This chain of diffusion of the solvent increases the homogeneity of the polymer. This process of inter-diffusion among polymer chains is called autohesion.

8.6.1.2. Film Solvent

A solvent's rate of evaporation depends on t1/2 in a process controlled or limited by diffusion. However, the removal of the final traces of solvent from solvent cast films is a problem attributed to the fact that the polymer may be plasticized by the solvent. Elevated temperatures (to help with diffusion of the solvent in the polymer), vacuum and long drying times are used to overcome the problem. The removal of final traces is important in respect to toxicity and also to the permeability properties of the film.

8.6.1.3. Volatile Organic Components in an Aqueous Solution

It is common to add volatile organic components to the aqueous solution to help with the coalescence lowering the elasticity and allowing movement in the polymer chain which provides a better film coating. The removal of these organic compounds depends on the size and polarity of the molecules. The more polar a compound, the easier it can free itself from the hydrophilic network of the evaporating water, and a lesser quantity of residual solvent will remain in the film. The initial rate of water evaporation is not dependent on the solvent, but it can be slowed if the additive is hygroscopic or interacts with the water, forming hydrogen bonds; in which case the evaporation will not be diffusion-controlled.[140]

8.6.1.4. Additives

Polymer latex films may contain a number of additives, ranging from stabilizing surfactants to plasticizers to help with film formation. They may also include colorants added to films to be used in paints.

The plasticizer or coalescing aid is added to help the deformation of latex particles, leading to a film formation at a given temperature. The plasticizers are generally used to harden the film, but in some cases, they can be used to control film permeability.[140]

During film drying, coalescence needs to provide contact between particles, facilitating film formation, giving a good appearance and high durability. These properties are related to coalescent characteristics in decreasing the glass transition temperature and the MFFT (Minimum Film Formation Temperature) (Figure 8.69).

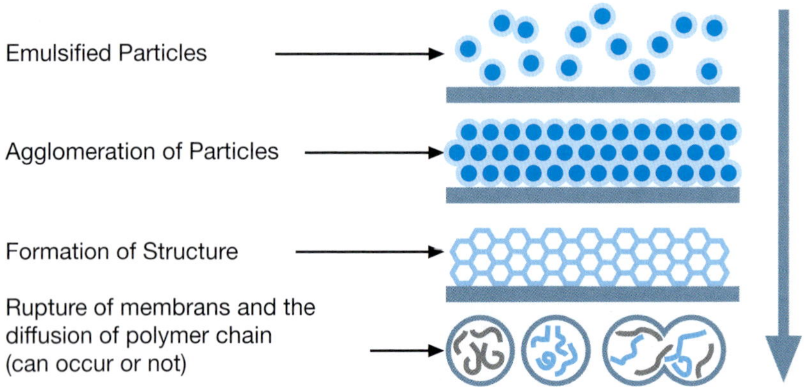

Figure 8.69. Film formation mechanism.

Coalescents are usually average and heavy solvents that improve film formation from dispersion, but they do not remain in the system after complete drying.

Decorative coatings are recognized by their good appearance after application, however, other factors also need to be considered, such as: durability, finishing, covering, easy application, spreading/thickness, among others. These come from coating composition which includes several raw materials, like fillers, thickeners, pigment, surface-active agents, and among them, the coalescing activity is very important to film end properties.

Coalescents need to give a high-performing latex coating characteristic to the formulation and are compatible with a wide variety of emulsion resins available on the market.

8.7. Solvents in Processes

In industrial chemical processes, solvents can be used in several operational stages such as separation (liquid extraction, extractive distillation, azeotropic distillation, and gas absorption), reactions, washing, and many others.

In solvent selection, the process as a whole must include: atom efficiency, energy use, the demand of renewable sources, transportation costs, solvent recovery, and environmental impact aspects (Figure 8.70).

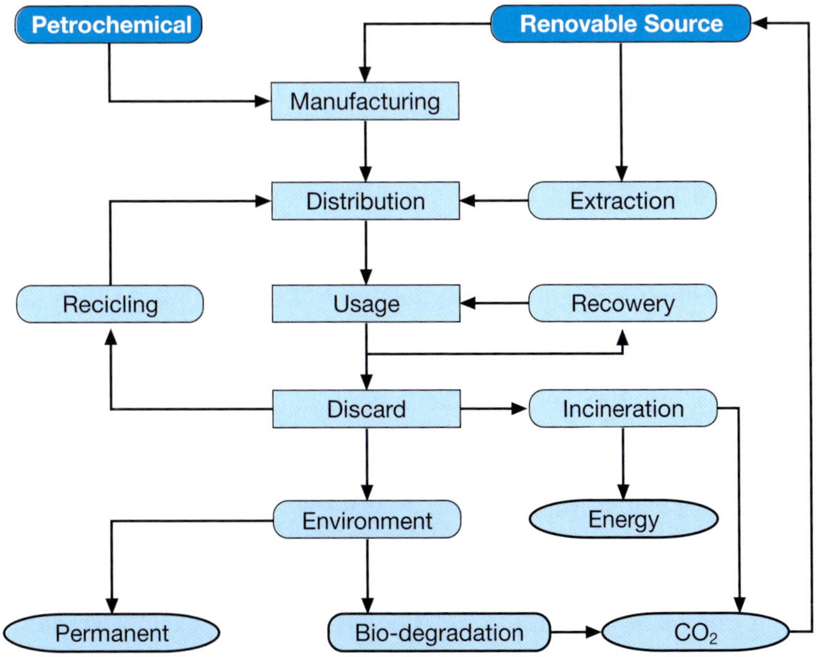

Figure 8.70. Life cycle of a solvent.

Commonly used solvents in chemical processes are petrochemical or are from renewable resources. With the increase in awareness of the need for more sustainable technologies for chemical and pharmaceutical production, the use of alternative means has been focused on some processes[1-3]. Among the alternative solvents are water and a group called neoteric solvents, such as supercritical fluids, perfluoride solvents, and ionic solvents (also called ionic liquid) at room temperature.

Ionic liquids have peculiar properties such as low vapor pressure; ability to dissolve organic materials, inorganics, and polymers; and high thermal stability. They can be used in organic and catalytic reactions and in separation and electrochemical processes. Supercritical fluids are solvents that can be used by all types of separation and as a reaction medium[4-8]. An advantage is that they can change their properties with pressure and temperature. When pressure is lessened, they can be easily removed and recycled, leading to the separation from the solute or reaction products. Among the supercritical fluids, carbon dioxide has been used in many industrial applications[4-8].

Perfluoride solvents[9 and 10] are non-polar, hydrophobic, chemically inert, non-toxic, and have a high density. They usually have limited miscibility, depending on temperature, with conventional organic solvents, forming bi-phasic solvent systems with such substances. With different solubilities for reagents, catalyzers and products like organic/fluoride bi-phasic solvents can facilitate product separation of the reaction mixture. These systems can become monophasic at higher temperatures, which can serve as a homogeneous reaction medium. After completing the reaction, the reaction mixture is cooled and again they form two phases[9-13].

Depending on the role of the solvent in the process, its evaluation can be made from a microscopic or macroscopic point of view. Operations involving heat transfer and in endothermic and exothermic reactions solvents are macroscopically treated as a continuous mass, where macroscopic physical constants, like boiling point, vapor pressure, density, cohesive pressure, refractory index, relative permissibility, thermal conductivity, surface tension, and others, must be considered. On the other hand, solvents are treated as individual molecules, i.e., from a microscopic point of view, when the solute-solvent interaction is important. This interaction can influence reaction rate and chemical equilibrium position[14 and 15]. Solvents from a molecular point of view can be characterized by the dipolar moment, electronic polarizability, hydrogen-bond donor (HBD) and hydrogen-bond acceptors (HBA), by the electron-pair donors (EPD), and the electron pair acceptors (EPA), and others. According to the extension of the solute-solvent intermolecular interaction, there can be highly structured solvents (water with strong hydrogen bonds driven to form an intermolecular network of cavities) and less-structured solvents (hydrocarbons with non-directed weak dispersion forces filling in the spaces in a more regular way)[16-20].

Solvent choice can allow the development of a more efficient process in terms of energy and chemical consumption. The solvation effect can change a chemical equilibrium or a mechanical path or decrease the activation energy which can lead to the process being more selective, having a lower temperature, or reducing the exothermicity, and consequently require less cooling. In the process of choosing, in addition to solvent properties being studied in several details[2, 14, 21] the processability, reactivity, and safety aspects must be considered. For optimization of all this information, algorithms for solvent selection have been proposed. Solvent selection in processes is presented in Item 8.7.6.

8.7.1. Extraction in Liquid Phase

8.7.1.1. Definitions

Extraction in liquid phase, sometimes called solvent extraction, is a process by which solution components are separated through contact with another insoluble or partially soluble liquid. The minimum requirement for liquid phase extraction is intimate contact between the immiscible or partially miscible liquids to allow mass transfer of constituents of a liquid (or phase) to another, following the physical separation of the phases. Any mechanism for which this process is carried out is a stage. The efficiency of this process depends on the unequal distribution of components between the two liquid phases and can be increased through the multiple contacts[22].

The feed to a liquid-liquid extraction process is the solution that contains the components to be separated. The liquid added to do the extracting is the solvent. When the solvent is mostly constituted of one substance, it is a simple solvent. When more than one substance makes up the composition for having special properties, it is a mixed solvent. The residual solution from the solvent-poor feed, with one or more solute removed by the extraction, is the raffinate phase. The solvent-rich solution containing the solute or the extracted solutes is the extract phase. Two immiscible solvents in which constituents of a solution are distributed is a double solvent; in this extraction the terms extract and raffinate are no longer appropriate.

Solvent recovery from the extract and the raffinate, even with solvent level relatively low, is almost always necessary in the extraction process because of economic and environmental conditions. Recovery is usually done through a distillation process (Figure 8.71A) that strongly depends on solvent characteristics (boiling point, solubility, relative volatility and azeotropic formation). Figure 8.71B shows the process using liquid phase extraction for solvent recovery. An example of liquid phase extraction is using lixiviation extraction in uranium recovery.

Exact knowledge of the equilibrium relationship between phases is vital to quantitative considerations of the extraction processes. The necessary quantities of solvent (and reflux when used) are determined by this data. Stable formation of two liquid phases in contact with each other is an essential condition of the process. When a solvent has a high selectivity for a component, it should be a substance in which other components have coefficients with high activity and where these solutes are excluded when there is an equilibrium between the phases.

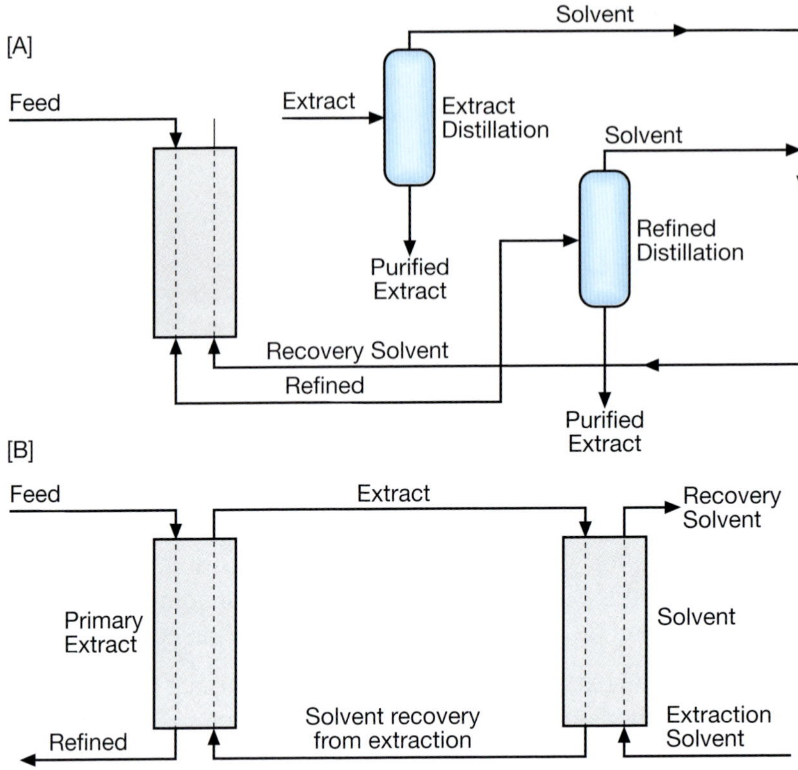

Figure 8.71. Flowchart of the liquid-phase extraction processes with solvent recovery by [A] distillation and [B] extraction.

Considering a simple system, the solution has three components: "A" called inert, "B" is the solute (these two make up the solution to be separated), and "S," the solvent. Phase equilibrium for this simple system can be represented by ternary equilibrium diagrams as seen in Figure 8.72. These graphs show isotherms in enough pressure to keep the system entirely liquid. The Type I diagram (Figure 8.72A), with an immiscible space, tangent only on one side is the most common. A-B and B-S liquid pairs are completely soluble: A and S dissolve limitedly, forming mutually saturated solutions represented by G and H. Adding B to the mixture tends to make A and S more soluble. At the critical point (the point where they interconnect) P, the two phases become one. The B solute distributes among the saturated phases forming solutions in equilibrium as shown as the tie line called M and N. In this tie line, the distribution coefficient K is shown in equation 4.

$$K = \frac{X_{BN}}{X_{BM}} \qquad \text{ou} \qquad K = \frac{x_{BN}}{x_{BM}} \tag{8.4}$$

where:, X_{BN} indicates the ponderal fraction, and x_{BN} is the molar fraction of B in solution N.

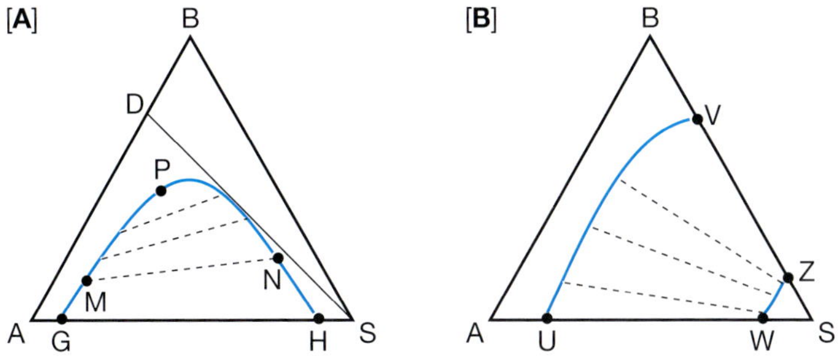

Figure 8.72. Extraction – triangular phase diagrams [A] Type I and [B] Type II.

An extraction by solvent can be carried out using any method in which mass transfer among different phases is done. Industrial processes normally use continuous contact in a countercurrent at various stages in a sequence of mixers and settlers, or in baffle trays; or differential contact in a continuous countercurrent in a filled baffle or with mechanical agitation. In laboratory studies, it is common to carry out the non-continuous operation with parallel currents, or with a simple contact in several stages.

8.7.1.2. Desirable Solvent Characteristics

Characteristics that should be used in choosing a solvent for liquid-phase extraction are as follows:[22]

1. **Selectivity.** One of the most important attributes of a good S solvent is its capacity to extract B, preferentially, from an A and B mixture, such that the relationship between A and B in the extract after the solvent is removed is different from the relationship between them in the solvent-free raffinate. Figure 8.73 illustrates the effect of an extraction in only one stage. The feed F in contact with solvent S, in which its quantity is enough to form a two-phase mixture Q, produces an equilibrium between the extract E (rich in S) and the raffinate R (rich in A), on the RE tie line. Point $E9$ represents the solvent-free raffinate with the same relationship between B and A that is in R. The $\beta_{B,A}$ selectivity of solvent S in the B and A separation, on the tie line is given in equation 8.2

$$\beta_{B,A} = \frac{x_{BS}/x_{AS}}{x_{BA}/x_{AA}} = \frac{X_{BS}/X_{AS}}{X_{BA}/X_{AA}} = \frac{K_B}{K_A} \tag{8.5}$$

where: x_{BS} is a molar fraction of B in solvent S, X_{BS} is the ponderal fraction and x_{BS}/x_{BA} is the relationship of B to A in E or in E'. K is the distribution coefficient.

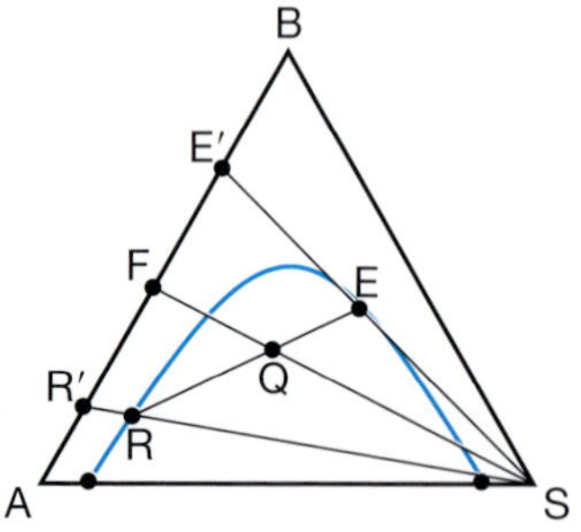

Figure 8.73. Selectivity of solvent *S* in the separation between *A* and *B*.

Considering that in a ternary system the three substances are present in both phases, the α activity coefficiency of each substance must be the same in the two phases, as long as the reference state (standard) is the same. This circumstance is guaranteed when choosing the pure solvent on system pressure and temperature as the reference state. In this case, phases rich in *A* and solvent *S*, there is: [22]

$$\alpha_{AA} = \alpha_{AS} \qquad \alpha_{BA} = \alpha_{BS} \qquad \alpha_{SA} = \alpha_{SS} \tag{8.6}$$

where: α_{BA} is the activity of *B* in *A*.

Taking this into consideration, the definition of the α activity coefficients (equation 6), equation 5 can be represented by equation 7.

$$\beta_{B,A} = \frac{\gamma_{AS}\,\gamma_{BA}}{\gamma_{BS}\,\gamma_{AA}} \tag{8.7}$$

Selectivity is a numerical measure of the degree of separation in the initial feed F. Separation will only occur when $\beta_{B,A}$ is different from the unit and it will be even better when the selectivity is greater than the unit.

2. **Recoverability**. Normally, the solvent must be recovered from the extract stream and the raffinate stream for reuse. Distillation is the usual medium for this. If distillation is used, the solvent should not form an azeotrope with the extracted solute and the mixture should have a relative high volatility to reduce recovery costs. The extract substance (solvent or solute) that is in a lesser quantity should be more volatile with the objective of reducing heating cost. If the solvent has to be evaporated, its latent evaporation heat must be low.

3. **Distribution Coefficient**. An extracted component's distribution coefficient is preferably high because the quantity of solvent and the number of extraction stages for a set relationship between the solvent and the feed is reduced. This coefficient directly influences selectivity (equation 4).

4. **Capacity**. This is the capacity of a solvent to dissolve the extracted solute. In Type I systems (Figure 8.72A), solvent S can dissolve an infinite quantity of B, and this has the least demands for the solvent than those in the Type II systems or in those systems in which B is a solid and its solubility in S is a limiting factor.

5. **Solubility of a Solvent**. In Type I systems it is better to have a higher insolubility grade between A and S. This leads to a large selectivity (K_A is little). In Figure 8.74, for both the systems shown, only AB solutions in the concentration range between D and A can be separated by the use of solvent S or S', since only they will form two immiscible phases with the solvent. The decrease in the solubility of A in the solvent, which can be seen by the amplification of the immiscibility interval in the composition triangle – Figure 8.74, provide, in general, an increase in the feed concentrations that can be used. Solvent recovery costs are less when insolubility is greater. In Type II systems there are no limitations, however solvent capacity is limited.

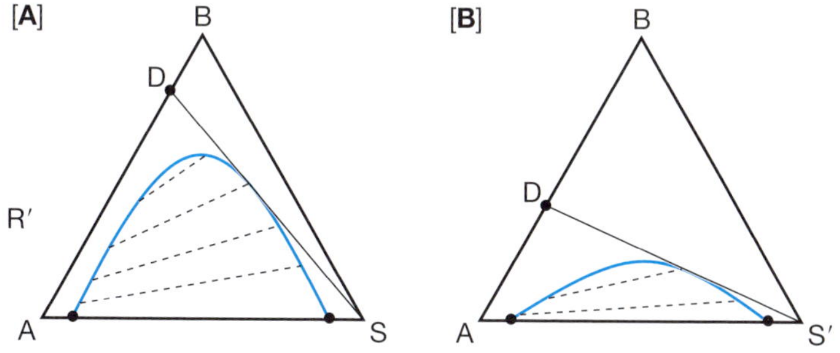

Figure 8.74. Influence of solvent solubility on extraction.

6. **Density**. Density difference between phases in equilibrium is essential because running speed of the liquids in contact, through the extractor, as well as the final separation between phases is directly affected by this factor. In Type I systems, density difference, in general, decreases monotonically with the growth of B concentration, reaching zero at the critical point. In some systems, the difference in density is nullified in an intermediary tie-line, (isopycnic line), and then is nullified. In these systems, extraction in differential extractors cannot be carried out in this line.

7. **Interfacial Tension**. This should be preferentially high, favoring the coalescence of the emulsion, thus making the dispersion of a liquid in another more difficult. The coalescence of the dispersed drops in the liquid is very important to the extraction process. The greater the miscibility between phases, the lesser the surface tension will be; in Type I systems this parameter falls to zero at the intertwining point.

8. **Chemical Reactivity**. The solvent must be chemically stable and inert in relation to the other system components, and in relation to the construction materials.

9. **Other Factors**. The following solvent properties are desirable: low viscosity (in order to have higher mass transfer speeds and dispersion separation), inflammability, low toxicity, and low cost.

8.7.1.3. Areas of Application

Liquid phase extraction is capable of separations that are impossible using ordinary methods of distillation because it primarily works according to chemical types of components. For example, aromatic hydrocarbons and paraffin with identical boiling intervals can be separated by liquid phase extraction using di-ethylene glycol as the solvent.

Some solutions can be fractionated by distillation, but for those in which treatment using this procedure is expensive, they can be separated more easily using a liquid extraction intermediary. As an example, distillation of a diluted solution of acetic acid in water can be cited; this is an expensive procedure because large quantities of water need to be evaporated with a high reflux rate and the latent evaporation heat of water is high. The extraction of acid using ethyl acetate, followed by the distillation of the new solution, is a cheaper process. Analogly, high quantities of water evaporation for recovering a non-volatile solute can be avoided through a solute extraction using a solvent that has less latent evaporation heat.

Fractionated crystallization, which is expensive, can be avoided, for example, in niobium separation in solution, using a liquid phase extraction.

In the case of thermosensitive substances, like penicillin, they can be separated from the mixtures in which they are formed through extraction using an appropriate solvent at a low temperature.

8.7.2. Extractive Distillation

8.7.2.1. Definitions

Extractive distillation is a distillation form involving the addition of a solvent which modifies the liquid-vapor equilibrium of the components to be separated thus making separation easier. The solvent modifies the volatility of each component, each one different from the other, by its effects on liquid phase properties. These effects can come from the formation of complexes associated with them or the existing associated structural changes in the presence of the solvent that is different from the solvent-free mixture.

The extractive distillation is applied to binary systems with close boiling points, which have the tendency to form azeotropes and multi-component systems, where the order of component volatilities do not correspond to the desirable distribution of the components among the products.

Figure 8.75 shows a typical flowchart of an extractive distillation process. The chosen solvent must be less volatile than either of the two components, *A* and *B*, and in order to maintain its high concentration throughout the greater part of the column, it is necessary to always introduce it above the stage of the virgin feed opening. Usually, some plates are introduced below the highest one, with the plate difference being dictated by the need to reduce solvent concentration at an insignificant level to the rising vapor before removing the product from the top.

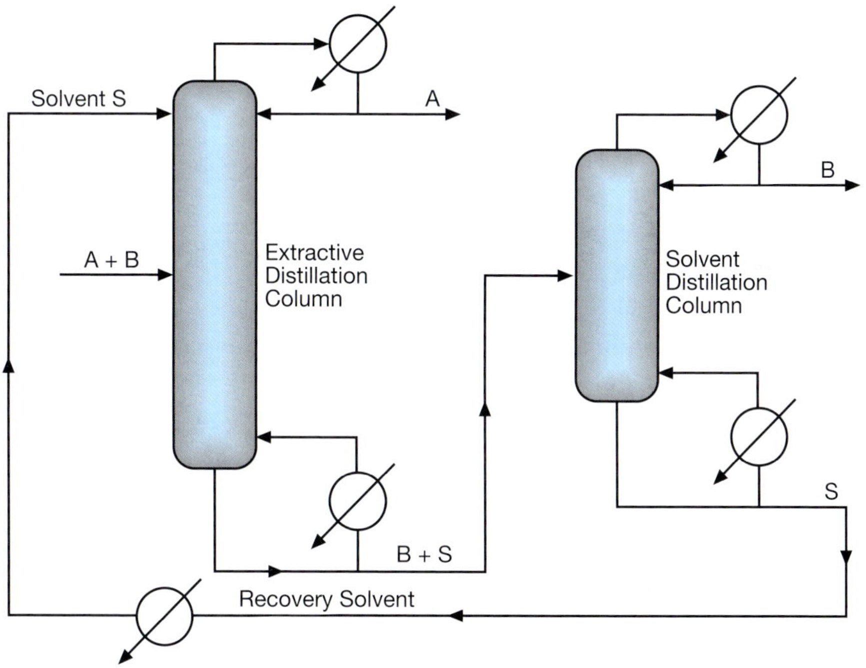

Figure 8.75. Simplified flowchart of the extractive distillation process.

The transfer rate of the liquid solvent from plate to plate is relatively constant in virtue of its low volatility. Solvent concentrations in the higher trays is desirable (typically 65/90 mol%) to get the greatest difference between the volatilities of the components to be separated. To ensure solvent concentration remains in the total solubility range, it is necessary to know the amount of the mutual solubilities. The solvent concentration profile in a column is controlled by manipulating the entering flow rate and solvent enthalpies of the virgin feed and reflux streams.

Solvent recovery is simpler with extractive distillation than azeotropic distillation. The chosen solvent does not form an azeotrope with other substances in the end product of the extractive distillation baffle and its recovery is possible using a fractionated distillation.

Phase Equilibrium

Non-idealic mixtures are the physical principle that allows extractive distillation to be a viable alternative to other separation methods. The non-ideality is expressed qualitatively by the activity coefficient of the liquid phase and by the fugacity coefficient of the vapor phase. At low pressures, fugacity coefficients are usually near 1 (ideal) and the non-ideality of the system is established by the activity coefficients of the liquid phase.

The molar fraction of i in the vapor phase (y_i) in equilibrium with x_i is given by:

$$y_i = \frac{\gamma_i f_i^0 x_i}{\phi_i P} \tag{8.8}$$

where:

y_i = molar fraction of i in vapor phase
x_i = molar fraction of i in liquid phase
f_i^0 = fugacity of i in the standard state
γ_i = activity coefficient of i
ϕ_i = fugacity coefficient of i
P = total pressure

If $f_i^0 = P_i^0$ where P_i^0 = pure component i vapor pressure, then:

$$y_i = \frac{\gamma_i P_i^0 x_i}{P} \tag{8.9}$$

Definition of relative volatility:

$$\alpha_{ij} = \frac{\dfrac{y_i}{x_i}}{\dfrac{y_i}{x_j}} \tag{8.10}$$

$$\alpha_{ij} = \frac{\gamma_i P_i^0}{y_i P_j^0} \tag{8.11}$$

$$\beta_{ij} \text{ ideal} = \frac{\gamma_i}{\gamma_j} \tag{8.12}$$

In some cases a significant change in operation pressure and therefore, in the temperature, α_{ij} changes enough to eliminate the azeotrope.

In addition to relative volatility, the selectivity β_{ij} parameter is used to characterize non-ideality of the mixtures:

$$\beta_{ij} \text{ ideal} = \frac{\gamma_i}{\gamma_j} \tag{8.13}$$

(activity coefficient of the key components in the presence of the solvent).

As an example, consider an acetone and methanol binary mixture. This mixture forms a binary azeotrope at 80 mol% of acetone (20 mol% of methanol).

The azeotrope composition is:

$$\alpha_{AM} = 1{,}0 \qquad \beta_{AM} = 0{,}7$$
$$\gamma_A = 1{,}1 \qquad \gamma_M = 1{,}5$$

If water is added to the mixture, there will be an equilibrium change in the phases of this binary mixture. The ternary system to the same relative composition as that of the binary azeotrope with the addition of H_20 at 60 mol% will have the following characteristics:

$$\alpha_{AM} = 3{,}3 \qquad \beta_{AM} = 2{,}4$$
$$\gamma_A = 1{,}8 \qquad \gamma_M = 0{,}7$$

Notice an increase of α and the substantial increase in β. Thus, the addition of water to the acetone/methanol mixture favors the separation of the azeotropic mixture.

Solvent Concentration Effect on Selectivity

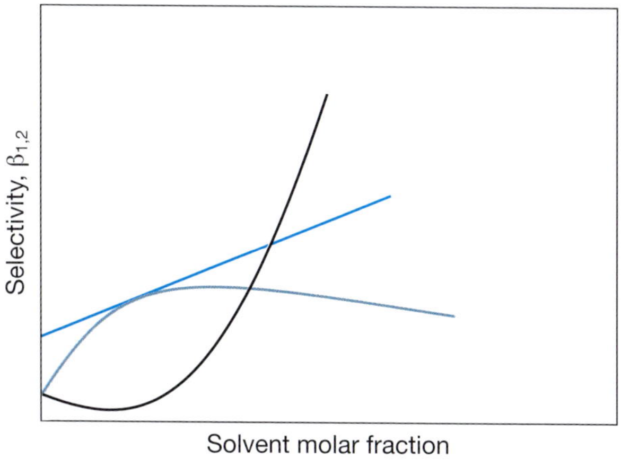

Figure 8.76. Influence of solvent solubility on selectivity.

1) Selectivity usually increases almost linearly with solvent concentration.
2) Selectivity increases more than linearly with the solvent concentration (high solvent concentration can cause immiscibility).
3) Selectivity has a maximum point (usual).

8.7.2.2. Desirable Solvent Characteristics

The number of possible solvents for an extractive distillation is generally greater than the convenient ones for azeotropic distillation because there are fewer restrictions on volatility. The following restriction must be considered in choosing the solvent:

- The solvent must boil at a temperature sufficiently higher than that of the feed components to be separated in order to impede azeotropic formation;
- Solvent boiling point must not be so high so that a high consumption of energy from the sensitive heat associated with the solvent cycle is avoided;
- High selectivity, or the ability to change liquid-vapor equilibrium of the original mixture enough to allow easy separation, committed to using a small quantity of solvent;
- High capacity or ability to dissolve components in the mixture to be separated;
- Recoverability: the solvent must be totally separated from the mixture;
- Other restrictions can be considered, like: cost, chemical stability, corrosion, and toxicity.

8.7.3. Azeotropic Distillation

8.7.3.1. Definitions

Like extractive distillation, azeotropic distillation involves adding a third component to a binary system to facilitate system separation using distillation. The component added modifies the activity coefficients of the liquid phase, and consequently, the liquid-vapor equilibrium of the other two components in a favorable direction.

In the extractive distillation solvent volatility must be lower than the feed components to be separated, but in the azeotropic distillation, solvent volatility (carrier) should be higher or intermediate to the feed components. In extractive distillation, the solvent cannot form an azeotrope with the mixture components to be separated. On the other hand, in azeotropic distillation, the solvent must form an azeotrope with the component to be removed at the top and this azeotrope must be heterogeneous.

Azeotropic distillation is recommended for mixtures that form azeotropes or for mixtures with close boiling points.

An azeotrope is a mixture of two or more volatile components having identical liquid and vapor equilibrium compositions and their composition does not change during distillation. The azeotropic mixture can boil at a higher or lower temperature than that of its components. They can also be classified as minimum or maximum azeotropes. Minimum boiling temperature azeotropes are the most common.

Another classification is related to the condensed phase. When the azeotrope is condensed and the liquid phase is homogeneous, or in other words, there is complete miscibility, the azeotrope is called homogeneous. However, when there is separation of liquid phases, the azeotrope is called heterogeneous.

The heterogeneous azeotrope can be easier to separate than many ideal mixtures. The high immiscibility of liquid phases from vapor allows complete separation of components that before formed a homogeneous azeotrope. This separation can be done through a second distillation column or using a settler.

To separate a homogeneous azeotrope, its properties need to be changed. There are four processes that can do this:

- Extractive Distillation;
- Distillation with pressure adjustment;
- Formation of a heterogeneous ternary azeotrope;
- Formation of a heterogeneous binary azeotrope.

During azeotropic distillation, a solvent (called the carrier) is added to the original mixture to form an azeotrope with one or more components in the mixture, and this azeotrope is removed, either as an overhead product or as a bottom product.

8.7.3.2. Desirable Solvent Characteristics

The carrier solvent must form an azeotrope with the following characteristics:

- Low boiling point;
- The new azeotrope must be sufficiently volatile to be totally separated from the remaining constituent;
- Allow a low amount of solvent to reduce the heat used in vaporization.
- It should be preferentially heterogeneous; this could simplify solvent recovery a lot.

Additionally, a satisfactory solvent must be::

- Cheap and available
- Chemically stable
- Non corrosive
- Non-toxic

It should also have:

- Low latent vaporization heat
- Low viscosity
- Low freezing point

8.7.3.3. Comparison between Extractive and Azeotropic Distillations

Extractive distillation is generally considered more interesting than azeotropic distillation mainly for two reasons: (i) there is a greater possibility of solvent choice because there is no requirement to form an azeotrope; and (ii) generally less solvent should be made volatile.

However, azeotropic distillation becomes more interesting when the volatized impurity is the smallest constituent in the feed and the azeotropic composition is favored. For example, in the dehydration of ethanol starting from an 85.6% ethanol-water molar solution, azeotropic distillation with n-pentane is more economical than extractive distillation using ethylene glycol[23]. The reverse can be true for a more diluted alcohol feed.

8.7.4. Absorption of Gases

8.7.4.1. Definitions

A gas absorption is a unitary operation in which one or more components in a gaseous mixture are dissolved in a liquid. Absorption can be a purely physical phenomenon or it can involve the solubilization of the substance in the liquid followed by a reaction with one or more the constituents.

Equipment used for continuous contact between a vapor and a liquid can be a column filled with a solid, an empty column in which a liquid is sprayed into it, a bubble tray column or with openings or appropriate valves. In general, gas and liquid run in a countercurrent in order to have the greatest difference of concentration, and therefore, greater absorption speed.

In an absorption column project, liquid-vapor equilibrium data are necessary to determine the amount of liquid to absorb a determined amount of the gas's soluble components.

8.7.4.2. Desirable Solvent Characteristics

If the main purpose of the absorption operation is to produce a specific solution, the solvent is chosen according to the product.

If the purpose of the absorption is to remove a gas constituent, there is the possibility solvent choice and the following properties to be considered:

1. **Gas Solubility.** Gas solubility should be high, increasing the absorption rate and decreasing the amount of necessary solvent. Generally, solvents of a chemical nature similar to the solute to be absorbed will provide good solubility. The solvent's chemical reaction with the solute normally will result in a high solubility of gas, but if the solvent is to be recovered, the reaction should be reversible.

2. **Volatility.** The solvent must have a low vapor pressure, or in other words, be a little volatile so the gas leaving the absorption column, normally saturated in the solvent, does not lead to a high loss of it. If necessary, a second, less volatile liquid can be used to recover an evaporated portion of the first.

3. **Low Corrosivity.**

4. **Low cost.**

5. **Viscosity.** Low viscosity is preferred because this results in a greater rate of absorption; it improves flooding characteristics in the absorption column, less load loss in the pump and good heat transfer characteristics.

6. **Other Factors.** Whenever possible, the solvent must be non-toxic, non-inflammable, chemically stable and have a low freezing point.

8.7.5. Reactions

Solvents can have multiple functions in reactions. They can be used as a reaction medium to keep reagents together, as a reagent to react with a solute when it cannot be dissolved, and as a carrier to keep chemical compounds in solution in required quantities until it is used.

As a reaction medium, a solvent can be used for some purposes. For example, in endothermic reactions, heat can be supplied through a heated inert solvent having a high heat capacity, while in exothermic reactions freed heat can be removed through boiling or absorption of the solvent. If reactions involve solid reagents, solvents can be used to create a solution (reaction medium) through which the solid reagents can be maintained in contact. Solvents can also be used to indirectly influence the reaction by removing of one or more reaction medium products.

8.7.5.1. Solvent Effect on the Reaction Rate

A solvent influences reaction rate and this effect depends on its characteristics. In thermally activated reactions, the solvent can decrease reaction rate by reducing reagent free energy in the transition state[14 and 24] or they can increase the rate stabilizing the energy in the transition state (Figure 8.77). These thermodynamic or static effects result from the interaction energy between solvent molecules and the reaction species. Charged species dissolved in a polar solvent is a characteristic example of how these energy stabilizations are generated. Charged species will polarize solvent molecules around them orienting the dipolar solvent molecules such that a decrease in the system's free energy is produced. This energy stabilization is frequently called solvation energy, the driving force for molecular dissolution. Solvents that do not have an appreciative potential for stabilizing solute molecules (non-polar solvents) will dissolve such species poorly, and in these cases, other competitive processes (ex. agglomeration and precipitation) will prevail over solvation[25].

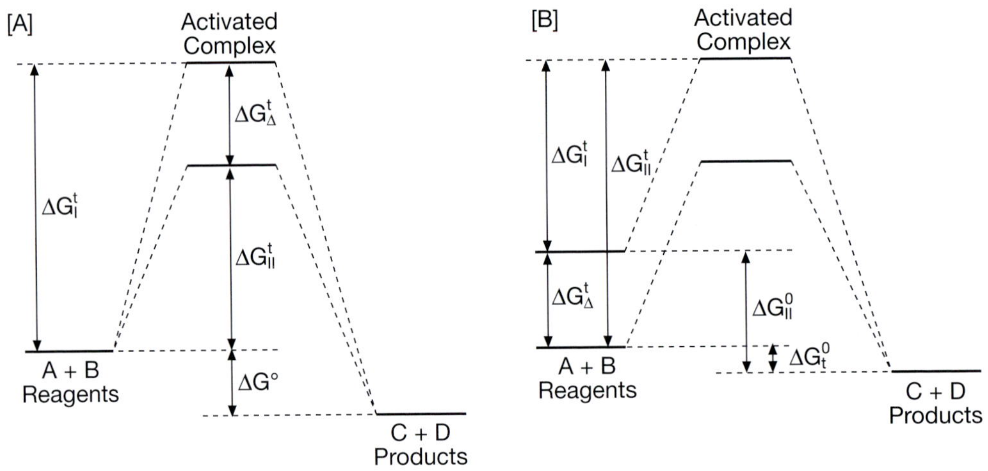

Figure 8.77. Static influence of the solvent at reaction rate. [A] Different reaction rate due to the preferential solvation of the activated complex. [B] Different reaction rate due to the preferential stabilization of the reagents. Gibbs free energy, G, is plotted on the *y* axis and the reaction (analyzed in three positions: reagent, products, and activated complex) is plotted on the *x* axis. The differences in free energy, ΔGI and ΔGII, are the activation energies for reactions in solvents I and II respectively. Standard Gibbs free energy, ΔGI e ΔGII, are the differences of free energy in the equilibrium for reagents and products in solvents I and II respectively.

Solvation is not limited to charged species. If a specie in reaction has a significant electrostatic moment (example: dipole, quadropole), the dissolution of these species in a polar solvent will also lead to a stabilization of species and the solvent molecules will adjust themselves to the species' electric field. Oppositely, a weak stabilization will occur if the species are dissolved in a non-polar solvent, where the interaction energy between the solvent and the species is minimum. Consequently, reaction rate can be experimentally adjusted by the selection of an appropriate solvent[14 and 24].

The effects of solvent interactions are important for many reactions, including electron transfer reactions. Figure 8.78 illustrates the effects of stabilization energy in an electron transfer reaction model: $D + A \rightarrow D^+ + A^-$. For both reagents and products, constructed free energy curves relate to system free energy (solute + solvent) versus reaction fluctuation[26]. Computational simulations suggest these free energy curves are better described as parabolas[27-30]. The reaction can then be seen as a population transfer of the reagent free energy curve base to the product curve base, on activation energy, $\Delta G\dagger$. Activation energy is determined by the difference in free energy of the reagent curve base to the intersection of the curves. The Marcus theory[26] is a commonly used model for describing electron transfer reactions, which relates reaction rate, k_{ET}, to the magnitude of activation energy that reagents must overcome.

$$k_{ET} \propto \exp \frac{-\ G^{\dagger}}{RT} \qquad (8.14)$$

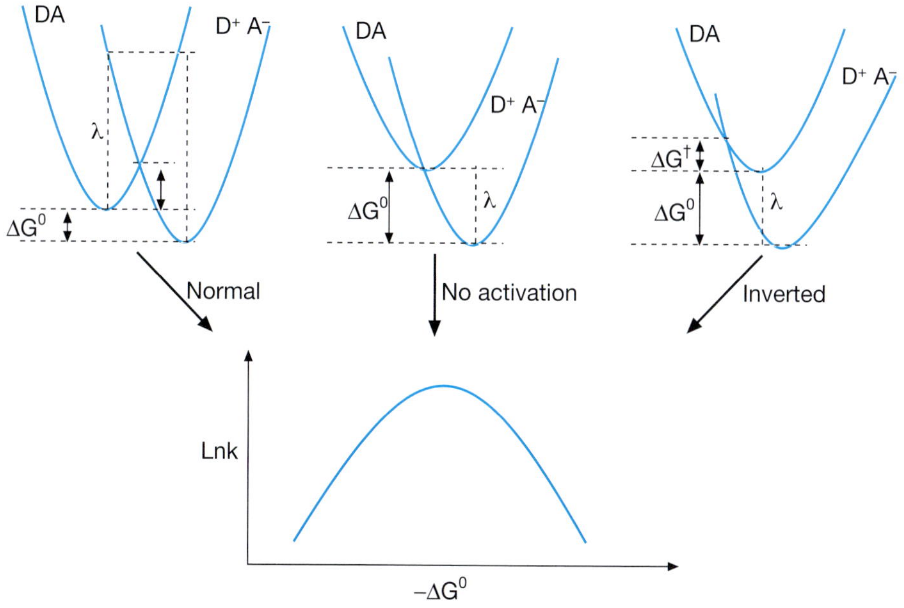

Figure 8.78. Static solvation effects on determining electron transfer speed according to the Marcus theory. Three examples of electron transfer systems are shone: normal, no activation, and inverted with their corresponding regions on reaction rate curves versus Gibbs free energy.

Based on the Marcus theory describing reaction rate as a thermally activated process on the activation energy (ΔG†), if this decreases, the reaction rate increases. One method to change ΔG† is to preferentially transfer the energy stabilization, ΔG°, of the products to the reagents (i.e. a curve is vertically dislocated). Through the free energy parabolic curves, ΔG† can be related to the ΔG° and to a reorganization energy, λ, that quantifies the magnitude of the stabilization required by the solvent to accommodate the reaction.

$$G^{\dagger} = \frac{(\ G^{0} + \lambda)^{2}}{4\lambda} \tag{8.15}$$

The curve ln kET vs. –ΔG° shows reaction regions controlled by effects from the solvent: a normal region where the increase in ΔG° increases the reaction rate, a region without activation where rate is maximized, and an inverted region where the increase of ΔG° will result in the decrease of reaction rate[31]. Therefore, energy stabilization from the solvent plays an important role in electron transfer reaction rate.

Treating energy transfer reactions using the Marcus theory presumes the solvent relaxes appreciatively faster than the reaction time, and then only solvent static properties affect reaction rate. With fast reactions or when refrigerated solvents are used (super cooled liquids), adiabatic considerations should be left out and non-adiabatic effects must be incorporated in the description of reaction rate[32 and 33]. These effects originating from the solvent dynamic can appreciatively affect reaction properties. For example, thermal fluctuations in the solvent can give the necessary thermal energy to reagents to overcome the big activation barrier, and also act as heat for product species that appear. These temporary properties affect reaction rate either by the probability to overcome the activation energy or inversely, by probability of product reduction pass the barrier to form the reagents.

Dynamic effects from the solvent are especially important in the description of reactions, in which time-scale are compared with the time-scale of solvents. Solvent dynamic has been a key factor in understanding many fast reactions including electron transfer reactions[32 and 33] and energy[34 and 35]. The Kramers study[36] showed that understanding the dynamic of solvents is important for understanding and optimizing chemical reaction rate in solutions.

8.7.5.2. The Role of Solvents in Biochemical Reactions

The biocatalyst has become an advantageous alternative for chemical transformation of a variety of compounds with the food industry, animal feed, chemical and pharmaceutical applications. However, it is not necessarily easy to get desired performance levels in rate, productivity, and selectivity of the reaction. One of the strategies for optimizing biocatalytic performance is the use of non-conventional mediums, i.e., non-aqueous or organic-aqueous systems.

The main advantages for using organic solvents in biocatalysis compared to the traditional enzymological aqueous system can be summarized as[37-39]:

(i) Increase in the solubility of nonpolar substrates and products;

(ii) Inversion of the thermodynamic equilibrium in favor of synthesis and not hydrolysis, allowing usually reactions not favored in aqueous systems (for example, transesterification, tioesterification, aminolyse);

(iii) Drastic changes in enantioselectivity of the reaction when an organic solvent is changed for another;

(iv) Suppression of undesirable lateral reactions depending on water;

(v) Elimination of microbial contamination in the reaction mixture [1,5-14].

Water-restricted organic systems used in biocatalysis can be classified as a homogeneous or heterogeneous system. A review of the homogeneous systems was written by Torres and Castro[38] and by Castro and Knubovets[40], and a review for heterogeneous systems by Nadia Krieger and collaborators[39]. The heterogeneous systems include liquid-liquid macro-heterogeneous systems in which the water represents a 1 to 5% reaction medium; liquid-solid macro-heterogeneous systems; and micro-heterogeneous systems.

Despite the advantages of the biocatalytic system in an organic medium, enzymes are denaturalized or inactive in the presence of organic solvents, and enzyme catalytic activity that is stable at non-aqueous ambients is generally less than the aqueous systems[41-43]. To solve this problem, many studies are being done to improve enzymatic activity and stability in a non-aqueous medium using several types of strategy. Examples include protein engineering[44-47], covalent bonding of amphipathic compounds (PEG, aldehydes, and imidoesters)[48-49], non-covalent interactions with lipids or surfactants[48, 50], water microemulsion in oil or reverse micelle[51], immobilization in appropriate insoluble supports (inorganic porous supports, polymers and molecular sieves)[52, 53], and the use of enzymes in lyophilized powder or bonded to suspended crystals in organic solvents[54].

Enzymatic stability is less in miscible solvents in water like acetone and ethers than in hydrophobic solvents like alkanes or haloalkanes. With reactions in an aqueous medium, reaction rate increase is seen mostly due to the stabilization of polarized activated complex by the hydrogen bound and a decrease in the hydrophobic surface of reagent molecules during the activation process[55]. With hydrophobic organic solvents there is a crucial water bond starting at the enzymatic surface[56-58], while this bond is difficult in hydrophilic organic solvents. As a result, acceptable enzyme stability in hydrophilic organic solvents is rare. Two enzymes are reasonably stable in these solvents: the lipase from Pseudomonas mendocina PK12CS[59], which maintains a good activity at 10% of ethanol, and the lipase from Bacillus megaterium

CCOC-P2637, which is stable at fractions greater than 80% ethanol and acetone, and 100% isopropanol[60].

In homogeneous non-aqueous biocatalytic systems, the solvent should solubilize the enzymes. Enzyme solubility dramatically depends on the solvent, principally on its hydrophobicity. A large number of enzymes is soluble in protic solvents, which suggests an importance on solvent propensity to form hydrogen bonds. Only a small number of solvents dissolve enzymes at a level of 1 mg/ml or more.

Some syntheses in a very successful non-aqueous biocatalytic system are ester and peptide syntheses, the production of cocoa butter substitutes[61]. Four of the 15 industrial processes using hydrolysis cited by Krishna[47] are driven in the presence of organic solvents: the synthesis of amina chiral and alcohols in MTBE-ethyl methoxy acetate, the synthesis of an anticholesteral medicine in toluene, the generation of an intermediary in the synthesis of Diltiazem in a toluene-water mixture, and the synthesis of palmitate and isopropyl myristate in 2-propanol.

Currently, principal industrial applications of non-aqueous enzymological systems are based on the use of suspended enzymes (liophilizated and bonded to crystals) and immobilized enzymes, both represent heterogeneous biocatalytic systems. However, heterogeneous biocatalyses in organic solvents have some disadvantages: low reaction rate which is difficult to control and large-scale reproducibility problems. In a homogeneous system, these problems do not occur because of the absence of organic-water solvent interface and the elimination of diffusion limitations for substrates and products. Additionally, substrate and product concentrations at the enzyme surface can easily be controlled in homogeneous systems, thus favoring reaction rate and prevents enzymatic inhibition. On the other hand, with homogeneous systems the separation and purification stages of products are more difficult.

Among all the developments to increase industrial use of biocatalytic systems in a non-aqueous system on a large scale, micro-homogeneous system developments can be cited. These combine homogeneous and heterogeneous biocatalysis as being one of the best approaches to robust enzymatic catalysis for industrial applications[40]. A notable system that includes this combination, the homogeneous and heterogeneous system, is an HIP system (hydrophobic ion pairing) plastic. Biocatalytic plastic is a biocatalysis obtained through incorporation of enzymes covered in surfactant ions in plastic material. These systems have excellent activity and stability that make them attractive for industries. Biocatalytic plastics can be compared to immobilized enzymes in an aqueous system, so the initial high rate in a micro-homogeneous ambient can combine with easy biocatalysis recycling and purification of reaction products due to the system's macro-heterogeneity. This system increases applications on a large scale.

8.7.5.3. The Role of Alternative Solvents in Reactions

Supercritical carbon dioxide ($scCO_2$) shows potentiality as a solvent in synthesis in high-value industries like pharmaceuticals and fine chemistry. In addition to environmental benefits, $scCO_2$ can cause an increase in diastereoselectivity and enantioselectivity compared to processes in conventional solvents[62-67]. It allows control of reagent and product solubility, which can lead to a selective separation of the product[68-70]. Further, reactions can be driven to subcritical pressures using CO_2 Lewis acidity or carbonic acid Bronsted acidity formed in aqueous solutions under CO_2 atmosphere[71-77].

Although the $scCO_2$ shows some advantages as a solvent in reactions, it presents some limitations as a solvent due to its low solubility in several organic compounds and in aqueous solutions. Another aspect that limits use is economics: the reactions require high-pressure reactors and high energy costs and the cost of CO_2 depends on the possibility of having a supply station near the reactor to make it economically feasible. In principle, continuous reactors are particularly attractive for these high-pressure reactions, which allow maximizing the intensification of processes and help minimize costs.

Polar molecule solubility problems in CO_2 have been minimized by adding a co-solvent (ex.: methanol) or surfactant[78]. Another possibility for favoring solubility is by using compound modification, normally by increasing lipophilicity or adding perfluorinated chains or silane groups[79-81].

Among some commercial examples where the CO_2 is used as a reaction solvent, fluorine polymer production can be cited[82]. Thomas Swan Ltd. in England has a plug-flow continuous reactor for hydrogenation alkylation reactions on a large scale[83], reactions involving light gases: hydrogen[84-86], CO, and oxygen[87]. A review of the perspectives on CO_2 use in the pharmaceutical industry was presented in 2004 by Beckman[82].

Ionic liquids have peculiar properties such as low vapor pressure, ability to dissolve organic materials, inorganics and polymers, and high thermal stability[19 and 89]. These characteristics have made the best known ionic liquids an alternative solvent[90-98].

Ionic liquids are salts with low melting temperatures (typically <100 °C) obtained by combining large organic cations with a variety of anions. There are numerous possible combinations which allow for a large potential for ionic liquids. Before these numerous combinations, functionalized ionic liquids have been developed by incorporating additional functional groups in cations and/or anions leading to ionic liquids with specific properties[99-101]. Interest in ionic liquid properties is expanding quickly. However, there are several studies concentrated in the preparation, physicochemical properties, and use as a reaction medium because there is little known about its properties and reactivity. An understanding of how chemical reactivity is influenced by different ionic liquid classes is probably key to obtaining safer technological improvement with significant environmental and economic benefits.

An important review was written by Chiape and Pieraccini[102], in which the most used ionic liquid properties and their reactivity in synthesis were discussed, emphasizing the effect of ionic liquids on mechanistic aspects of some organic reactions.

Some ionic liquids are commercially evaluated and find several industrial applications[103-106].

The first large-scale use of an ionic liquid was introduced by BASF AG, Ludwigshafen/Germany, in 2002, using 1-methylimidazone to remove hydrogen chloride that formed on the alcoxiphenylphosphane (precursor to photoinhibitor)[107-108].

Perfluorine solvents, because of their good solubility of gases (e.g.: up to 57 ml of O_2 can be dissolved in 100 ml of perfluoro (methyl-cycle-hexane), compared to only 3ml of O_2 in 100ml of water in normal temperature and pressure conditions) can be used as good mediums for aerobic oxidation reactions and have been tested as artificial blood[109]. Studies with biphasic fluorinated systems show many applications in organic syntheses[9-12].

Perfluorate solvents show a limited use because of their solubility attributed to their non-polar characteristic (showing better use in biphasic systems). Although chemically inert and nontoxic, they present characteristics of being bio-accumulative.

8.7.6. Method for Solvent Selection in Processes

In this section, a bibliographic review of methodologies on solvent selection in processes is discussed, as well as some tools and methods for helping to respond to selection criteria are presented.

Important considerations should be included in the process development in the chemical industry: chemical and physicochemical characteristics, reactivity, processability recovery, supply resources, and environmental aspects, as graphically illustrated in Figure 8.79.

Several methodologies and approaches to facilitate process solvent selection have been developed[110-128]. Many of them focus performance optimization of the solvent based only on physicochemical properties of which it is hoped they positively impact the process performance where it will be used. Some methodologies conciliate both criteria of properties and cost into the optimization. However, some studies have demonstrated that the use of methodologies of solvent selection through solvent-process integration is efficient enough[114, 115, 118-121, 123, 124 and 127]. In these methodologies, the best candidate selection includes, in addition to chemical and physicochemical properties, process and environmental restrictions.

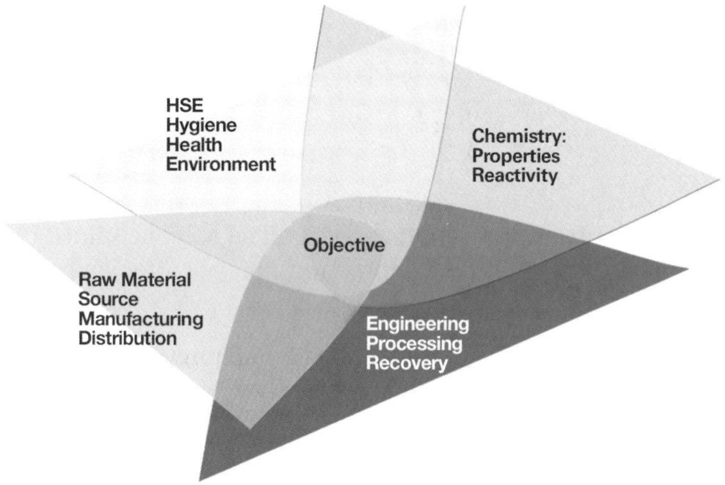

Figure 8.79. Solvent selection objective: a solvent that promotes chemical reactivity and benefits the process, minimizing the use of material and energy, and also minimizes HSE impacts.

We can generically group solvent selection into four stages:

1. Definition of the problem;
2. Definition of selection criteria;
3. Selection of solvent candidates;
4. Validation of results.

Stage 1. Definition of the Problem

The first stage of solvent selection is to define what the problem is and identify the direction needed to solve it. An additional question at this point is whether solvent use is necessary, i.e., if the objective can be reached by another means, for example, like a physical separation or a reaction without a solvent.

An example of this first stage can be that the problem is defined as a product outside specification. The solution could be the extraction of product contaminants.

A second example could be the process involving a reaction with the solvent is not meeting environmental restrictions. A possible solution could be solvent substitution.

Stage 2. Definition of Selection Criteria

At this stage, requirements that need to be met by the solvent are defined. The following criteria can be used:

- Physical and chemical properties
- HSE characteristics
- Operational properties
- Functional restrictions
- Economic assessment

Pure solvent properties and functions as well as solvent-solute interaction properties have an important role in selection and analysis of solvents to be added to the process.

Pure component properties, for example, like boiling point, melting point, vapor pressure, and the process components will allow verification of the solvent's physical state in process conditions.

Pure component properties which depend on temperature and/or pressure, like viscosity, vapor pressure, enthalpy, and vaporization heat will be important in process condition definitions.

HSE properties of pure components are used for a preliminary selection of solvents that meet hygiene, health, and environmental restrictions because they do not directly affect solvent need.

Solvent-solute interaction properties are the solution properties where the amount of solute and solvent in a mixture plays a part along with pressure and/or temperature in operational conditions.

The most common property is solute solubility in relation to mixture composition as well as solvent selectivity in process component solubilization. In reactions, reagent and product solubility is important. In some cases, the solvent should be more selective for products than for reagents, as is the case of a non-isolated reaction and is followed by other additional reactions in the same reaction recipient. In other situations, it is better to make the product completely immiscible or partially immiscible in the solvent system. For example, the solvent dissolves reagents and provides product precipitation.

Other useful properties of solvent-solute interaction in selection are solute saturation point in relation to temperature and pressure, solution viscosity, and diffusion coefficient.

Functional properties are those that define solvent performance, like the amount of solute removed from or dissolved in the solvent, solvent loss during the extraction process, reaction rate and yield, solute behavior in relation to evaporation rate, and others.

Using the first example in describing stage 1, in stage 2 the following requirements are defined: the solvent should dissolve high amounts of contaminant, but little product; it must be immiscible with the product; liquid in operational conditions; non-toxic; it should be available at an acceptable cost; and if possible, it should already be used in the factory/unit.

Once necessary solvent requirements are defined, they have to be quantified. They can be obtained experimentally, using a database or using estimates from models.

Some better-known databases are shown in Table 8.26. Pure compound properties are easy to get, but solute-solvent interaction properties and functional properties are not so simple. It is difficult to find data that was obtained for the same solute-solvent system which is being used and the same process conditions of the same functional properties. The majority of solute-solvent properties and solvent functional properties can be predicted through the models. Table 8.27 shows some tools for solvent property prediction.

Table 8.26. List of the more known data bases

Name	Address	Comment
CambridgeSoft ChemFinder	http://chemfinder.cambridge-satt.com	Database and hyperlink index for thousands of compounds
CRC Handbook of Chemistry and Physics	http://www.hbcpnetbase.com	Database of thousands of organic, inorganic, and polymer compounds and HSE data
DECHEMA Chemistry Data Series	http://www.dechema.de/CDS.html	Collection of solubility data
SOLV-DB	http://solvdb.ncms.org	Collection of solvent property data and their respective HSE properties
Integrated Solvent Substitution Data System (ISSDS)	http://es.epa.gov/issds/	Database system for acessing alternative solvent properties
Solvents Database	http://chemtec.org/cd/ct_23.html	Solvent database containing data for many applications
DIPPR	http://www.aiche.org/TechnicalSocieties/DIPPR/About/Mission.aspx	Thermophsical data
Knovel Solvents – A Properties Database	http://www.knovel.com/knovel2/Toc.jsp?BookID=635	Database including physical and chemical properties of more than 1100 solvents

Table 8.26. List of the more known data bases (*Continuation*)

Name	Address	Comment
Solvents Database	http://www.williamandrew.com/titles/1463.htm	Database including physical and chemical properties of more than 1100 solvents
TAPP - Thermochemical and Physical Properties Database	http://www.chempute.com/tapp.htm	Database of physical and thermo-chemical properties of pure chemical compounds
The NIST Webbook	http://webbook.nist.gov	Thermophsycial and thermochemical data source
SciGlass 6.5	http://www.esm-software.com/sciglass/	Database of physicochemical and thermodynamic properties

Table 8.27. List of tools for solvent property prediction

Name	Main Characteristics	Address or Reference
Group Contribution Methods	Prediction of physical and physico-chemical properties	[128-129]
ASPEN PLUS	Software that allows physicochemical and thermodynamic properties of pure and mixed compounds	http://www.aspentech.com
COSMO-RS	Software that allows physicochemical and thermodynamic properties of pure and mixed compounds	COSMO-RS is implemented in the COSMOtherm available in http://www.cosmologic.de
PREDICT Plus 2000	Software for prediction of physicochemical and thermodynamic properties	http://www.mwsoftware.com/dragon/brftour.html
SciGlass 6.5	Software for prediction of physicochemical and thermodynamic properties	http://www.esm-software.com/sciglass/
Quantitative structure-property relationships (QSPR)	Software for predicting polymer properties	http://www.polymerexpert.biz/products.html

Stage 3. Selection of Solvent Candidates

In the third stage, possible solvent candidates are selected. This selection can be based on three paths.

Benchmarking: Sometimes it is possible to select solvent candidates by analogies to similar processes, or based on chemical intuition and technical experiences.

Database Research: These are used to search for compounds that meet set requirements. This search should primarily start with pure compound properties, then by HSE properties, and finally by solvent functional properties.

Simulation: This can be used to calculate compounds that the criteria find. There are two distinct computational approaches:

- It generates a predefined list of candidates and classifies them according to criteria.
- It uses molecular modeling. The use of molecular modeling methods to generate and analyze chemical structures that meet desired properties.

At this stage of solvent candidate selection, algorithms can be used with the degree of varied complexity depending on the process and its testing. Some algorithms are based only on solvent properties that are to meet process needs, other algorithms with more accurate results are based on the integration of solvent properties with the process.

Stage 4. Result Verification

The final stage for solvent selection is to verify that the candidates really act as expected. A computational validation can be done, for example by the simulation stage of the extraction process. Experimental validation of the solvent candidate is required for all process development stages, from the laboratory and pilot trials to industrial tests.

Note that for a given problem not all stages are necessary. The database might be necessary for all four stages. In selecting solvent candidates, if many candidates are found, selection criteria must be refined and stage 3 repeated to reduce the number of alternatives. On the other hand, if few or no alternatives are found, the defining selection criteria can be relaxed.

Bibliographical References

1. Knochel, P. *Modern Solvents in Organic Synthesis*. Topics in Current Chemistry, Springer Berlin/Heidelberg, v. 206, pages 1-152, 1999.

2. Adams, D. J.; Dyson, P. J.; Tavener, S. J. *Chemistry in Alternative Reaction Media*. November 2003. Chichester: John Wiley. 268 pages.

3. Mikami, K. *Green Reaction Media in Organic Synthesis*. November 2005. Oxford, U.K.: Blackwell Publishing. 200 pages.

4. Jessop, P. G.; Leitner, W. *Chemical Synthesis Using Supercritical Fluids*. Weinheim: Wiley-VCH, 1999. 500 pages.

5. Noyori, R. Supercritical Fluids: Introduction. *Chemical Reviews*, USA: American Chemical Society, v. 99, issue 2, pages 353-634, february 1999.

6. Wells, S. L.; DeSimone, J. M. *CO_2 Technology Platform*: An Important Tool for Environmental Problem Solving. Angewandte Chemie, Wiley: InterScience, v. 113, issue 3, pages 534-544, february 2001.

7. Leitner, W. *Supercritical Carbon Dioxide as a Green Reaction Medium for Catalysis*. Accounts of Chemical Research, ACS Publications, v. 35, issue 9, pages 746-756, july 2002.

8. Eckert, C. A.; Liotta, C. L.; Bush, D.; Brown, J. S.; Hallett, J. P. Sustainable. Reactions in Tunable Solvents. *The Journal of Physical Chemistry B*, ACS Publications, v. 108, issue 47, pages 18108-18118, November 2004.

9. Horva´th, I. T.; Ra'bai, J. Facile *Catalyst Separation without Water*: Fluorous Biphase Hydroformylation of Olefins. Science, AAAS: American Association for the Advancement of Science, v. 266, issue 7, pages 72-75, october 1994.

10. Horva´th, I. T. *Fluorous Biphase Chemistry*. Accounts of Chemical Research, ACS Publications, v. 31, issue 10, pages 641-650, october 1998.

11. Betzemeier, B.; Knochel, P. *Perfluorinated Solvents* – a Novel Reaction Medium. Topics in Current Chemistry, Springer Berlin/Heidelberg, v. 206, pages 60-78, 1999.

12. Gladysz, J. A.; Curran, D. P.; Horva'th, I. T. *Handbook of Fluorous Chemistry*. 2004. Weinheim: Wiley-VCH. Pages 624.

13. Riess, J. G.; Le Banc, M. *Solubility and Transport Phenomena in Perfluorochemicals Relevant to Blood Substitution and Other Biomedical Applications*. Pure Applied Chemistry, v. 54, issue 12, pages 2383-2406, july 1982.

14. Reichardt, C. *Solvents and Solvent Effects in Organic Chemistry*. 3rd updated and enlarged edition, 2003. Weinheim: Wiley-VCH. Pages 653.

15. Buncel, E.; Stairs, R.; Wilson, H. *The Role of the Solvent in Chemical Reactions*. 2003. Oxford, New York : Oxford University Press. Pages 176.

16. Fawcett, W. R. Liquids, *Solutions, and Interfaces* – from Classical Macroscopic Descriptions to Modern Microscopic Details. 2004. Oxford, New York: Oxford University Press. Pages 640.

17. Marcus, Y. *The Properties of Solvents*. 1998. Chichester, New York: John Wiley. Pages 254.

18. Marcus, Y. *Solvent Mixtures* – Properties and Selective Solvation. 2002. New York, Basel: Marcel Dekker. Pages 288.

19. Wypych, G. *Handbook of Solvents* (+ Solvent Data Base on CDROM). 2001. Toronto: ChemTec Publishing, and New York :William Andrew Publishing. Pages 1727.

20. Cheremisinoff, N. P. *Industrial Solvents Handbook*. 2nd edition, 2003. New York: M. Dekker. Pages 346.

21. Nelson W. M. *Green Solvents for Chemistry*: Perspectives and Practice. march 2003. New York, USA: Oxford University Press. Pages 400.

22. Robert H. Perry; Secil H. Chilton. *Manual de Engenharia Química*. 5. Ed., 1980. Rio de Janeiro: Editora Guanabara II.

23. Robert E. Treybal. *Mass Transfer Operation*. 3rd Edition, 1981. Singapure: MacGraw Hill Book Company. Pages 800.

24. Reichardt, C. *Solvents effects in Organic Chemistry*. 3rd. Edition, 2003. Weinheim, New York: Verlag Chemie. Pages 653.

25. J. Linnanto; V. M. Helenius; J. A. I. Oksanen; T. Peltola; J. L. Garaud; J. E. I. Tommola. Exciton Interactions and Femtosecond Relaxation in Chlorophyll α-water and Chlorophyll α-dioxane aggregates. *The Journal of Physical Chemistry A*, ACS Publications, v. 102, issue 23, pages 4337-4349, june 1998.

26. R. A. Marcus. *Journal of Chemical Physics*. 24, 1956.

27. J. S. Bader; R. A. Kuharski; D. Chandler. Role of nuclear tunneling in aqueous ferrous–ferric electron transfer. *The Journal of Chemical Physics*, v. 93, pages 230-236, july 1990.

28. J. S. Bader; D. Chandler. Computer simulation study of the mean forces between ferrous and ferric ions in water. *Journal of Physical Chemistry*, ACS Publications, v. 96, issue 15, pages 6423-6427, july 1992.

29. P. J. Rossky; J. Schnitker; R. A. Kuharski. Quantum simulations of aqueous systems. *Journal of Statistical Physics*, Springer, v. 43, issue 5-6, pages 949-965, june 1986.

30. R. A. Kuharski; J. S. Bader; D. Chandler; M. Sprik; M. L. Klein; R. W. Impey. Molecular model for aqueous ferrous–ferric electron transfer. *The Journal of Chemical Physics*. v. 89, pages 3248-3257, september 1988.

31. G. L. Closs; J. R. Miller. *Intramolecular Long-Distance Electron Transfer in Organic Molecules*. Science, AAAS: American Association for the Advancement of Science, v. 240, pages 440-447, 1988.

32. P. F. Barbara; G. C. Walker; T. P. Smith.*Vibrational Modes and the dynamic solvent effect in electron and proton transfer*. Science, AAAS: American Association for the Advancement of Science, v. 256, pages 975-981, may 1992.

33. P. F. Barbara; T. J. Meyer; M. A. Ratner. Contemporary Issues in Electron Transfer Research. *The Journal of Physical Chemistry*, ACS Publications, v. 100, issue 31, pages 13148-13168, august 1996.

34. S. Speiser. Photophysics and mechanisms of Intramolecular electronic energy transfer In bichromophoric molecular system: solution and supersonic jet studies. *Chemical Reviews*, ACS Publications, v. 96, issue 6, pages 1953-1976, october 1996.

35. V. Sundstrom; T. Pullerits; R. van Grondelle. Photosynthetic light-harvesting: Reconciling dynamics and structure of purple bacterial LH2 reveals function of photosynthetic unit. *The Journal of Physical Chemistry B*, ACS Publications, v. 103, issue 13, pages 2327-2346, march 1999.

36. H. A. Kramers. *Brownian motion in a field of force and the diffusion model of chemical reactions*. Physica A, Elsevier, v. 7, issue 4, pages 284-304, april 1940.

37. G. R. Castro; T. Knubovets. Homogeneous Biocatalysis in Organic Solvents and Water-Organic Mixtures. *Critical Reviews in Biotechnology*, Informa Healthcare, v. 23, issue 3, pages 195-231, 2003.

38. S. Torres; G. R. Castro. *Non-Aqueous Biocatalysis in Homogeneous Solvent Systems*. Non-Aqueous Biocatalysis. Food Technology and Biotechnology, Croatia, v. 42, issue 4, pages 271-277, october-december 2004.

39. N. Krieger; T. Bhatnagar; J. C. Baratti; A. M. Baron; V. M. de Lima; David Mitchell. Non-Aqueous Biocatalysis in Heterogeneous Solvent Systems. *Food Technology and Biotechnology*, Croatia, v. 42, issue 4, pages 279-286, october-december 2004.

40. G. R. Castro; T. Knubovets. Homogeneous Biocatalysis in Organic Solvents and Water-Organic Mixtures. *Critical Reviews in Biotechnology*, Informa Healthcare, v. 23, Issue3, pages 195-231, 2003.

41. H. Ogino; H. Ishikawa. Enzymes whichh are stable In the presence of organic solvents. *Journal of Bioscience and Bioengineering*, Elsevier, v. 91, issue 2, pages 109-116, 2001.

42. M. T. Ru; J. S. Dordick; J. A. Reimer; D. S. Clark. Optimizing the salt-induced ac-

tivation of enzymes in organic solvents: Effects of lyophilization time and water content. *Biotechnology and Bioengineering*, Wiley: Knowledge for generations, v. 63, issue 2, pages 233–241, april 1999.

43. G. Pencreac'h; J. C. Baratti. Comparation of hydrolytic activity in water and heptane for thirty-two commercial lipase preparations. *Enzyme and Microbial Technology*, Elsevier, v. 28, issue 4-5, pages 473-479, march 2001.

44. F. H. Arnold. Directed evolution: Creating biocatalysts for the future. *Chemical Engineering Science*, Elsevier, v. 51, issue 23, pages 5091-5102, december 1996.

45. F. H. Arnold; J. C. Moore. Optimizing industrial enzymes by directed evolution. *Advances in Biochemical Engineering/Biotechnology*, Springer, v. 58, pages 1-14, 1997.

46. F. H. Arnold. *Combination and computational challenges for biocatalyst design*. Nature, NPG: Nature Publishing Group, v. 409, pages 253-257, january 2001.

47. S. H. Krishna. *Developments and trends in enzime catalysis in nonconventional media*. Biotechnology Advances, Elsevier, v. 20, issue 3-4, pages 239-267, November 2002.

48. H. Ogino; H. Ishikawa. Enzymes whichh are stable In the presence of organic solvents. *Journal of Bioscience and Bioengineering*, Elsevier, v. 91, issue 2, pages 109-116, 2001.

49. A. B. Salleh; M. Basri; M. Taib; H. Jasmani; R. N. Rahman; M. B. Rahman; C. N. Razak. *Modified enzymes for reactions in organic solvents*. Applied Biochemistry and Biotechnology, Humana Press, v. 102, Issues 1-6, pages 349-357, july 2002.

50. N. Kamiya; M. Inoue; M. Goto; N. Nakamura; Y. Naruta. *Catalytic and Structural Properties of surfactant-horseradish peroxidase complex in organic media. Biotechnology Progress*, Wiley InterScience, v. 16, issue 1, pages 52-58, 2000.

51. L. M. Pera; M. D. Baigori; G. R. Castro. Enzyme behaviour in non-conventional system. *Indian Journal of Biotechnology*, CSIR: National Institute of Science Communication and Information Resources, v. 2, issue 3, pages 356-361, july 2003.

52. M. Persson; E. Wehtje; P. Adlercreutz. Factors governing the activity of lyophilised and immobilised lipase preparations in organic solvents. *A European Journal of Chemical and Biology*: ChemBioChem, Wiley InterScience, v. 3, issue 6, pages 566-571, june 2002.

53. A. X. Yan; X. W. Li; Y. H. Ye. Recent progress on immobilization of enzymes on molecular sieves for reactions in organic solvents. *Applied Biochemistry Biotechnology*, Humana Press Inc., v. 101, issue 2, pages 113-129, may 2002.

54. V. P. Torchilin; K. Martinek. Enzyme stabilization without carriers. *Enzyme and Microbial Technology*, Elsevier, v. 1, issue 2, pages 74–82, april 1979.

55. W. Blokzijl; J. B. F. N. Engberts; Hydrophobic Effects – Opinions and Facts. *Angewandte Chemie International Edition*, Wiley InterScience, v. 32, issue 11, pages 1545-1579, November 1993.

56. A. Zaks; A. M. Klibanov. Enzymatic catalysis in nonaqueous solvents. *Journal of Biological Chemistry*, The American Society for Biochemistry and Molecular Biology: ASBMB, v. 263, pages 3194-3201, march 1988.

57. H. Sztajer; H. Lünsdorf; H. Erdmann; U. Menge; R. Schmid. *Biochimica et Biophysica Acta*, Elsevier, v. 1124, pages 253-261, 1992.

58. S. Hazarika; P. Goswami; N. N. Dutta; A. K. Hazarika. Ethyl oelate synthesis by Porcine pancreatic lipase in organic solvents. *Chemical Engineering Journal*, Elsevier, v. 85, issue 1, pages 61-68, january 2002.

59. U. K. Jinwal; U. Roy; A. R. Chowdhury; A. P. Bhaduri; P.K. Roy. Purification and characterization of an alkaline lipase from a newly isolated Pseudomonas mendocina PK-12CS and chemoselective hydrolysis of fatty acid ester. *Bioorganic & Medicinal Chemistry*, Elsevier, v. 11, issue 6, pages 1041-1046, march 2003.

60. V. M. G. Lima; N. Krieger; D. A. Mitchell; J. C. Baratti; I. Filippis; J. D. Fontana. Evaluation of the potential for use in biocatalysis of a lipase from a wild strain of Bacillus megaterium. *Journal of Molecular Catalysis B: Enzymatic*, Elsevier, v. 31, Issues 1-3, pages 53-61, october 2004.

61. P. Adlercreutz. Biocatalysis in Non-Conventional Media. In: Applied Biocatalysis, A. J. J. Straathof; P. Adlercreutz (Eds.), Harwood Academic Publishers, Amsterdam. 2000. p. 295-316.

62. R. S. Oakes; A. A. Clifford; K. D. Bartle; M. T. Pett; C. M. Rayner. *Sulfur oxidation in supercritical carbon dioxide*: dramatic pressure dependant enhancement of diastereoselectivity for sulfoxidation of cysteine derivates. Chemical Communication, RSC Publishing, issue 3, pages 247-248, 1999.

63. F. A. Luzzio. *The Henry reaction*: recent examples. Tetrahedron, Elsevier, v. 57, issue 6, pages 915-945, february 2001.

64. A. J. Parratt; D. J. Adams; A. A. Clifford; C. M. Rayner. *Manipulation of the stereochemical outcome and product distribution in the Henry reaction using* CO_2 *pressure*. Chemical Communication, RSC Publishing, issue 23, pages 2720-2721, 2004.

65. D. A. Evans; K. A. Woerpel; M. M. Hinman; M. M. Faul. Bis(oxazolines) as chiral ligands in metal-catalyzed asymmetric reactions. Catalytic, asymmetric cyclopropanation of olefins. *Journal of The American Chemical Society*, ACS Publications, v. 113, issue 2, pages 726-728, january 1991.

66. A.A. Clifford; K. Pople; W. J. Gaskill; K. D. Bartle; C. M. Rayner. Potential turning and reaction control in the Diels-Alder reaction between cyclopentadiene and methyl acrylate in supercritical carbon dioxide. *Journal of the Chemical Society*, Faraday Transactions, RSC Publishing, v. 94, issue 10, pages 1451-1456, 1998.

67. R. S. Oakes; T. J. Heppenstall; N. Shezad; A. A. Clifford; C. M. Rayner. *Use of scadium tris(trifluoromethanesulfonate) as a Lewis acid catalyst in supercritical carbon dioxide*: efficient Diels-Alder reactions and pressure dependent enhancement of endo:exo stereoselectivity. Chemical Communication, RSC Publishing, issue 16, pages 1459-1460, 1999.

68. D. Basavaiah; A. J. Rao; T. Satyanarayana. Recent advances In the Baylis-Hillman reaction and applications. *Chemical Reviews*, ACS Publications, v. 103, issue 3, pages 811-892, march 2003.

69. I. E. Marko; P. R. Giles; Hindley, N. J. *Catalytic enantioselective Baylis-Hillman reactions*. Correlation between pressure and enantiomeric excess. Tetrahedron, Elsevier, v. 53, issue 3, pages 1015-1024, january 1997.

70. T. Oishi; H. Oguri; M. Hirama. *Asymmetric Baylis-Hillman reactions using chiral 2,3-disubstituted 1,4-diazabicyclo[2.2.2]octanes catalysts umder high pressure comditions*. Tetrahedron: Asymmetry, Elsevier, v. 6, issue 6, pages 1241-1244, june 1995.

71. P. Raveendran; Y. Ikushima; S. L. Wallen. Polar attributes of supercritical carbon dioxide. *Accounts of Chemical Research*, ACS Publications, v. 38, issue 6, pages 478-485, june 2005.

72. P. W. Bell; A. J. Thote; Y. Park; R. B. Gupta; C. B. Roberts. Strong Lewis acid-Lewis base interactions between supercritical carbon dioxide and carboxylic acids: Effects on self-association. *Industrial & Engineering Chemistry Research*, ACS Publications, v. 42, issue 25, pages 6280-6289, december 2003.

73. P. Raveendran; S. L. Wallen. Cooperative C-H···O hydrogen bonding in CO_2-Lewis base complexes: Implications for solvation in supercritical CO_2. *Journal of the American Chemical Society*, ACS Publications, v. 124, issue 42, pages 12590-12599, october 2002.

74. E. J. Beckman; T. Sarbu; T. Styranec. Polymeric Materials: Science and Engineering, *Chinese Electronic Periodical Services*, v. 84, pages 269, 2001.

75. M. R. Nelson; F. B. Borkman. Ab initio calculations on CO_2 binding to barbonyl groups. *The Journal of Physical Chemistry A*, ACS Publications, v. 102, issue 40, pages 7860-7863, october 1998.

76. J. C. Meredith; K. P. Johnston; J. M. Seminario; S. G. Kazarian; C. A. Eckert. Quantitative equilibrium constants between CO_2 and Lewis bases from FTIR spectroscopy. *The Journal of Physical Chemistry*, v. 100, issue 26, pages 10837-10848, june 1996.

77. Y. Ikushima; N. Saito; K. Hatakeda; O. Sato. Promotion of a lipase-catalyzed esterfication in supercritical carbon dioxide in the near-critical region. *Chemical Engineering Science*, Elsevier, v. 51, issue 11, pages 2817-2822, june 1996.

78. M. Sagisaka; S. Yoda; Y. Takebayashi; K. Otake; Y. Kondo; N. Yoshino; H. Sakai; M. Abe. *Effects of CO_2-philic Tail structure on phase behavior of fluorinated aerosol-OT analogue surfactant/water/supercritical CO_2 systems*. Langmuir, ACS Publications, v. 19, issue 20, pages 8161-8167, september 2003.

79. D. Clarke; M. A. Ali; A. A. Clifford; A. Parratt; P. Rose; D. Schwinn; W. Bannwarth; C. M. Rayner. *Current Topics in Medicinal Chemistry*, Bentham Science Publishers, v. 4, pages 729-771, 2004.

80. D. Clarke; M. A. Ali.; A. A. Clifford; A. Parratt; P. Rose; D. Schwinn; W. Bannwarth; C. M. Rayner. *Chemical Society*, pages 2964, 1962.

81. S. Saffarzadeh-Matin; C. J. Chuck; F. M. Kerton; C. M. Rayner. *Poly (dimethylsiloxane)-derived phosphine and phosphinite ligands synthesis, characterization, solubility in supercritical carbon dioxide, and sequestration on silica. Organometallics*, ACS Publications, v. 23, issue 22, pages 5176-5181, october 2004.

82. E. J. Beckman. Supercritical and near-critical CO2 in green chemical synthesis and processing. *The Journal of Supercritcal Fluids*, Elsevier, v. 28, Issues 2-3, pages 121-191, march 2004.

83. R. Amandi; J. R. Hyde; S. K. Ross; T. J. Lotz; M. Poliakoff. Continuos reactions in supercritical fluids; a cleaner, more selective synthesis of thymol in supercritical CO2. *Green Chemistry*, RSC Publishing, v. 7, issue 5, pages 288-293, 2005.

84. P. Cramers; C. Selinger. Advanced hydrogenation technology for fine chemical and pharmaceutical applications. *PharmaChem*, v. 1, pages 7-9, june 2002.

85. D. Clarke; M. A. Ali; A. A. Clifford; A. Parratt; P. M. Rose; D. Schwinn; W. Bannwarth; C. M. Rayner. *Current Topics in Medicinal Chemistry*, v. 4, pages 729-71, 2004.

86. P. G. Jessop; T. Ikariya; R. Noyori. Homogeneous catalysis in supercritical flu-

ids. *Chemical Reviews*, ACS Publications, v. 99, issue 2, pages 475-494, february 1999.

87. C. Bianchini; G. Giambastiani. Chemtracts, v. 16, pages 301-309, 2003.

88. J. L. Kuiper; P. A. Shapley; C. M. Rayner. Synthesis, structure, and reactivity of the ruthenium (VI)-Nickel(II) complex (dppe)Ni(μ3-S)2{Ru(N)M2}2. Organometallics, ACS Publications, v. 23, issue 16, pages 3814-3818, august 2004.

89. P. Wasserscheid; T. Welton. *Ionic Liquids in Synthesis*. 2nd Edition, 2003. Weinheim: Wiley-VCH, pages 364.

90. R. D. Rogers. Ionic Liquids: Industrial Applications to Green Chemistry. K. R. Seddon editor, august 2002, pages 488.

91. J. D. Holbrey; K. R. Seddon KR. *Ionic Liquids*. Clean Products and Processes, Springer Berlin/Heidelberg, v. 1, issue 4, pages 223-236, december 1999.

92. M. J. Earle; K. R. Seddon. Ionic liquids. *Green solvents for the future*. Pure and Applied Chemistry, IUPAC, v. 72, issue 7, pages 1391-1398, 2000.

93. T. Welton. Room-temperature ionic liquids. Solvents for synthesis and catalysis. *Chemical Reviews*, ACS Publications, v. 99, issue 8, pages 2071-2083, august 1999.

94. P. Wasserscheid; M. Keim. Ionic liquid – New "soluctions" for transition metal catalysis. *Angewandte Chemie Interanational Edition*, Wiley InterScience, v. 39, issue 21, pages 3772-3789, November 2000.

95. R. Sheldon. Catalytic reactions in ionic liquids. *Chemical Communication*, RSC Publishing, issue 23, pages 2399-2407, 2001.

96. H. Olivier-Bourbigou; L. Magna. Ionic liquids: perspectives for organic and catalytic reactions. *Journal of Molecular Catalysis A*: Chemical, Elsevier, v. 182-183, pages 419-437, may 2002.

97. J. Dupont; R. F. de Souza; P. A. Z. Suarez. Ionic liquid (molten aslt) phase organometallic catalysis. *Chemical Reviews*, ACS Publications, v. 102, issue 10, pages 3667-3692, october 2002.

98. J. S. Wilkes. Properties of ionic liquid solvents for catalysis. *Journal of Molecular Catalysis A*: Chemical, Elsevier, v. 214, issue 1, pages 11-17, may 2004.

99. J. H. Davis, Jr. Task-specific ionic liquids. *Chemistry Letters*, CSJ Publications, v. 33, issue 9, pages 1072-1077, september 2004.

100. S. Lee. *Functionalized imidazolium salts for task-specific ionic liquids and their applications*. Chemical Communications, RSC Publications, issue 10, pages 1049-1063, 2006.

101. Z. Fei; T. J. Geldbach; D. Zhao; P. J. Dyson. From Dysfunction to Bis-function:

On the Design and Applications of Functionalised Ionic Liquids. *Chemistry A European Journal*, Wiley InterScience, v. 12, issue 8, pages 2122-2130, march 2006.

102. C. Chiappe; D. Pieraccini. Ionic liquids: solvent properties and organic reactivity. *Journal of Physical Organic Chemistry*, Wiley InterScience, v. 18, issue 4, pages 275-297, april 2005.

103. R. D. Rogers; K. R. Seddon; S. Volkov. *Green Industrial Applications of Ionic Liquids* (NATO Science Series II: Mathematics, Physics and Chemistry). First edition, january 2003. Dordrecht: Springer, pages 580.

104. R. D. Rogers; K. R. Seddon. *Ionic Liquids*: Industrial Applications for Green Chemistry (ACS Symposium Series). Revised edition, august 2002. Washington, DC: American Chemical Society Publication. pages 488.

105. R. D. Rogers; K. R.Seddon. *Ionic Liquids as Green Solvents*: Progress and Prospects (ACS Symposium Series). First edition, september 2003. Oxford, U.K.: An American Chemical Society Publication, pages 616.

106. P. L. Short. IUPAC project aims to plug ionic liquids data gap. *Chemical Engineering News*, American Chemical Society, v. 84, Issue17, pages 15-21, april 2006.

107. K. R. Seddon. Ionic liquids – A taste of the future. *Nature Materials*, NPG: Nature Publishing Group, v. 2, pages 363-365, june 2003.

108. M. Freemantle. BASF's smart ionic liquid. *Chemical Engineering News*, American Chemical Society, v. 81, issue 13, march 2003.

109. J. G. Riess; M. Le Banc. *Solubility and Transport Phenomena in Perfluorochemicals Relevant to Blood Substitution and Other Biomedical Applications*. Pure and Applied Chemistry, IUPAC, v. 54, issue 12, pages 2383-2406, july 1982

110. E. A. Brignole; S. Botini; R. Gani. *A strategy for the design and selection of solvents for separation processes*. Fluid Phase Equilibria, Elsevier, v. 29, pages 125-132, october 1986.

111. M. Cockrem; J. Flatt; E. Lightfoot. *Solvent selection for extraction from diluite solution*. Separation Science and Technology, Ingenta Connect Plubication, v. 24, pages 769-807, 1989.

112. S. Macchieto; O. Odele; O. Omatsone. Design of optimal solvents for liquid-liquid extraction and gas absorption processes. *Chemical Engineering Research and Design*, Elsevier, v. 68a, pages 429-433, 1990.

113. A. Buxton; A. G. Livingston; E. N. Pistikopoulos. Reaction path synthesis for environmental impact minimization. *Computers and Chemical Engineering*, Elsevier, v. 21, Supplement 1, pages S959-S964, may 1997.

114. E. N. Pistikopoulos; S. K. Stefanis. Optimal solvent design for environmental impact minimization. *Computers & Chemical Engineering*, Elsevier, v. 22, Isuue 6, pages 717-733, june 1998.

115. A. Buxton; A. G. Livingston; E. Pistikopoulos. Optimal design of solvent mixtures for environmental impact minimization. *AIChE Journal*, Wiley Inter-Science, v. 45, issue 4, pages 817-843, april 1999.

116. A. D. Curzons; D. J. C. Constable. Solvent selection guide: A guide to the integration of environmental, health and safety criteria into the selection of solvents. *Clean Products and Processes*, Springer Berlin/Heidelberg, v. 1, pages 82-90, 1999.

117. E. C. Marcoulaki; A. C. Kokossis. On the development of novel chemicals using a systematic optimization approach. Part II. Solvent design. *Chemical Engineering Science*, Elsevier, v. 55, issue 13, pages 2547-2561, july 2000.

118. K. J. Kim; U. M. Diwekar. Integrated solvent selection and recycling for continous processes. *Industrial and Engineering Chemistry Research*, ACS Publications, v. 41, issue 18, pages 4479-4488, september 2002.

119. Y. Wang; L. E. K. Achenie. Computer aided solvent design for extrative fermentation. *Fluid Phase Equilibria*, Elsevier, v. 201, issue 1, Page 1-18, august 2002.

120. Y. Wang; L. E. K. Achenie. A hybrid global optimization approach for solvent design. *Computers & Chemical Engineering*, Elsevier, v. 26, issue 10, pages 1415-1425, october 2002.

121. A. Giovanoglou; J. Barlatier; C. S. Adjiman; E. N. Pistikopoulos; J. L. Cordiner. *Optimal solvent design for batch separation based on economic performance*. A.I.Ch.E. Journal, Wiley InterScience, v. 49, issue 12, pages 3095-3109, december 2003.

122. M. Folic; C. S. Adjiman, E. N. Pistikopoulos. *The design of solvents for optimal reaction rates*. In14th European Symposium on Computer Aided Process Engineering (ESCAPE-14) proceeding, pages 175-180, 2004.

123. M. R. Eden; S. B. Jorgensen; R. Gani; M. M. El-Halwagi. A novel framework for simulation separation process and product design. *Chemical Engineering and Processing*, Elsevier, v. 43, issue 5, pages 595-608, june 2004.

124. W. Y. Xu; U. M. Diwekar. Environmentally friendly heterogeneous azeotropic distillation system design: integration of EBS selection and IPS recycling. *Industrial and Engineering Chemistry Research*, ACS Publications, v. 44, issue 11, pages 4061-4067, may 2005.

125. C. Jiménez-González; A. D. Curzons; D. J. C. Constable; V. L. Cunningham. Expanding GSK's solvent selection guide – application of life cycle assessment to

enhance solvent selections. *Journal of Clean Technologies and Environmental Policy*, Springer Berlin/Heidelberg, v. 7, issue 1, pages 42-50, december 2004.

126. R. Gani; C. Jiménez-González; D. J. C. Constable. Method for selection of solvents for promotion of organic reactions. *Computers and Chemical Engineering*, Elsevier, v. 29, issue 7, pages 1661-1676, june 2005.

127. A. I. Papadopoulos; P. Linke. Efficient integration of optimal solvent and process design using molecular clustering. *Chemical Engineering Science*, Elsevier, v. 61, issue 19, pages 6316-6336, october 2006.

128. R. Gani; C. Jiménez-González; A. ten Kate; P. A. Crafts; J. H. Atherton; J. L. Cordiner. A Modern Approach to Solvent Selection. *Chemical Engineering*: Technical & Pratical, pages 30-42, march 2006.

129. J. Murrero; R. Gani. Group-contribution based estimation of pure component properties. *Fluid Phase Equilibria*, Elsevier, v. 183-208, july 2001.

130. J. W. Kang; J. Abildskov; R Gani; J. Cobas. Estimulation of Mixture Properties from First- and Second-Order Group Contributions with the UNIFAC Model. *Industrial & Engineering Chemistry Research*, ACS Publications, v. 41, issue 13, pages 3260-3273, june 2002.

131. http://www.icell.com.br/infotech/infotecnicas2.htm

132. http//www.abtg.or.br

133. http//www.pneac.org/printprocesses/gravure

134. http//desktoppub.about.com

135. http//pt.wikipedia.org/wiki/Flexografia

136. http//pt.wikipedia.org/wiki/rotogravura

137. http//www.pneac.org/printprocesses/screen

138. http//en.wikipedia.org/wiki/Offset printing

139. http//www.planetaplastico.com.br

140. Steward, Paul A.; Literature Review of Polymer Latex Film Formation and Particle Coalescence; 1995, www. initium.demon.co.uk/filmform.htm

141. P. Swaraj (ed.), *Surface Coatings*, Science and Technology; Second Edition; John Wiley & Sons, 1996.

142. T. Paul. *Waterborne and Solvent-based Surface Coatings, Resins, and Their Applications*; v. 3 Polyurethanes 1998; John Wiley and Sons.

9 Methods for Analysing Solvents

In this chapter, the most used analytical methods for specifying solvents in industries will be presented. Practical aspects and how some analytical strategies can be used in investigation and solving problems related to coloration, odor, contamination, and methodologies for identification and quantification of solvents retained in films, packaging, raw material, etc., will also be presented.

Cristina Maria Schuch

Solvents can be analyzed using a series of parameters selected based on chemical class, impurities, and final application requirements.

Determining the degree of purity stands out from among the analytical methods normally used in solvent specification, and for this, gas chromatography is used. This will be discussed in detail later in this chapter. In addition, physical and chemical tests are also a part of final product specification. Among them, the most used are density, color, acidity or alkalinity, distillation range, non-volatile material, water content, miscibility in water, permanganate resistance, and residual odor.

Analytical requirements are chosen in relation to the type of analyzed solvent. For example, one of the important specification parameters for acetone is chemical testing for potassium permanganate resistance. This parameter is not analyzed when the solvent is ethyl acetate. However, in this case, residual odor is an important parameter required by the market.

Overall, the test that determines purity of solvents is gas chromatography analysis. The degree of purity here is calculated based on impurity and water content, using the following formula:

$$\% \text{ purity} = 100 - (\Sigma A + B)$$

where:

A = impurities determined by gas chromatography

B = water content determined by the Karl Fischer technique.

In many cases, solvent specification shows compound that can impact final solvent application separately. For example, in the butyl acetate specification one of the reported parameters is the butyl alcohol content.

A summary of the main parameters used in some commercial solvents and respective norms is presented in Table 9.1.

9.1. Gas Chromatography

Chromatography is a physicochemical separation method based on the differential migration of the mixture components using two immiscible phases: a mobile phase and a stationary phase. Possible differences between mobile and stationary phases make chromatography a very versatile technique allowing the development of different analytical techniques, like Gas Chromatography (or HRGC – High Resolution Gas Chromatography) and Liquid Chromatography (or HPLC – High Performance Liquid Chromatography), which are widely applied in the academic and industrial medium.[1,2]

When speaking of solvent analysis, gas chromatography is the most applied technique. Especially useful for analyzing mixture components, it allows separation, characterization, and quantification of compounds and organic impurities present in available solvents on the market and accompanies the industrial synthesis process, leading to optimization of production parameters carried out in real time.

Tabla 9.1. Comercial solvents and its standards for analysis

Analysis	Reference	Methodology
Purity	—	Chromatography
Density	ASTM D-4052	Physical test
Color (Pt-Co)	ASTM D-1209	Physical test
Acidity (as acetic acid)	ASTM D-1613	Chemical test
Distillation range (760 mm Hg)	ASTM D-1078	Physical test
Non volatile material	ASTM D-1353	Physical test
Water content	ASTM D-1364	Chemical test
Residual odor	ASTM D-1296	Physical test
Water miscibility	ASTM D-1722	Physical test
Resistance to $KMnO_4$	NBR-5824	Chemical test

HRGC uses 10-100 m-length columns with internal diameters of 0.10-0.75 mm and a stationary phase with a variable polarity and film thickness. Hydrogen and helium are the preferred carrier gases.

The most used detectors are the Flame Ionization Detector (FID) and the Thermal Conductivity Detector (TCD). Another commonly used type of detector is the Mass Spectrometry Detector (MSD). In addition to these, there is the Electron Capture Detector (ECD); Flame Photometric Detector (FPD); specific detectors for nitrogenated species and phosphorates like Nitrogen Phosphorus Detectors (NPD), Photo Ionization Detector (PID); and coupled detectors (FID/PID, TCD/FID, and MS/FID), where detection principles are used simultaneously.

Gas Chromatography equipment coupled with Mass Spectrometry (GC/MS) joins the chromatography column exit to a system that can generate molecular fragments that, once they are selected, detected, and recorded, they determine the component's mass spectrum. Generally, the component leaving the column in gas form enters the ionization chamber (operating at reduced pressures of 10^{-5} to 10^{-6} torr), where it is bombarded by a stream of electrons from a heated filament. Through a sequence of stages of positive ion acceleration generated by the electron impact, the collector system captures a group of ions, using collimator openings, that are then detected and amplified in an electron multiplier.[3]

The result is recorded in a mass spectrum characterizing a compound correspon-ding to a determined peak. With the use of traditional gas chromatography (FID, TCD or other detector), component identification is made by comparing retention time for a determined compound against its pure standard. For very complex mixtures, it is almost impossible to check all compounds against known standards.[4]

Therefore, the advantage of the GC/MS technique is in the qualitative analysis of mixture or complex mixtures of components, using the mass fragmentation spectrum obtained for each compound.

Ethyl acetate mass spectrum and the comparison with the available spec-trum in the library of the equipment are shown in Figure 9.1. Here, the typical spectrum of each peak in the chromatogram can be compared to the library ac-companying the equipment or by the interpretation of the specialist responsible for the analysis.

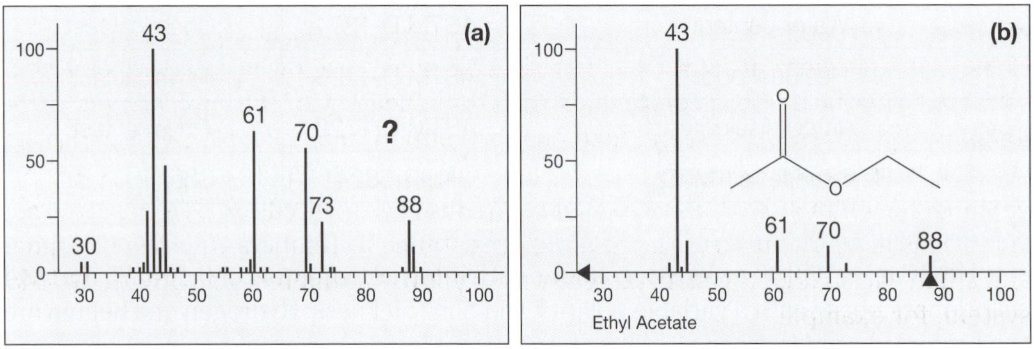

Figure 9.1 (a) Component mass fragmentation spectrum with retention time at 2.10 min; (b) comparison with the ethyl acetate spectrum available in the equipment library.

9.1.1. Choosing Chromatography Columns

Choosing the type of column for analysis will depend on a number of factors, such as polarity and boiling point of the solvent system to be analyzed. In general, capillary column polarity follows the polarity tendency of the major component in the mixture to be chromatographed. In other words, when the system is highly polar, a high-polarity column, available on the market, will probably provide a good separation of all components. What is important here is to guarantee that all mixture components are capable of eluting in the chromatographic conditions chosen. The same happens when the column has a low polarity. However, when chromatographic separation is critical, in addition to modifying column polarity, it is possible to change film thi-ckness and length to increase component retention and consequently improve the separation and/or selectivity of the total mixture separation.

The Comprehensive Two-Dimensional Gas Chromatography (GC × GC), used mostly for hydrocarbons, was developed to improve separation efficiency. It uses two columns connected sequentially, generally a conventional one and a short one, so that all effluent or a large part of it from the first is directed to the second one using a modulator. With this technique, sensitivity is significantly increased and resolution increases expressively, as compared to traditional gas chromatography. This technique is mainly used for the separation of oil derivatives where, depending on the fraction being analyzed, the number of components can reach one million. The classes typically analyzed using this technique are n-alkanes or paraffins, branched alkanes or isoparaffins, alkenes or olefins, branched alkenes or isoolefins, cyclic alkanes or naphtalenes, cyclic alkenes, aromatics, monocyclic aromatics, bi-cyclic aromatics, tri-cyclic aromatics, among others.[5]

For solvent analysis, PLOT (Porous Layer Open Tubular) columns can be used especially to improve the separation performance of high volatile and gas compounds. These columns were developed to solve analytical problems related to low retention compounds since the other columns used very fine films giving poor resolution and separation efficiency with high volatile and gas compound analyses. These columns were developed in 1988 by Chrompack, allowing users to analyze a variety of volatile and gas compounds with the efficiency of the capillary gas chromatography, offering speed and high peak resolution, thus making it an interesting alternative in solvent analysis. The technology used in manufacturing these columns includes film growth in-situ, chemically fused to melted silica of the external layer. This gives the internal layer a high resistance to temperature, low bleeding, and a film with a high mechanical resistance, allowing an interface with the GC/MS system, for example.

Recently, with the increasing need for fast high-performance analyses, fast gas chromatography (or fast-GC) was introduced with columns that have smaller internal dimensions (100 μm or 0.10 mm).

The main differences between traditional GC and fast-GC are:

- the sample volume injected in the fast-GC is less;
- fast-GC works with smaller column diameter, length, and film thickness than traditional GC;
- fast-GC uses greater carrier gas pressure than traditional GC and the preferred order is $H_2 > He >>> N_2$;
- oven rate temperatures are higher than the traditional method.

There are translation programs that allow an easy adaptation of the traditional method to the fast-GC method and help the technician to determine best analysis conditions quickly.

An example of changes in injection conditions using the translation is shown in Table 9.2.

Table 9.2. Results from a translation – traditional gas chromatography to Fast-GC

Column	Traditional	Fast-GC
Length (m)	60,0	10,0
Internal Diameter (μm)	250	100
Film Thickness (μm)	1,00	0,40
Carrier Gas	H_2	H_2
Column Head Pressure (psi)	11,4	26,6
Gas Flow (ml/min)	1,0	0,5
Average Gas Velocity (cm/s)	36	115
Initial Oven Temperature (°C)	45	45
Initial time (min)	0	0
Oven Rate (°C/min) – Ramp 1	4	52,5
Oven Rate (°C/min) – Ramp 2	12	157,5
Final Oven Temperature (°C)	180	180
Final Time (min)	0	0
Total Run Time (min)	27,8	2,1
Injected Sample Volume (μl)	1,0	0,4

In Table 9.2., injection time was reduced by nearly 13 times, making the methodology very interesting if applied in accompanying industrial processes, for example.

In transforming traditional method conditions to fast-GC, especially with low volatile components such as solvents, some care must be taken in relation to flow and split values. However, the main advantage for using the fast-GC technique remains with the fact that analysis time is drastically reduced.

Figure 9.2. presents a typical chromatogram obtained in fast-GC conditions. Retention times for components in the mixture are recorded for each case.[6]

Figure 9.3 shows a second example using the fast-GC methodology for analyzing a solvent mixture with the major compound being MIBK (methyl isobutyl ketone) and the total chromatographic run time is 3.0 min.[6]

As in all methodology implementation, the modified method should be carefully verified case-by-case, using statistic parameters.

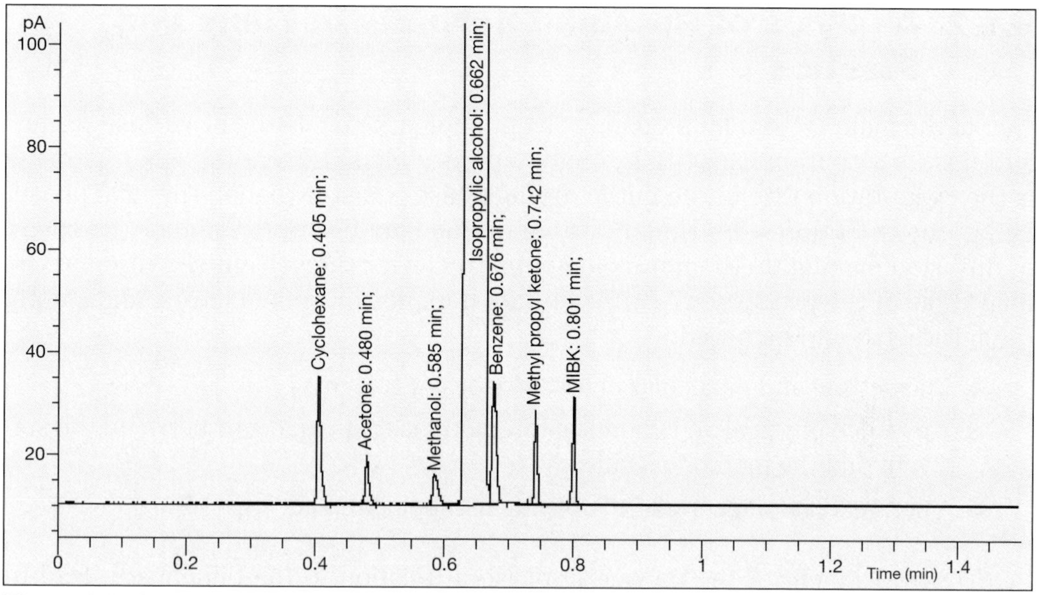

Figure 9.2. Analytical Conditions: Column DB-WAX; 10 m; 100 μm; 0.2 μm; Temperature Limit: 20-250 °C (Serial #: US1434617A, Agilent); Phase: Polyethylene Glycol. Split: 700:1; Flow: 0.7 ml · min^{-1}; Initial Oven Temperature: 45 °C; Heating Ramp: 52.5 °C · min^{-1}; 100 °C (0.3 min); 40 °C · min^{-1}; 140 °C (0.3 min); 30 °C · min^{-1}; 180 °C (1 min); Equilibration Time: 0.5 min; Injector Temperature: 200 °C; Detector Temperature (FID): 220°C; Carrier Gas: H_2.

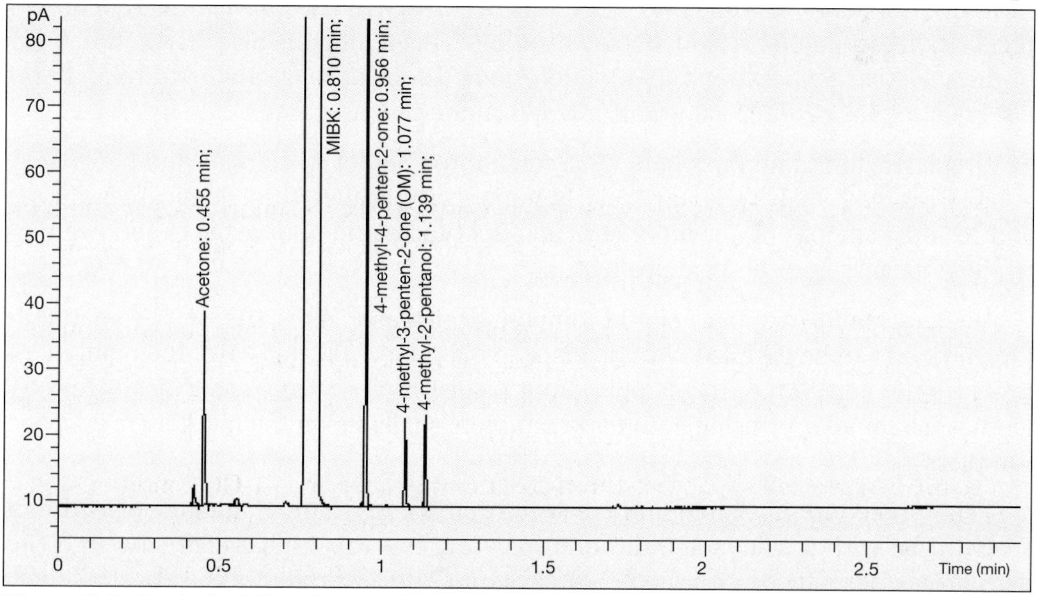

Figure 9.3. Analytical Conditions: Column DB-WAX; 10 m; 100 μm; 0,2 μm; Temperature Limit: 20-250 °C (Serial #: US1434617A, Agilent); Phase: Polyethylene Glycol. *Split*: 700:1; Flow: 0,7 ml · min^{-1}; Oven Initial Temperature: 45 °C; Temperature Programation: 52,5 °C · min^{-1}; 100 °C (0,3 min); 40 °C · min^{-1}; 140 °C (0,3 min); 30 °C · min^{-1}; 180 °C (1 min); Equilibration Time: 0,5 min; Injector Temperature: 200 °C; Detector Temperature (FID): 220 °C. Carrier Gas: H_2.

9.1.2. Analysis of Retained Solvents in Films and Plastics

One of the main factors influencing solvent choice for a coating or varnish formulation is its retention capacity in humid and dry films. In general, what happens is the evaporation rate in the initial drying phase is faster than in the final phase, i.e., in the retention step. Under certain drying conditions, the higher the speed of diffusion making up the film's elements, the lower the solvent retention in the final stage of the process is.7 Diffusion rate depends on some factors, and among them these should be noted:

- structural and molecular characteristics of solvents
- possible physical and chemical interactions that can occur between the solvent and the macromolecule
- permeability capacity of the plastic film being tested.

The enrichment of the polymeric phase in relation to the diluent can lead to precipitation or coagulation of the polymer, and this needs to be avoided. If it does occur, it becomes necessary to study drying phenomena using precise analytical methods. At the start of the process when the solvent has high evaporation rates, it is possible to measure the solvent content by using thermobalance. However, when the solvent amount becomes very low, it is necessary to use chromatography in gas phase to determine its residual concentration. This technique also allows the study of the selective evaporation rate of a determined solvent in a total mixture and precisely quantifies residual values after a determined testing time.

In plastic films and packaging that come into contact with foods and beverages, volatile residual analysis is particularly important because the chemical class type and residual amount of a component can affect taste and odor aspects and even introduce toxicity into the final product.

In determining residual volatile compounds in plastic films, the gas chromatography method can be coupled with headspace injection.[8] In this case, the equipment provides the injection of the sample vapor phase using a syringe specifically for collecting gas samples or using automatic samplers available on the market (headspace samplers).

The procedure is simple and consists of heating a sample amount placed in a sealed tube with a silicon septum and injecting the vapor phase directly into the gas chromatography or into the GC/MS system. The latter allows the spectroscopic characterization of the mixture released in the vapor phase, making it an important analytical strategy when a difference in odor or sample contamination is suspected. Quantifying is a result of adding the standard of the component to be quantified directly in the sample, resulting in a calibration curve that allows quantifying solvent concentration retained in the packaging or plastic film.

Figure 9.4. shows a chromatogram obtained from a GC/MS system coupled with headspace injection that allowed volatile compounds retained in polypropylene samples to be characterized and quantified.

The chromatogram in Figure 9.4.a. shows the initial peak at a retention time (RT) of 1.46 min corresponding to injected air that, in this case, is detected as carbon dioxide. An example of how it is characterized is shown in Figure 9.4.b. where the retention time peak at 5.64 min (compound A) was characterized as 2,4-di-methyl heptane or C9 isomer derived from polypropylene chain degradation.

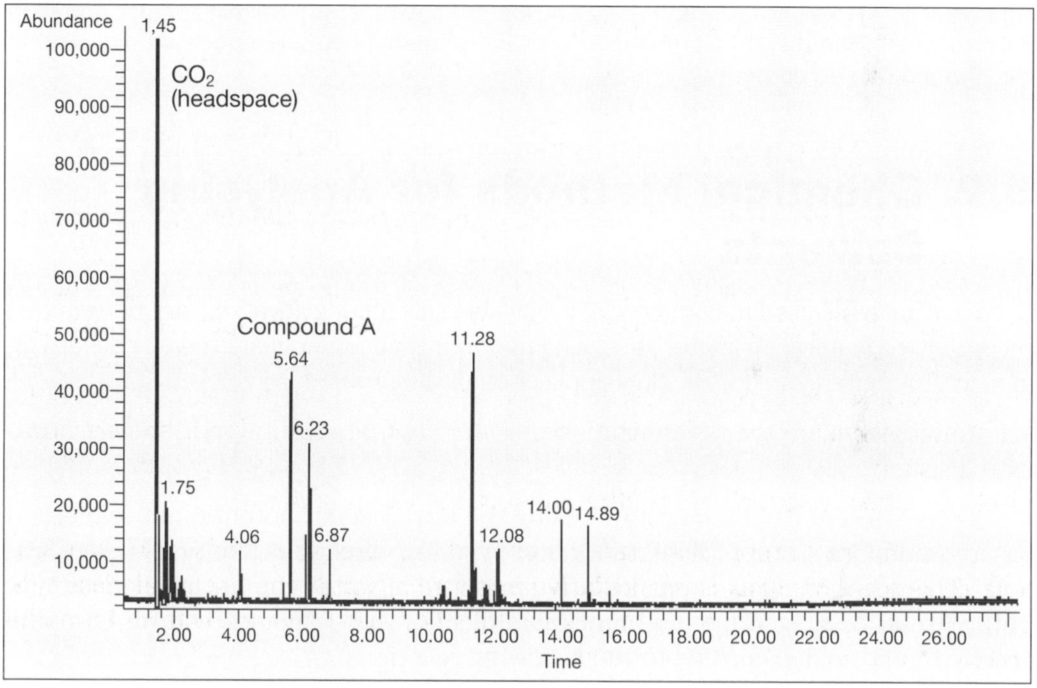

Figure 9.4.a. Agilent Equipment 6890/MSD 5973N, Headspace Sampler; Agilent 7694; Column HP-5MS; 30 m; 250 μm; 0.25 μm; Phase: Sulphenylene Siloxane; Initial Oven Temperature: 45 °C; Heating Ramp: 45 °C (5 min); 8 °C · min^{-1}; 200 °C (5 min); Injector Temperature: 200 °C; Detector Temperature (MSD): 230 °C. Mass Range: 30-400 (scan mode); Gas Carrier: He; Headspace Conditions: 2.0g of sample; Vial Temperature: 90 °C (30 min); Transfer Line Temperature: 100 °C; Pressurization Time: 0.5 min; Injection Time: 0.5 min

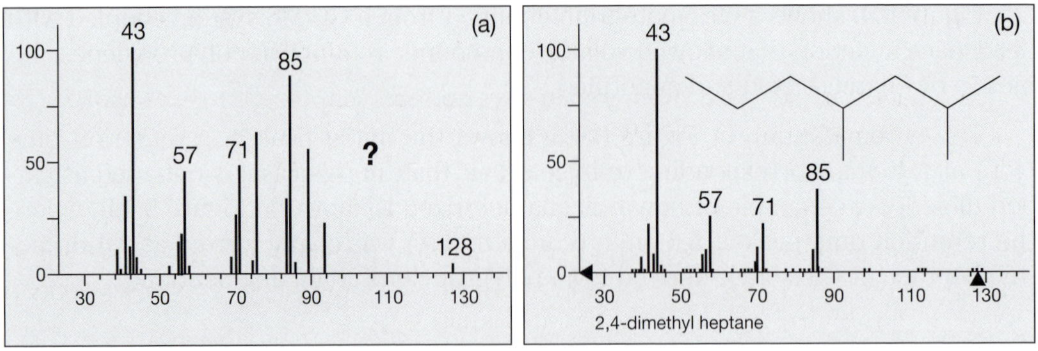

Figure 9.4.b. (a) Peak Mass Fragmentation Spectrum with RT = 5.64 min (Compound A) in Polypropylene Sample; (b) Comparison with 2.4-dimethyl heptane spectrum available in the library of the equipment

9.2. Chemical Methods for Analysing Solvents

Chemical analyses used in commercially available solvent specifications are determined according to chemical characteristics of compounds that contaminate them or according to market requirements for the final application of the product.

Among them are the determination of water content, acidity, and the permanganate resistance test.

Water content can be measured using the Karl Fischer potentiometric method based on the reaction of iodine reduction by sulfur dioxide in the presence of water.[9] This reaction allows a quantitative measure of water content in the medium, in the presence of pyridine and a primary alcohol that react with sulfur tri-oxide and hydriodic acid according to the following reactions:

$$H_2O + I_2 + 3C_5H_5N \rightarrow 2C_5H_5N \cdot HI + C_5H_5N \cdot SO_3$$

$$C_5H_5N \cdot SO_3 + R\text{—}OH \rightarrow C_5H_5N \cdot HSO_4\text{—}R$$

Karl Fischer solutions are commercial and include iodine, pyridine, and sulfur dioxide dissolved in 2-methoxyethanol. The reagent is standardized immediately before being used. Due to required care in handling, there are pyridine-free Karl Fischer reagent solutions available on the market.

The method described in ASTM D1364 is indicated for volatile solvents and chemical intermediaries used in coatings, varnishes, and similar products and can be used in any range of water content from very low quantities (given in ppm) to large amounts (given in percentages). Solvents that absorb water easily from moisture, like ketones, acetates, and glycol ethers, should be handled with care to avoid this effect, especially when the water content to be measured is very low.

The method is not sensitive to the presence of mercaptans, peroxides, or appreciative amounts of aldehydes or amines, having a wide range of applicability.

In total acidity determination, results are normally expressed in acetic acid content. In industrial solvents, acidity can originate from contamination of the productive process itself or from decomposition during storage or distribution.

The ASTM D1613 method[10] is applicable for acidity content below 0.05% (500 ppm) in organic compounds, hydrocarbon mixtures, solvents and diluents used in coatings, varnishes, and chemical intermediaries. Among the compounds that can be analyzed using this method are low molecular weight saturated and unsaturated alcohols, ketones, ethers, esters, and hydrocarbons.

The method's principle is based on titration with a sodium hydroxide aqueous solution in the presence of phenolphthalein or alternatively employing a potentiometric titration with pH electrode. Acetic acidity content is calculated based on the amount of NaOH used for neutralizing the total medium acidity.

The permanganate resistance test in ethanol, methanol, and acetone is described in NBR 5824 and is also known as the Barbet method.[11] The test consists of a qualitative trial to determine solvent impurities, indicative of the presence of substances that in a neutral medium reduce the permanganate to manganese dioxide. The necessary time for discoloration of the potassium permanganate solution to occur is compared to that of a standard solution of uranyl nitrate and cobaltous chloride. The test temperature varies as a function of the solvent type to be tested. For example, for methyl and ethyl alcohols, compounds capable of oxidizing in test conditions, the temperature is maintained at 15 °C. For acetone, the test is run at 25 °C.

In general, substances that react with the potassium permanganate, reducing it, correspond to:

- double and triple bond compounds, except aromatic rings;
- aldehydes;
- primary and secondary alcohols;
- phenols;
- nitrogen compounds.

9.3. Physical Methods for Analysing Solvents

Physical analyses used in commercially available solvent specifications are determined according to physical characteristics necessary for final application of the product.

Among the most used physical methods are non-volatile material determination, water miscibility, volatile solvent odor, density and relative density, and distillation range. The color test is also considered a physical test although it cannot be completely disassociated from chemical trials that correlate possible contaminants (organic, inorganic, etc.) with factors contributing to solvent color specification.

The color test is based on the visual comparison or spectrophotometric determination of color of a known standard concentration with the sample color.[12] In the method where a scale is used based on Pt-Co standards, the comparison is visual and the sample should present only a weak color, as is the case for the majority of commercially available solvents. The presence or absence of color is an indication of the degree of material purification, of the storage conditions or both. The ASTM method for determining solvent color is D1209 and is known by many as APHA color specification. Preparation on a comparison scale is made using concentrated solutions known as potassium hexachloroplatinate (K_2PtCl_6) and cobalt chloride hexahydrate ($CoCl_2 \cdot 6H_2O$).

The nonvolatile material content or dry extract is an applicable method for solvents used in coating, varnish or chemical intermediary production. ASTM D1353 describes the methodology,[13] based on the gravimetric method for determining maximum nonvolatile material content necessary for final application of the solvent. This test is of fundamental importance since the presence of any residual after solvent evaporation can affect coating, varnish, and other product quality and performance. To run the test, some safety procedures are noted in the standard as some solvents can form peroxides and become potentially explosive when they are concentrated and heated under test conditions. Another safety precaution mentioned in the regulation refers to care with solvents with low self-ignition temperatures, like aliphatic hydrocarbonates.

Another physical test used to control industrial solvent quality is distillation range determination. ASTM D1078 is applicable to liquids that boil between 30 and 350 °C and are chemically stable during the distillation process, which can be done in a manual or automatic system.[14] It can be applied to organic liquid compounds like hydrocarbonates, oxygenated solvents, and chemical intermediaries including mixtures.

The test result is obtained from a temperature range indicating relative volatility of the analyzed organic liquid used for identification and as an indicator of an analyzed solvent or mixture quality (degree of purity).

Density or relative density is a physical magnitude that indicates, along with other physical properties, final solvent application characteristics. ASTM D4052 deals with analytical requirements of testing for liquid compounds in the range of 15 to 35 °C, with a vapor pressure less than 600mm Hg and viscosity below approximately 15000cSt (mm^2/s) in the test temperature, which should be indicated.[15]

Density is defined as mass per volume unit at a determined temperature. Relative density is the quotient between product density at a determined temperature and the density of water under the same condition.

For density tests carried out using digital densimeters, amounts are usually expressed as follows:

density at 20 °C = value in g/ml;

relative density, 20/20 °C = o,xxxx (adimensional value).

The analytical method used to analyze solvent odor and volatile diluents is explained in ASTM D1296.[16] This is a comparative method in which characteristics and volatile organic solvent residual odor are observed to determine its applicability and tolerance in the solvent system of the final application. The method is not used to determine odor difference or its intensity.

In this test, solvents are classified in relation to characteristic odor. A sample placed on a filter paper is evaporated. The odor from the moist paper is compared to a dry paper after solvent evaporation.

The water miscibility test for solvents is described in ASTM D1722. The main characteristic is to determine immiscible contaminants in water qualitatively.[17] This is a good indicator for the presence or absence of substances that cannot solubilize completely in water, resulting in cloudiness when the test is carried out. Among the substances that can cause cloudiness are paraffins, olefins, aromatic compounds, alcohols or ketones with high molecular weight, among others. After the trial, the solvent that does not pass the test will be analyzed carefully to identify substances present that can interfere with the test.

9.4. Spectroscopic Characterization of Solvents

Many solvents require tests using spectroscopic methods that unequivocally characterize them when they arrive at the industrial unit. The importance of the receiving test varies in function of the chemical intolerance type between the material to which the solvent will be added and the product's chemical characteristic. For example, when a substrate (material to receive the solvent dilution) can react easily, it may be necessary to maintain this type of control. In this case, if there is an error in receiving, it can result in undesirable reactions in the product, reacting with the solvent used in the dilution (material cure, etc.). The test is also run for safety reasons since some reactions can involve chemical incompatibility and lead to an undesirable risk situation in the industrial process.

One of the tests often used in receiving and releasing of lots is the infrared spectroscopic analysis or Fourier Transform Infrared (FTIR).

Infrared radiation between 10,000 and 100 cm^{-1} when absorbed by an organic molecule is transformed into molecular vibration energy. The vibrational spectrum usually appears as a series of bands from the overlapping of multiple rotational energy lines, denominated by vibration-rotation bands. The interaction of infrared radiation with an organic molecule, as simple as it may be, gives a high complexity spectrum, in the range between 4,000 and 400 cm^{-1}. High complexity in the obtained spectrum allows the analyst to compare characteristic bands of functional groups to standard absorption data from literature, using important information to identify its structure.

An infrared spectrum gives two ranges that help in interpretation: the range between 4,000 and 1,300 cm^{-1} and between 900 and 650 cm^{-1}. The intermediary range between 1,300 and 900 cm^{-1} is known as the fingerprinting region. Here, the associated bands that are specific for the different chemical compounds will be looked for.[18]

The main functional groups present in the organic compounds absorb at higher frequency range (4,000 to 1,300 cm^{-1}).

For example, characteristic bands of absorption of aromatic and heteroaromatic groups appear in the range of 1,600 to 1,300 cm^{-1}. These groups strongly absorb in the range of 900 to 650 cm^{-1}. These bands are from angular deformation in the aromatic rings and indicate their substitution pattern. The range between 1,300 and 900 cm^{-1} should also be observed giving some particular information of the molecular structure being analyzed.

Figure 9.5. shows an FTIR spectrum in absorbance obtained as the result of the reception test of xylene (mixture of o, m, and p-dimethylbenzene) used in dilution of a compound sensitive to the presence of protic functional groups (hydroxyls, amines, and amides, etc.).

In the spectrum in Figure 9.5., the aromatic substitution pattern is in the range of 697 to 795 cm^{-1}, the aromatic C—H axis deformation band from 3027-3084 cm^{-1}, the methyl C—H deformation from 2873-2996 cm^{-1}, the combination frequency or harmonic bands from 1941-1742 cm^{-1}, and the aromatic ring C=C axis deformation band from 1606 to 1496 cm^{-1} characterizing the compound as a mixture of *orto*, *meta* and *para*-xylene unequivocally. In addition, bands that could suggest the presence of alcohol compounds, carbonyl compounds, amines, etc., do not appear in the solvent received, thus, it is released for use in the industrial plant.

Nuclear Magnetic Resonance is used more in academic settings than in industrial ones. However, it is a powerful tool for analyzing solvent molecular structure and allowing the unequivocal characterization and quantification of the isomer structure present, contaminant structure, etc.

Information from a Nuclear Magnetic Resonance can be used along with that from the infrared and the mass spectrometry, explained in this chapter, to interpret components present through spectroscopic techniques. Samples are prepared in commercially available deuterated solvents. The analysis can be done at different temperatures depending on the deuterated solvent boiling temperature in which the sample was prepared and different nuclei can be monitored, with hydrogen and carbon being the most used.[19]

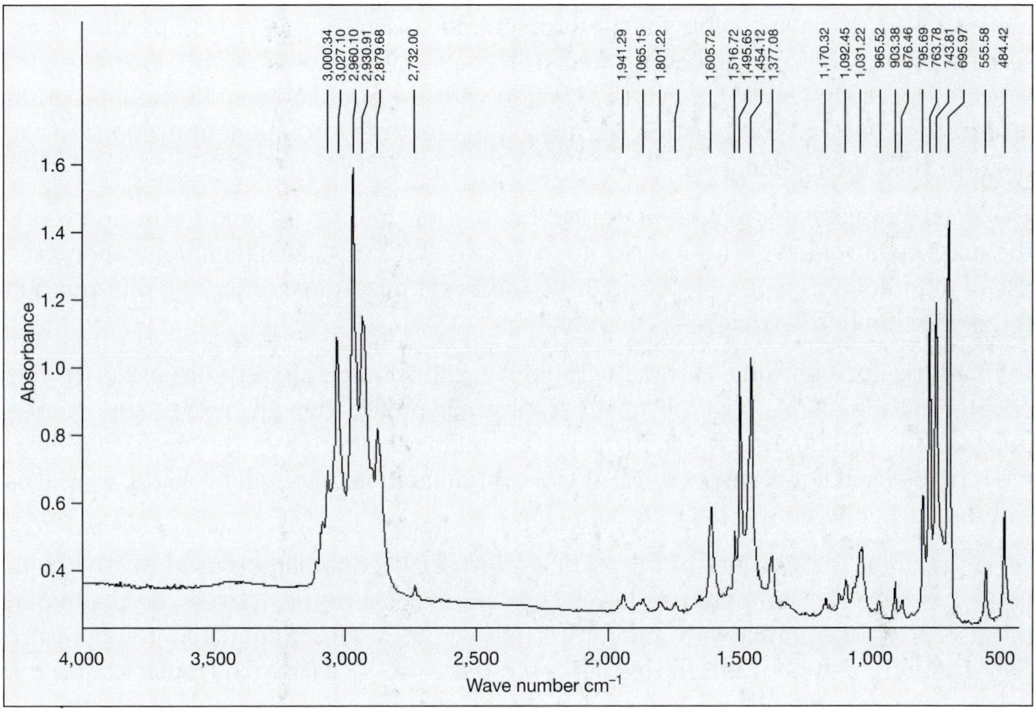

Figure 9.5. FTIR Spectrum (absorbance), from a Bruker Equinox 55, in a KBr cell for liquids of 0.025 mm of optic path.

9.5. Use of Coupled Analytical Techniques for Problem Solving in the Solvent Industry

In many cases, it is necessary to use more than one analytical technique to solve industrial problems. For example, when differences in odor or color are observed in the final product, techniques that allow a detailed investigation of possible components present from organic or inorganic origins are necessary. The sequence of techniques used in a final analytical approach is defined as an analytical strategy, in which its dimension varies in relation to size, degree, and difficulty of the problem. The investigative approach extrapolates normal analytical control limits and becomes the responsibility of laboratories specializing in research and development, and will give necessary technical support to define the best strategy for each case.

In general, with problems related to coloring or odor differences that do not meet specifications, an investigative strategy for organic and inorganic compounds can be used since the contamination level can be in extremely low concentrations. One possible approach is to use a reference lot, one with no problem, to compare with the problematic lot. In situations like these, many times it is important to confirm what compound(s) are responsible for the problem. In this case, it is very important to assess the qualitative differences as much as the quantitative ones. Many times the presence of a certain component in the sample with the problem can be the answer. In others, the chemical composition is similar but a mixture's component proportion can possible hold the explanation.

For an organic compound investigation, one of the first techniques to be used is the mass spectrometry with gas chromatography (GC/MS). This technique allows the use of different chromatographic columns, different injections forms, and different ion fragmentation to help diagnose the problem.

Criteria for choosing chromatographic columns were already discussed in this chapter in the section on gas chromatography. In general, they follow the same criteria in which the compound type being investigated is considered. In this case, Plot columns can be used if it is suspected that the contaminant is extremely volatile, as well as whether the columns are polar or nonpolar.

In relation to the types of injectors, a normal injection can be used for the component, where the temperature lower than normal for the injector to be made and the use of a programmable temperature injector, PTV (Programmable Temperature Vaporization), can be used. In this case, the risk of loss or thermal transformation of contaminant compounds is reduced during the injections process in chromatography.

Thermal degradation is very common for alcohols where eliminating water leads to joined or unjoined unsaturated compounds, which in turn can lead to erroneous amounts or conclusions about the solvent composition being analyzed.

Another possible form of injection, especially when analyzing odor caused by residual solvents (in plastic, rubber, solids, films, etc.) is the injection using the static or dynamic headspace technique. Here, only the gas phase (headspace) of the sample is injected, having a qualitative sweeping of compounds present and will have the residual odor found in the sample. This was discussed in 9.1.2. in relation to the residual solvent analysis in plastic and packaging films.

Figure 9.6. shows an example where the residual odor from an inorganic support sample (inorganic filler) was examined.

Here, headspace injection analysis allowed the identification and quantification of the compounds present in the vapor phase of the sample with the odor, making it possible to characterize them as ethanol, limonene and dodecane. As a comparison, a sample's witness lot with no residual odor was used. Quantification was made using an addition of standard on the sample with no contamination.

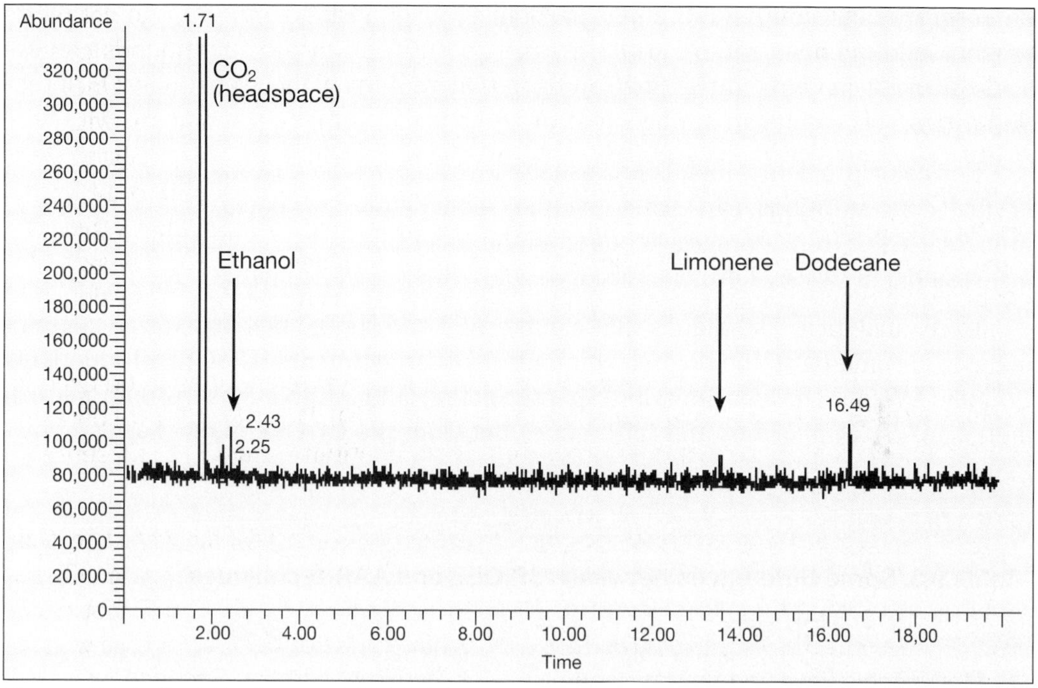

Figure 9.6. Agilent Equipment 5890/MSD 5971, Headspace Sampler Agilent 7694E. Innowax Column: 60 m; 250 µm; 0.25 µm; Phase: polyethylene glycol; Oven Starting Temperature: 45 °C (5 min); 8 °C · min^{-1}; 200 °C (5 min); Injector Temperature: 200 °C; Detector Temperature (MSD): 230 °C. Mass Range: 30-400 (scan mode); Gas Carrier: He; Headspace Conditions: 2.0 g of sample; Vial Temperature: 90 °C (30 min); Transfer Line Temperature: 100 °C; Pressurization Time: 0.5 min; Injection Time: 0.5 min

The injection technique known as SPME (Solid Phase Micro Extraction) can be used to detect residual solvents in pharmaceutical, food, environmental matrix and other products.

The SPME technique allows the analysis of a greater variety of compounds with different boiling points, and shows high sensitivity in detecting volatile compounds. Additionally, this technique is costs less than the static headspace system. Fiber choice for SPME injection depends on the type of analysis to be done. Some technical notes mention that for some matrixes, fibers responding to the adsorption mechanism are more appropriate for identifying volatility than for quantifying, where the absorption mechanism is more efficient.

Forms of detection can also be chosen in relation to the time of contaminant being investigated. In general, EI (Electronic Impact) detectors are used in investigating organic contaminants in solvents. The ion sweep can be made using the total selection technique, choosing a range based on total mass (SCAN mode) or selectivity, choosing some typical fragments of a component (SIM – Selective Ion Monitoring mode). The selection using SIM mode injection considerably increases sensitivity in detecting organic compounds and is particularly useful for quantifying contaminants present in low concentrations. Since impurity identification and quantification analyses can result in color in the final product that require higher sensitivity to detect impurities, this can be an important strategy to use.

For color problems, metals are usually investigated because metal cations, even in small concentrations, can change sample color. Metals can come from the industrial process itself or from storage systems, and their identification and quantification become essential in cases where final product color differences or changes are detected.

Here, analytical techniques that can be used and should be treated as a complement to chromatographic organic compound investigation are ICP OES (Inductively Coupled Plasma Optical Emission Spectrometry) and Atomic Absorption Spectrometry (AAS). Similarities and differences between the two techniques are summarized in Table 3.

Table 9.3. Some Differences between ICP OES and AAS Techniques

ICP OES	AAS
Atomic/Ionic Emission Spectroscopy Technique	Atomic Absorption Spectroscopy Technique
Atomic or ionic specie is generated by plasma	Atomic specie is generated by flame or graphite oven
Argon plasma temperature can be higher than 10,000 K	Acetylene/air flame temperature reaches a maximum of 2300-2900 °C
Best limits for detecting refractory elements (B, Ti, V, Al, etc.)	Allows determining high content of alkaline and alkaline soils

Commercially available ICP OES equipment can be sequential (determining one element at a time) or simultaneous (determining several elements at once). The advantage of the latter is that it reduces analysis time since many elements can be done at the same time.

Torch position allowed the development of radial type equipment, and most recently, axial type. The main characteristic of the radial equipment is the greater range of linearity, which allows the highest content of alkaline elements to be determined. The equipment with the torch placed axially is more modern and has a wider emission line range. Additionally, the equipment is more sensitive and therefore, is more subject to interferences from other elements and/or sample matrix.

In addition, a series of other advantages or disadvantages can be considered in choosing a metal analysis technique, especially when required levels for detection are extremely low.

Figure 9.7. shows an example of emission spectra from Fe and Al analysis using simultaneous ICP OES equipment. In this case, two distinct wavelengths for each metal are being used to improve possible interference evaluations due to other element emission and/or sample matrix, thus increasing selectivity from the measurement.

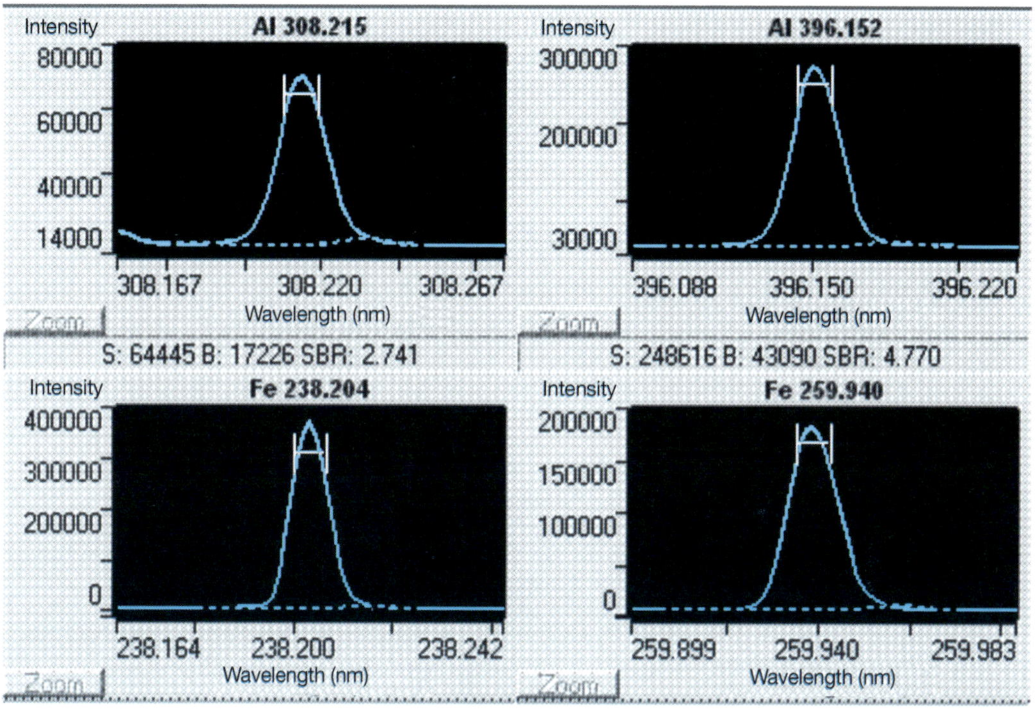

Figure 9.7. Optic emission spectra by plasma inductively coupled for aluminum (308.215 and 396.152 nm) and iron (238.204 and 259.940 nm) obtained using ICP-OES Varian-ES equipment, simultaneous, axial vision.

Currently, there are some accessories on the market that can be coupled to the ICP OES equipment that introduce small quantities of oxygen in plasma, facilitating the volatile compound analysis. In solvents that are high vapor pressure organic solutions, oxygen, stabilizes the plasma at the moment of aspiration. This allows the direct analysis of metal content in organic solvents (without preparing a sample). When there is no equipment for direct analysis, it is more common to evaporate the solvent completely, preferentially in mild conditions, using an inert gas atmosphere, to avoid the loss of possible volatile metal species.

9.6. Final Considerations

To analyze industrial solvents, different analytical methods can be used for final product or raw material specification. However, one of the main challenges is choosing the analytical strategy most appropriate in cases where specification is not reached or when product final application problems appear or even in the development of new products and/or applications. In these cases, many times there is no one possible unique strategy for approaching the analytical source and interaction between process specialists and analysis specialists to define together which is the best path to follow for investigating problem solutions is a fundamental importance.

Bibliographical References

1. W. Bertsch; W. G. Jennigs; R. E. Kaiser. *Recent Advances in Capilary Gas Chromatography*. Alemanha: Hutchig Verlag Heidelberg, 592 p., 1981.

2 W. Bertsch; W. G. Jennigs; R. E. Kaiser. *Recent Advances in Capilary Gas Chromatography*, v. 2. Alemanha: Hutchig Verlag Heidelberg, 557 p., 1981.

3 W. McFadden. *Techniques of Combined Gas Chromatography/Mass Spectrometry*: Applications in Organic Analysis. Bristol: John Wiley & Sons, 463 p., 1973.

4 J. R. Chapman. *Practical Organic Mass Spectrometry* – A Guide for Chemical and Biochemical Analysis, 2. ed. Chichester: John Wiley & Sons, 339 p., 1995.

5 C. Von Muhlen; C. A. Zini; E. B. Caramão; *et al*. Characterization of petrochemical samples and their derivatives by comprehensive two-dimensional gas chromatography. *Química Nova*, Brasil, v. 29, no.4, p.765-775, Jul/Ago, 2006.

6 A. O. Maldaner; M. A. Guilherme. *Cromatografia gasosa ultra-rápida – Implementação de metodologia*. Documento 2002-19-C02, Centro de Pesquisas de Paulínia, Rhodia Poliamida e Especialidades Ltda., 2002.

7 H. Verneret. *Solventes Industriais*: Propriedades e Aplicações. São Paulo: Toledo, 145 p., 1984.

8 B. Kolb. *Applied Headspace Gas Chromatography*. Bristol: Heyden & Son, 185 p., 1982.

9 ASTM. *Standard Test Method for Water in Volatile Solvents* (Karl Fischer Reagent Titration Method), D 1364. USA, 2002.

10 ASTM. *Standard Test Method for Acidity in Volatile Solvents and Chemical Intermediates Used in Paint*, Varnish, Lacquer, and Related Products, D 1613. USA, 2006.

11 ABNT. *Acetona, Álcoois etílico e metílico*. Determinação do Tempo de redução de permanganato. Método de Barbet, NBR 5824. Brasil, 1986.

12 ASTM. *Standard Test Method for Color of Clear Liquids* (Platinum-Cobalt Scale), D 1209. USA, 1997.

13 ASTM. *Standard Test Method for Nonvolatile Matter in Volatile Solvents for Use in Paint*, Varnish, Lacquer, and Related Products, D 1353. USA, 2003.

14 ASTM. *Standard Test Method for Distillation Range of Volatile Organic Liquids*, D 1078. USA, 2003.

15 ASTM. *Standard Test Method for Density and Relative Density of Liquids by Digital Density Meter*, D 4052. USA, 2002.

16 ASTM. *Standard Test Method for Odor of Volatile Solvents and Diluents*, D 1296. USA, 2001.

17 ASTM. *Standard Test Method for Water Miscibility of Water-Soluble Solvents*, D 1722. USA, 2004.

18 R. M. Silverstein; F. X. Webster. *Identificação Espectrométrica de Compostos Orgânicos*. 6. ed. Rio de Janeiro: Livros Técnicos e Científicos, 460 p., 2000.

19 T. W. Solomons; C. B. Fryhle. *Organic Chemistry*. 7. ed. New York: John Wiley & Sons, 1258 p., 2000.

10 A Segment Committed to Sustainability

Relevant aspects in the management of health, safety, and the environment when using industrial solvents are presented in this chapter.

Hidejal Santos
Maria Luiza Teixeira Couto
Fernando Zanatta
Rosmary De Nadai

In response to growing demands from society, customers, and the sector itself, the ICCA (*International Council of Chemical Associations*) implemented and maintains the *Responsible Care*® program as one of its main initiatives. Originally conceived in Canada in 1985, the principle objective of the program is to deal with target audience concerns about production, distribution, and the use of chemicals, focusing on continual improvement of performance, communication, and responsibilities by implementing and maintaining management practices. These good management practices were organized under six codes of practices: **Process Safety, Transportation and Distribution, Environmental Protection, Employee Health and Safety at Work, Community Awareness and Emergency Response, and Product Stewardship.**

The *Responsible Care*® program was adopted in 1992 by the Brazilian Chemical Industry Association (ABIQUIM) under the name of *Atuação Responsável*®. In 1998, the program became mandatory for associated companies.

With the publication of the *Responsible Care Global Charter* in 2005, *Responsible Care*® became a global chemical industry initiative in which companies, through national associations, commit themselves to work together to continually improve product and process performance in terms of safety, industrial hygiene, health, and environment and to contribute to sustainable development in local communities and society as a whole.

10.1. Product Stewardship

Product Stewardship may be described as the ethical and responsible management of health, safety, and environmental aspects of chemicals throughout their life cycle. Its practice results in continual improvement in reducing impacts from products on the market by selling safer and environmentally acceptable products, improving emergency controls, reducing accidents or decreasing the severity of known hazards, and managing risks associated with the use of chemicals.

In 2006, the *International Conference on Chemical Management* (ICCM) adopted the *United Nations Environment Programme* (UNEP) *Strategic Approach to International Chemicals Management* (SAICM). This is a policy for international action on chemical product hazards. SAICM supports objectives ratified in 2002 during the *Johannesburg World Summit on Sustainable Development* ensuring that by 2020 chemicals produced and used will continue to reduce adverse health and environmental impacts.

At the beginning of 2006, in response to the SAICM, the worldwide chemical industry, represented by the ICCA, introduced the *Global Product Strategy* (GPS) to intensify application of *Product Stewardship* best practices throughout the supply chain. The GPS is an important part of the *Responsible Care*® program because it brings together several chemical management initiatives by developing a basis for continuous improvement, transparency, and principally, the first global effort for leveraging *Product Stewardship* from industry to customers.

Product Stewardship management practices can be grouped as:

- *Product Stewardship* Management Principles
  - Defining policy and objectives, and establishing action plans for implementation and continual improvement;
  - Defining accountability, authorities, and interaction between concerned groups;
  - Managing resources;
  - Monitoring global process and individual performances.

- Identification Classification of Hazards
  - Identifying and classifying hazards to health, safety and the environment, based on internationally recognized hazard classification systems.

- Risk Assessment
  - Defining tools for analyzing risks;
  - Identifying indicators for application and recognition in chemical use;
  - Identifying legislation applicable to where products are produced and sold;
  - Characterizing product risk – application – applicable regulation.

- Risk Management
  - Defining management strategies and actions for identified risks;
  - Systematically analyzing accidents and complaints involving health, safety, and environmental aspects;
  - Managing crises;
  - Managing the distribution chain.

- Communication
  - Communicating hazards;
  - Communicating cooperatively with suppliers, subcontractors, customers, and authorities;
  - Communicating crises.

The *United Nations Recommendations on the Transport of Dangerous Goods* and the GHS are the main *Product Stewardship* tools for classifying and communicating chemical hazards.

10.2. Hazard Classification and Communication – GHS

In 1989, at the International Labor Organization (ILO) General Assembly during the first discussion on ILO rules for safety in the use of hazardous chemicals in the workplace, the ILO adopted the resolution presented by India's government. This resolution is on a Harmonization System of Classification and Labelling for use of hazardous chemicals in the workplace. Following the Chemical Substance Convention model, adopted in 1990, the ILO started a project for harmonizing existing systems for classification and labelling of chemicals.

In 1992, supported by the Conference on the Environment and the Development of the Organization for Economic Cooperation and Development (OECD) – Rio 92, the objective moved on and became one of the six areas of action identified in Chapter 19, Agenda 21, on the environmental control of toxic chemicals. The OECD recommended that a globally harmonized hazard classification and a compatible labelling system, including Safety Data Sheets (SDS), containing relevant safety data be developed.

> "UNCED – Agenda 21 – Chapter 19 – GHS Objectives
>
> 19.27. A globally harmonized hazard clasification and compatible labeling system, including material safety data sheets and easily understandable symbols, should be available, if feasible, by the year 2000."

This commitment was reaffirmed at the World Summit on Sustainable Development in Johannesburg in 2002 (Rio+10):

> "Plan of Implementation Paragraph 23 (c):
>
> Encourage countries to implement the new globally harmonized system for the clasification and labeling of chemicals as soon as posible with a view to having the system fully operational by 2008."

GHS is the Globally Harmonized System of Classification and Labelling of Chemicals. Its first edition – also known as the Purple Book – was published in 2003 and can be found with its subsequent revisions and amendments at:

http://www.unece.org/trans/danger/publi/ghs/ghs_welcome_e.html.

The GHS establishes a wide range of logical criteria for classifying physical hazards to health and the environment and for chemicals and mixtures, as well as establish the structured communication of these hazards using safety data sheet labels.

- **Physical Hazards**

  For physical hazards, the GHS includes the following classes for both substances and mixtures: explosives, flammable substances (gases, aerosols, liquids, and solids); oxidizing gases, solids, and liquids; organic peroxides, gases under pressure; self-reactive substances and mixtures; pyrophoric liquids and solids; self-heating substances and mixtures; substances and mixtures that emit flammable gases when they come into contact with water; and substances corrosive to metals. For solvents discussed in this book, flammability is probably the most relevant physical hazard.

- **Health Hazards**
  - Acute Toxicity;
  - Skin Corrosion and Irritation;
  - Serious Eye Damage/Eye Irritation;
  - Respiratory or Skin Sensitization;
  - Germ Cell Mutagenicity;
  - Carcinogenicity;
  - Reproductive Toxicity;
  - Specific Target Organ Toxicity (single and repeated exposure);
  - Aspiration Hazard.

- **Environmental Hazards**

  For environmental hazards, the GHS includes acute aquatic toxicity, potential for or actual bioaccumulation, degradation (biotic or abiotic) for organic chemicals, and chronic aquatic toxicity.

During the development of the GHS, there was an effort to approach harmonization with the existing system for risk classification in transporting hazardous products. However, for transportation, recommendations established by the *United Nations Committee of Experts for the Transport of Dangerous Goods* still prevail.

As with all UN recommendations for the transportation of dangerous goods, GHS adoption, form, and the extent of implementation are subordinate to legal requirements of each country.

10.2.1. Hazard Classification for Solvents

In relation to health hazards, common solvent characteristics include local dermal effects from lipid removal from skin (degreasing), central nervous system depression, neurotoxicity, hepatoxicity, nephrotoxicity, and in some cases carcinogenicity with variable risk, for example:

Substance	Potential Signals/Symptoms
Ethyl Acetate	Eye, skin, nose, throat irritation; narcosis, dermatitis
Acetone	Eye, nose, throat irritation; headache, dizziness, central nervous system depression; dermatitis
Toluene	Eye irritation, lethargy/fatigue (lacking energy, tiredness), confusion, euphoria, dizziness, headache, pupil dilation, tearing,; anxiety, muscle fatigue, insomnia, paresthesis, dermatitis, hepatic and renal injuries
Methyl Ethyl Ketone	Eye, skin, nose irritation; headache; dizziness; vomiting; dermatitis
n-Hexane	Eye, nose irritation; headache; peripheral neuropathy; sleeping of extremities, muscle weakness; dermatitis; dizziness; chemical pneumonia (liquid aspiration)
Cyclohexane	Eye, skin, respiratory system irritation; drowsiness; dermatitis; narcosis, coma
Ethylene Glycol	Eye, skin, nose, and throat irritation; lethargy/fatigue (lack of energy, tiredness); headache, dizziness, central nervous system depression; abnormal eye movement (nystagmus); skin sensitivity Acute ingestion: Abdominal pain, nausea, vomiting, loss of coordination, stupor, convulsion, unconsciousness; cardiac rhythm acceleration, congestive cardiac failure; damage to kidneys, and kidney failure

10.2.2. GHS Use for Classification and Communication of Hazards

Criteria established by the GHS are applicable for pure substances and mixtures, including the definition of the methods for obtaining data when necessary.

Accurate classification for a substance or mixture depends on the interpretation of criteria established by the GHS and the reliability of available data. Hazard classification should be done by trained specialists because in addition to the existing reliable data, evidence of the effects on humans and animals should also be taken into account in the assessment of hazards presented (*in the sense of to be offered*) by a substance or mixture to human health or the environment. For certain properties, classification can be directly obtained when data satisfies criteria. For others, classification can be based entirely on evidence. Another important aspect is the calculation formulas for mixtures are based on the additivity principle. However, antagonistic and synergistic interaction must be considered by experts for classification.

One of the general principles established by the GHS determines that experimental data already available for classification of chemicals in other existing systems should be accepted to avoid test repetition and the unnecessary use of laboratory animals. It is recommended that where possible, partnerships development and cooperative actions between companies and entities be taken in order to share data and avoid duplicating tests and unnecessary animal testing. Afterwards, if tests are still necessary, it is recommended that valid laboratories strictly committed to humane treatment of animals be used.

Considering a hypothetical homogeneous solvent mixture containing 40% of substance A, 32% substance B, 25% substance C, and 3% of substance D, classification can be made as follows:

	C_i%	Flash Point °C	Boiling Point °C	Vapor Pressure mm Hg ≅ 0 °C	$\Delta_{VAP}H_0$ kJ/mol
Substance A	40	–22.0	55.6	187.5	31.3
Substance B	32	–4.0	77.0	76	35.1
Substance C	25	92.8	196.0	0.1	48.4
Substance D	3	4.4	114.1	21.8	38.1

	C_i%	Exposure Route		
		Oral LD_{50}	Dermal LD_{50}	Inhalation LC_{50}
Substance A	40	5,800 mg/kg	20,000 mg/kg	50,100 mg/m^3
Substance B	32	5,620 mg/kg	20,000 mg/kg	200 mg/m^3
Substance C	25	2,080 mg/kg	3,000 mg/kg	4,000 mg/m^3
Substance D	3	5,000 mg/kg	14,000 mg/kg	10,640 mg/m^3

LD_{50} (Lethal Dose 50%): chemical dosage causing death in 50% (half) of a group of animals submitted to tests.
LC_{50} (Lethal Concentration 50%): chemical concentration in the air or in water causing death in 50% (half) of a group of animals submitted to tests.
C_i%: quantity or concentration for component i in mixture.

Because substance A is absorbed through the respiratory tract, when it is in high concentrations, it has a narcotic effect and is able to induce a coma.

Substance D is recognized as a neurotoxin.

10.2.3. Classification Example of Physical Hazards

10.2.3.1. Flammability of Liquids

For mixtures containing flammable liquids and in known concentrations, flash point and initial boiling point are used as classification criteria. In the above example, flash point (closed cup) can be experimentally determined, or according to the GHS, can be calculated using the method described by Gmehling and Rasmussen [Ind. Eng. Chem. Fundament, 21, 186 (1982)]. However, to use the latter method, some necessary data, like the activity coefficient and its dependence on temperature for each component of the mixture, would probably be difficult to find. For this reason, the experimental method can be faster.

Considering that for this hypothetical mixture, experimentally obtained data of 5.6°C and 99.2°C for flash point and boiling point, respectively, according to the above table in which cut-off values are given for flammability of liquids, this mixture should be classified as **flammable – category** 2.

Criteria	Categories			
	1	2	3	4
Flash Point	< 23 °C	< 23 °C	≥ 23 °C e ≤ 60 °C	> 60 °C e ≤ 93 °C
Initial Boiling Point	≤ 35 °C	> 35 °C	–	–
Labelling Elements				
Pictogram				Not Applicable
Signal Word	Danger	Danger	Warning	Warning
Hazard Statement	Extremely Flammable Liquid and Vapor	Highly Flammable Liquid and Vapor	Flammable Liquid and Vapor	Combustible Liquid

10.2.4. Classification Example of a Health Hazard

10.2.4.1. Acute Toxicity (AT)

Chemicals and mixtures are classified under five categories for acute toxicity based on possible oral, dermal, or inhalation exposure routes.

Exposure Route	Categories				
	1	2	3	4	5
Oral (mg/kg)	$0 < AT \leq 5$	$5 < AT \leq 50$	$50 < AT \leq 300$	$300 < AT \leq 2000$	$2000 < AT \leq 5000$
Dermal (mg/kg)	$0 < AT \leq 50$	$50 < AT \leq 200$	$200 < AT \leq 1000$	$1000 < AT \leq 2000$	$2000 < AT \leq 5000$
Gases (mg/m^3)	$0 < AT \leq 100$	$100 < AT \leq 500$	$500 < AT \leq 2500$	$2500 < AT \leq 5000$	Note 1
Vapour (mg/m^3)	$0 < AT \leq 0.5$	$0.5 < AT \leq 2.0$	$2.0 < AT \leq 10.0$	$10.0 < AT \leq 20.0$	Note 1
Dust/Mist (mg/m^3)	$0 < AT \leq 0.05$	$0.05 < AT \leq 0.5$	$0.5 < AT \leq 1.0$	$1.0 < AT \leq 5.0$	Note 1
Elementos de la Etiqueta					
Pictogram					Not Applicable
Signal Word	Danger	Danger	Danger	Warning	Warning
Frase de Perigo					
Oral	Fatal if swallowed (oral)	Fatal if swallowed (oral)	Toxic if swallowed (oral)	Harmful if swallowed (oral)	May be harmful if swallowed (oral)
Dermal	Fatal in contact with skin (dermal)	Fatal in contact with skin (dermal)	Fatal in contact with skin (dermal)	Fatal in contact with skin (dermal)	Harmful in contact with skin (dermal)
Inhalation	Fatal if inhaled (gas, vapor, dust, mist)	Fatal if inhaled (gas, vapor, dust, mist)	Fatal if inhaled (gas, vapor, dust, mist)	Fatal if inhaled (gas, vapor, dust, mist)	May be harmful if inhaled (gas, vapor, dust, mist)

Note 1 – GHS criteria for category 5 allow the identification of chemical substances or mixtures with relatively low acute toxicity, based on reliable information, indicate toxic effects for humans or may be hazardous to vulnerable populations according to expert judgement.

10.2.4.2. Classification Based on Available Data for All Components

Knowing the inhalation route is the most probable type of exposure and acute toxicity values for each component are known, it is possible to calculate an Acute Toxicity Estimate (ATE) for our mixture using the additivity formula:

$$\frac{100}{ETA_m} = \sum_n \frac{C_i}{ETA_i}$$

where: C_i = Concentration of ingredient i
n = Quantity of ingredients from 1 to n
ATE_i = Acute Toxicity Estimate of ingredient i
ATE_m = Acute Toxicity Estimate of mixture

To apply the above formula, it is necessary to:

- Include ingredients with known acute toxicity that fall under one of the GHS acute toxicity categories;
- Ignore known ingredients like non-toxic ones at the acute level (for example, water, sugar);
- Ignore ingredients for which the oral test does not show acute toxicity below 2,000mg/kg/body weight);
- Consider that there is no evidence of antagonistic or enhancing synergy among mixture components.

If we consider all mixture ingredients:

$$\frac{100}{ATE_m} = \frac{40}{50{,}1} + \frac{32}{0{,}2} + \frac{25}{4{,}0} + \frac{3}{10{,}64}$$

ATE_m = 0,6 mg/L

So a mixture would be classified under **category 2 for acute toxicity**.

However, if we consider that, based on available data, substance A does not contribute to the mixture's acute toxicity, we would have:

$$\frac{60}{ATE_m} = \frac{32}{0{,}2} + \frac{25}{4{,}0} + \frac{3}{10{,}64}$$

ATE_m = 0,36 mg/L

This mixture would be classified under **category 1 for acute toxicity**.

This example reinforces what has already been presented. Hazard classification should be done by trained specialists because, in addition to the simple existence of reliable data, evidence from human beings and animals should be taken into consideration in the assessment of hazards to human health mainly for cases with mixtures, as well as possible antagonistic or enhancing synergy among its ingredients.

10.2.5. Specific Target-Organ Toxicity

One of the ingredients in the example of the hypothetical mixture is a recognized neurotoxin and the other ingredient is a central nervous system depressor that, in high concentrations, can lead to a coma.

For classification of this hazard subclass, reliable evidence associated with a single or repeated exposure with identified, consistent effects must be used as support. Effects that should be considered are, for example:

a) Morbidity or death resulting from repeated or long-term exposure. Morbidity or death may be a result of repeated exposure, even in low concentrations/doses due to a substance or its metabolite bioaccumulation;

b) Significant functional changes in the peripheral or central nervous system or other organic system, including signs of CNS depression and changes in the senses (sight, hearing, sense of smell);

c) Any consistent and significant adverse change in biochemical, hematological or urinalysis parameters;

d) Significant organ damage that may be noted in necropsy and/or histologically confirmed;

e) Multifocal or diffuse necrosis, fibrosis, or granuloma formation in vital organs with regenerative capacity;

f) Morphological changes that are potentially reversible but provide clear evidence of marked organ dysfunction (ex.: hepatic esteatosis);

g) Evidence of appreciable cell death (including cell degeneration and reduced cell number) in vital organs incapable of regeneration.

The neurotoxic action of chemicals is difficult to characterize due to:

- Definition of non-specific cases;
- Possibility of prevalence rate duplication;
- Variability in neurobehavioral tests;
- Non-specificity of physiological measurements;
- Confusion with ethanol (habits and life style), trauma, and other factors;
- Multiple exposures;
- Nutritional deficiencies, metabolic deficiencies, hereditary deficiencies;
- Demyelinating conditions, paraneoplastic conditions.

The definition of the central nervous system is not functional: it is characterized anatomically by structures located in the axial skeleton (cranial cavity and vertebral column), i.e., brain, cerebellum, cerebral trunk, and spinal cord. However, some neurotoxic symptoms may be associated with these structures:

Central Acute	Overall activity like anesthesia or selective inhibition; narcosis; euphoria; disinhibited behavior; irritation; lack of coordination, ataxia, dysarthria
Central Chronic	Controversial – difficult to characterize Painter's Syndrome: depression; slow psychomotor performance; personality changes; short-term memory deficiencies Neurobehavioral test results from the World Health Organization (WHO)
Peripheral Neurotoxicity	Axonal distal neuropathy generally present in lower extremities primarily: decrease in sensory capacity – paresthesia (numbness, loss of proprioception), tingling; later on: decrease in motor capacity (motor weak and atrophy because of denervation)

10.2.5.1. Classification of Systemic Toxicity in Specific Target-Organs

In the following table, classification criteria for both one-time and repeated exposure are presented.

Categories	Single Exposure	Repeated Exposure
1	Substances that have produced significant toxicity in humans, or that, on the basis of evidence from studies in experimental animals can be presumed to have potential to produce significant toxicity in humans following single exposure.	Substances that have produced significant toxicity in humans, or that, on the basis of evidence from studies in experimental animals can be presumed to have the potential to produce significant toxicity in humans following repeated exposure.
2	Substances that, on the basis of evidence from studies in experimental animals can be presumed to have the potential to be harmful to human health following single exposure.	Substances that, on the basis of evidence from studies in experimental animals can be presumed to have the potential to be harmful to human health following repeated exposure.
3	Transient target-organ effects. Effects that adversely change human function, but do not meet category 1 or 2 may be classified in this category	

Labelling Elements			
	Categories		
	1	2	3
Pictogram			
Signal Word	Danger	Warning	Warning
Hazard Statement			
	Categories		
	1	2	3
Single Exposure	Causes damage to organs (or state all organs affected, if known) (state route of exposure if it is conclusively proven that no other routes of exposure cause the hazard)	May cause damage to organs (or state all organs affected, if known) (state route of exposure if it is conclusively proven that no other routes of exposure cause the hazard)	May cause respiratory irritation; or May cause drowsiness or dizziness
Repeated Exposure	Causes damage to organs (state all organs affected, if known) through prolonged or repeated exposure (state route of exposure if it is conclusively proven that no other routes of exposure cause the hazard)	May cause damage to organs (state all organs affected, if known) through prolonged or repeated exposure (state route of exposure if it is conclusively proven that no other routes of exposure cause the hazard)	

The hazard communication on the MSDS (Material Safety Data Sheet) and labels for mixtures, however, depend on hazard ingredient concentration. The GHS suggests cut-off values or concentration limits, shown in the table below, but the authorities of each country can establish their criteria.

Ingredient classified as:	Cut-off values/concentration limits triggering classification of a mixture as:	
	Category 1	Category 2
Category 1 Target organ toxicant	≥ 1.0 %(note 1)	1.0 ≤ ingredient < 10%(note 3)
	≥ 10 %(note 2)	1.0 ≤ ingredient < 10%(note 3)
Category 2 Target organ toxicant		≥ 1.0 %(note 1)
		≥ 10 %(note 2)

Note 1: If a category 1 or 2 specific target organ toxicant is present in the mixture as an ingredient in a concentration between 1.0% and 10.0%, the GHS places the responsibility with the authorities of each country to require information and to use the respective symbol.

Note 2: If a category 1 or 2 specific target organ toxicant component is present in the mixture as an ingredient in a concentration of ≥ 10%, it is expected that the authorities of each country require its identification on the label.

Note 3: If a category 1 specific target organ toxicant is present in the mixture as an ingredient in a concentration between 1.0% and 10.0%, some regulatory authorities may classify it as a category 2 target organ toxicant agent, but others may not.

An aspect to be considered is the existence of other internationally recognized classification systems.

Based on the above information, the classification of the hypothetical mixture following GHS criteria would be:

- Hypothetical mixture containing ingredient A characterized as a CNS depressor and may induce a coma in high concentrations, as a substance would be classified as **Category 2 – Target Organ Toxicant**.
- Ingredient D, a recognized neurotoxin would be classified as **Category 1 – Target Organ Toxicant**.
- In this case, regarding ingredient D concentration in the mixture and because of the potential synergistic interaction with ingredient A, the mixture would be hypothetically classified as **Category 1 – Target Organ Toxicant**.

10.2.6. Hazard Communication

The purpose of this section is to present the use of the information from the classification examples. Safety data sheet and labelling preparation is for experts that, in addition to GHS recommendations, should know and understand applicable legal requirements for labels in countries where a dangerous chemical is commercialized. GHS hazard communication is done using two principal ways:

- The safety data sheets are internationally known as MSDS – Material Safety Data Sheet, or SDS – Safety Data Sheet, or HDS – Hoja de Datos de Seguridad, in some Spanish-speaking countries; and as Ficha de Informação de Segurança de Produto Químico (FISPQ) in Brazil.
- Label. An important aspect for labelling is that the GHS has as one of its goals the act of establishing criteria for adding safety information to labels, mainly using pictograms, signal words, hazard statements, which should be located near each other on the label. Therefore, each country must make all needed arrangements to adapt these criteria according to their respective legal standards.
  - Pictogram: it is a graphic composition result of a combination of a hazard symbol and other graphic elements including borders, background pattern, and color;
  - Signal Word: these are used on the label to indicate the relative level of severity of hazard and as an alert on the label to the potential hazard. The signal words used in the GHS are "Danger" and "Warning." "Danger" is used for more severe hazard categories;
  - Hazard Statements: these are phrases assigned to classes and categories describing the nature of the hazards in a hazardous product, and where appropriate, the degree of hazard.

10.2.6.1. Multiple Hazards and Precedence of Hazard Information

- Pictogram

  For substances and mixtures covered by the *UN Recommendations on the Transport of Dangerous Goods*, the precedence of pictograms for physical risks should follow the rules established by this recommendation. When a risk pictogram for transport is shown on the label, the GHS pictogram for the same hazard should be omitted.

  For health hazards, the following principles of precedence apply:

  a) If the "skull and crossbones" pictogram is applicable, the "exclamation mark" pictogram should be omitted;

b) If the "corrosive" pictogram is applicable, the "exclamation mark" pictogram should be omitted when it is used for skin or eye irritation;

c) If the "health hazard" pictogram is applicable for respiratory sensitization, the "exclamation mark" pictogram should be omitted when it is used for skin sensitization or for skin or eye irritation.

- **Signal Word**

  If the signal word "Danger" is applicable, the signal word "Warning" should be omitted.

- **Hazard Statements**

  All assigned hazard statements related to classification should appear on the label. However, the authorities for each country may choose to specify the order in which they appear.

  For our mixture example, the following elements should be placed on the label:

Pictogram	Signal Word	Hazard Statements
	Danger	Highly Flammable Liquid and Vapor
		Fatal if swallowed. Fatal in contact with skin. Fatal if inhaled
		May cause damage to the Central Nervous System through prolonged or repetitive respiratory exposure.

10.3. Management of Hygiene, Health, and Environmental Aspects

10.3.1. Solvents in an Industrial Hygiene Context

In the industrial hygiene field, we look to establish a connection with chemical compound toxicology; in particular, we deal with the relationship to solvents. Man is an integral part of the environment and maintains a close relationship with consumption and contributions

When we speak of toxicity, we are talking about the characteristic of a molecule or compound to produce an adverse effect on an individual, both on the inside and outside of the body. Within the context of solvents, this susceptibility is increased mainly through dermal and respiratory paths.

10.3.1.1. Toxicology as an Information and Planning Tool in the Occupational Hygiene Practice

Classification of solvents according to their toxic potential leads to establishing a product's hazard as the probability for which an illness may be caused due to product use.

With this scenario, it can be concluded that an evaluation of solvent effects on users, whether in an industry or any other application segment, only has a comparative value if interpreted as a cross-matrix of toxicological data (*relationship of impact on the human organism*) and ecotoxicological data (*relationship of impact on the environment*). Therefore each one has a potential negative or positive characteristic.

Within this concept, the interpretative interdependence for the health, safety, and environmental areas, globally known as *Health, Safety, and Environment* (HSE) is reinforced.

In the world of solvents, human exposure can occur during manufacturing, handling, and use in work environments as well as residential environments so it is important to understand the relationship between the terms used in classifying products. Acute toxicity is a short-term effect, and chronic toxicity id when the effect is long-term exposure, either by inhalation or absorption through the skin.

The following table helps explain information and expert judgement in solvent hazard classification:

Danger		Expert Judgement
Classification	Description	
Unknown	Substances in which toxicological information could not be found in literature or other sources as experiments, where results obtained with inferior animals, are not recognized by the scientific community as information applied to human exposure.	The role of occupational hygiene is in recognizing the production route, information on raw material used in manufacturing the products under evaluation.
Not classified by the GHS – Non-Toxic	Substances that do not cause any risk under any condition of use or for those which cause toxic effects on humans only under very unusual conditions or through excessively high dosages.	The hygienist has the job of perception and background research of events collected in the production areas.
GHS Category 5 – Slightly Toxic Substances that produce changes in the human body but are immediately reversible and will disappear at the end of exposure, with or without medical treatment.	Products that have low toxicity may present a localized acute effect, i.e., mild effect on the skin or mucous membrane, independent of exposure time. When the cause is systemic acuteness, there may be absorption by the body through inhalation, ingestion, or through the skin, producing mild effects, i.e., exposure lasting seconds, minutes, or hours, if by ingestion of a single dose, independent of quantity. Localized chronic exposure may be continuous or repetitive, lasting days, months, or years, but sustaining light injuries to the skin or mucous membrane. Still within the classification of low toxic effect, there is the systemic chronic effect, a condition where the body absorbs a substance through inhalation, ingestion, or through the skin, producing mild effects, even though exposure is continuous and repetitive for days, months, or years.	Interaction of the hygienist with the site doctor brings added value in the identification of human toxicity potential, on health evaluations at site medical stations, and with information released on the commercialized Product Data Sheets.

Danger		Expert Judgement
Classification	Description	
GHS Category 3 and 4 – Moderately Toxic Substances may produce irreversible changes but not as severe as to be life-threatening or produce permanent physical incapacity.	Solvents are also classified as moderately toxic. In this category, *localized acuteness*; *systemic acuteness*; *localized chronic effects*; and *systemic chronic effects* can be found. The effects are moderate in situations of body absorption, whether via inhalation, ingestion, or through the skin, with different intensities that can last seconds, minutes, hours, days, months, or years.	The comparative study of: exposure time vs. entry possibilities vs. intensity of effect allows the best procedures for evaluation to be decided for assessing, quantifying, and applying preventive measures to lessen or eliminate toxic potential of a product or substance.
Category 1 and 2 – Severely Toxic	The substances may cause *localized acute* effects in a single exposure lasting seconds or minutes, causing injury to the skin or mucous membrane that is severe enough to threaten the life or cause permanent physical damage. Systemic acute effects are absorbed into the body through inhalation, ingestion, or through the skin, with severe injury for exposure lasting seconds, minutes, or hours, or even with a single ingested dose. In continuous or repetitive exposure situations, lasting days, months, or years, where the skin or mucous membrane suffers irreversible injury, with a threat to life, these substances are classified as *localized chronic* effects. The same irreversible effect or one that causes death occurs when substances are absorbed into the body by inhalation, ingestion, or through the skin, with exposure lasting days, months, or years. In this case, they are classified as *systemic chronic* effects.	Highly reliable and constantly updated data banks are available at research sites with subscriptions. They help in the search for information on effects for classifying toxicity potential of solvents and their raw materials.

Source: Data Banks - http://www.portaldapesquisa.com.br/databases/sites[1]

The table gives a summary of the most important information in relation to the evaluation of product hazard characteristics and can be applied to any solvent or solvent system on the market. If the solvent shows a hazard potential, there is a risk in its usage; and that is why hazard data knowledge is central to the risk evaluation process in a solvent production or application area. This concept has a fundamental importance in field evaluation planning, which will be discussed in detail later in the chapter.

A controversial consideration, but important in the solvents segment, is how results from toxicity cause-effect of products based on lab species experiments is extrapolated to human beings. For many years, toxicity of different chemical compound classes has been the object of exhaustive studies resulting in a large database. These databanks of information that indicate concentration limit levels, based on which of those substances will produce toxic effects on human beings known as Exposure Limit Values (ELV). At this point similarities between published scientific information in the databanks by different scientists and/or organizations of worldwide recognition should be considered. Publications on field studies carried out for years on employees directly in the production areas of solvents strongly add to the information obtained from live experiments of other species as comparison data sources. With all this information, in order to conclude toxic potential of a solvent and its effect, it is necessary to consider the *severity* of the effect, which can vary from contact dermatitis (simple skin irritation) to cancer. Additionally, data on the reversibility or irreversibility of the effect needs to be considered because in many cases a toxic effect may be severe and totally reversible. Additionally, there are substances known for being severe only in high concentrations of exposure, but with totally irreversible effects. Thus, these concepts are important for evaluation of a scientific publication on the toxic level of a product and its effects on humans.

10.3.2. Field Evaluations – Concepts and Practices

10.3.2.1. Fundamental Considerations

A hygienist or expert in the specialized area of HSE planning field work in which the objective is to evaluate solvent impacts in the production and application areas in any segment should remember that knowledge and exposure control is the first rule of occupational hygiene, according to Paul Hewett, (Ph.D. and Certified Industrial Hygienist) who formulates the following considerations in order to guarantee an evaluation using a sustainable approach:

- Monitoring Exposure;
- Management of Hazardous Material;
- Engineering Controls;
- Administrative Controls;
- Individual Protective Equipment;

- Medical Alert/Standby;
- Epidemiology
- Education and Training.

In addition to these important aspects of field planning, those that form the level of knowledge of the product being analyzed are added since this quantification has a systemic dimension that includes the direct effect on the exposed individual and the environmental conditions in the area of the product being evaluated. For this reason, field work requires an expert to first look at data on:

i. Valid toxicity for human health;
ii. Potential environmental impact.

Test procedures for toxicity need to follow protocols found in the OECD (Organization for Economic Cooperation and Development) Guide for Product Testing. Below are some examples of tests described in this protocol:

- Acute Oral Toxicity Test (TG 401);
- Acute Inhalation Toxicity Test (TG 403);
- Acute Dermatological Toxicity Test.

If there is enough information on the toxicological properties of individual mixture components, it is assumed that toxicity for the mixture is given by the total of the toxicities of the individual compounds.

Sources that can reliably give this data are:

- HSDB – Hazardous Substances Data Bank;
- CHRIS – CHemical Hazard Response Information System;
- IRIS – Integrated Risk Information System;
- NIOSH – National Institute for Occupational Safety and Health;
- MEDITEXT® – Medical Management;
- OHM/TADS – Oil and Hazardous Materials/Technical Assistence Data System;
- RTECS – Registry of Toxic Effects of Chemical Substances;
- REPROTEXT® System;
- TOMES PLUS;
- GUIDE TO MANAGING SOLVENT EXPOSURE (ESIG) – European Solvents Industry Group.

From an occupational point of view, employees in the industry sector are the most vulnerable to chemical product exposure, and solvent industry employees are especially vulnerable to contamination through skin absorption and respiratory track. To protect them from occupational risk, the American Conference of Governmental and Industrial Hygienists (ACGIH) annually publishes Exposure Limit Values (ELV), giving permitted maximum concentrations exposure for an employee in the work area, and specific amount of time in a workday. The ACGIH is a nongovernmental scientific association that establishes standards, organized by committees analyzing and compiling published data from scientific literature and publishing orientation guides called TLVs® (Threshold Limit Values) and BEIs® (Biological Exposure Indices). These are used by industrial hygienists when making decisions for safe exposure levels for various chemical products, including solvents and physical agents found in the work environment. Using these guidelines, hygienists should be aware of limit values used with other multiple factors to be taken into account when evaluating a work area and its inherent conditions. A positive measure for multidisciplinary experts in Brazil, including the area of occupational hygiene, toxicology, medicine, and epidemiology, is that the Brazilian Association of Occupational Hygienists (ABHO) promotes the translation and publication in current annual editions of the TLV® and BEI® guides.

TLV® substance concentrations are expressed in parts per million or milligrams per cubic meter in which the worker can be exposed without health risk in a workday. These values are only applicable in work spaces and cannot be used as air quality standards for the public. To interpret the information in the orientation guides correctly, as an ACGIH reference for exposure limits, the three different TLV forms need to be explained:

- TLV-TWA (TLV Time Weighted Average): the average concentration of a chemical product that a worker can be exposed to safely during an 8-hour workday, 5 days a week;
- TLV-STEL (TLV Short Term Exposure Limit): the permitted exposure for an individual for 15 minutes, and a maximum 4 times a day, with a minimum interval between exposures of 60 minutes;
- TLV-C (TLV Ceiling): the concentrations than can never be exceeded.

The BEIs correspond to another form of evaluating chemical compound exposure limits and are related to the permitted maximum quantities of a chemical product in the blood, urine, or air breathed by exposed employees.

The acceptable concentration standards published by the American Standards Association (ASA) are almost the same as the TLVs®. According to the ASA, these standards are designed for prevention:

i Undesirable changes in the structural concept or biochemistry of the body;
ii Undesirable functional reactions, many times without perceptible effect on health;
iii Irritation or other adverse sensory effects.

For gas and vapors, the Tolerance Limit Value is also expressed in parts per million (ppm) – parts of gas or vapor per million parts of air.

10.3.2.2. Field Measurements – Considerations for a Sustainable Evaluation

Independent of the standard being used, the ACGIH (European), or the ASA (American), the Brazilian Occupational Hygienist expert may adopt the internationally applied guidelines in its sample and evaluation planning, however, Brazilian Work Safety and Medicine legislation, which orients and establishes the NRs (Normas Regulamentadoras – Regulatory Rules), must be followed. These are current regulations found in Law No. 6.514 of December 22, 1977 and are available in regularly published and updated manuals.

The main premise for field planning is to know and control all exposure and for this, there needs to be a review of all processes and materials that are going to be involved in real exposure conditions based on systematic documentation of data, at least qualitative, given by the production areas involved in the measurement. The workforce as well as job positions, functions, and attributions in the location being studied also need to be known. When planning field work, most important is managing uncertainties of unexpected changes at the level of studied product effects on human health. Characteristics of the evaluated work area in which the procedures are directly related to those executing them are referred to, should be observed to establish, in justified situations, Homogeneous Groups of Exposure. The Homogeneous Groups of Exposure are groups with the same exposure profile based on similarities and frequency of work they execute, as well as the materials and processes with which they work. In this context, the analysis laboratory is at the same level of importance, where sample procedures and applied analytical methods should be validated and reliability proven using official quality control programs. World-recognized agencies that publish air sample and analysis methods are:

i *National Institute for Occupational Safety and Health* (NIOSH);
ii *Occupational Safety and Health Administration* (OSHA);
iii *Environmental Protection Agency* (EPA).

In other words, to guarantee a sustainable approach and success for characterization of field work, integrated groups of actions are summarized by what consultants and experts from the AIHA [American Industrial Hygiene Association (*Drs. Hewett, Paul e Mulhausen, John R.*)][2] have called Factors. The following table lists these factors in successive steps taken by the hygienist in the field.

Table 10.1. Exposure Evaluation Strategy

Factors Related to Work Areas	Objectives	Understand process and material flowchart Obtain chemical and physical process information inherent to the locale being evaluated (process description/chemical) Understand process flow arrangement Identify potential emission and control points Locate the work that handles materials Work practices and procedures (interview and observe employees/complexity/workday/routines/assistance/shift rotation) Transporting/moving materials
	Other Considerations	Types of building structures (plants) Energy sources (equipment/evaluation) Technical support (engineers/chemicals, etc.)
Workforce Factors	Objectives	Understand work division and practices Exposure standard profile (positions and work) Exposure frequency and duration (routine vs. sporadic work) Skin contact and respiratory tract potential (aspersions/depressurizations/projections)
	Other Considerations	Know factory personnel/organizational flowchart/work position description Management-level interviews Detailed review of work locale/critical procedures (job analysis)
Environmental Agent Factors	Objectives	Identification of agents (chemical/physical/biological – inventory review) Potential effects on health (MSDS) Manipulated quantities
	Other Considerations	Raw materials Intermediary products and additives Laboratory chemical substances (reagents) Harmful residuals Physical agents (noise/vibration/radiation/extreme temperatures) Biological agents (pathogens) Use of background records before exposure
	Establishment of Exposure limits	Occupational Exposure Limits (LEs) Regulation of Exposure Limit (established and regulated by governmental agencies – $NRs_{(BRAZIL)}/OSHA_{(EUROPE)}/EPA_{(USA)}$) Licensed Exposure Limit (defined by scientific organizations – NIOSH and $ACGIH_{(GLOBAL)}/AIHA_{(USA)}$) Internal Exposure Limit (defined by private entities) Occupational Exposure Limits (defined by occupational hygienist)

Table 10.1. (continued)		
General Information Sources	Objectives	Interview with medical and safety teams Knowledge of current rules and standards
	Initial Prioritization	Resource used but optional for choosing areas or agents for additional evaluation. Useful when there is an evaluation resource limitation or when the evaluation for operations is done for the first time. In these cases, the following should be considered: Agent type (e.g.: only chemical agents) Is there a regulated exposure limit? Is there a consensus rule (e.g.: ACGIH TLV)? Agent toxicity (e.g.: level of influence on health stipulated in specialized literature) Previous sample data
	Documentation Approaches	Informal (experiences/ daily living situations/related factors) Formal (documented studies and measurements) Relationship (verification/questionnaire lists) Databanks (available spreadsheets and documents) Videos (audio-visual documentation)

Source: Strategy for Assessing and Managing Occupational Exposures; Mulhausen, J.R Ph D/CIH) and Damiano, J.-MS/ CIH).Second Edition AIHA-Press.[2]

At this point of the description on the practice for a reliable and sustainable field approach, there are important elements to be noted in the occupational hygiene practice, both in obtaining data as well as its correct interpretation. From the operational perspective, there needs to be some definitions established of the activity in the field.

- **Exposure profile**: the situation of the work location being evaluated using all available information at the qualitative, semi-quantitative, and quantitative level.
- **Exposure classification**: the mathematical means for the Exposure Profile. It is classified as the greatest level of exposure; the value that is above the long-term average, followed by those between 50 and 100% of the long-term average, then the range of 10 to 50% exposure risk long term average, and finally, those that are below 10% long-term average as the lowest exposure risk.
- **Potential effects on health**: adverse effects that chemical, physical, or biological substances may cause on a live organism. This is a discussion focusing on occupational medicine questions

In environments where solvents are handled, it is important to consider basic principles for air samples. The active sample is a collection process of the contamination of interest in an appropriate system. A sampling pump aspirates the air, allowing the concentration or collection of the chemical agent to be maintained in the collected sample. This procedure is accepted and practiced throughout the world. The range usually used is from 1 to 5,000ml/min which is sufficient to be used as a an individual (person) sample and a set environment/point sample. The most commonly used configuration in gas, vapor, and aerodispersoid collections is the active collection using adsorption where the air pump is connected to small glass tubes usually filled with an adsorbing material, in which its methodology specifies the type to be used for each chemical agent of interest. The most common are: *Activated Charcoal; Silica Gel; Tenas; XAD-2 Ionic Resin; Chromosorb*.

There are several sample methodologies, with the most important being selection criteria based on knowledge about the chemical agent and its relation to the environment that is going to be evaluated.

- **Human Factor**: the interaction of the operator with the product, no matter the level of contact at the location, involves knowing the risks, level of demand, and behavior as an inherent factor of each individual;
- **Environmental Factor**: location conditions like ventilation, temperature, and air humidity; equipment conditions and preventative measures;
- **Technical Factor**: normalized practices found in procedures characterize homogeneous groups using exposure risks, guarantee speed, collection capacity, and sample stability;
- **Analytical Credibility**: methodologies that guarantee exact, precise recovery of the chemical agent(s) of interest and dominion of the interferences.

10.3.2.3. Field Results – Interpretation and Acceptability

Each measurement where the human factor has a relevant potential, as well as instrumental quality and reliability, presents a challenge. The success of the results requires competencies and transparencies on the part of the occupational hygiene professional in relation to the areas involved. This premise supports the requirement of selective judgement at the time of exposure classification since data obtained from the field should be grouped as *Acceptable, Uncertain, or Unacceptable*.

Basic considerations for this are grouped for a better understanding as follows:

- Acceptable Exposure
    - When exposure and variability are low enough for the risks associated with the exposure profile to be low;
    - When obtained values are higher than the exposure range and below the EL (Exposure Limit) uncertainty range, noting the specialized literature;
    - When high reliability of the exposure profile data allows greater proximity of the EL. The inverse is true when there is lower reliability in the field data, the distance to be covered will be higher for it to be considered an acceptable evaluation.

- Unacceptable Exposure
  - Generally, unacceptable exposure is the average exposure or a set of data in the range higher than the exposures, which is no more than the group of data that is found in the range below the EL uncertainty;
  - When there is evidence of adverse effects for health associated with an agent (this judgement is common in cases of thermal stress or dermatitis);
  - Data obtained from measurements where other data appear to be associated with community complaints, legal responsibility, regulatory observance requirements, or pure expert perception (ethics and sensitivity);
  - The control observation of unacceptable exposure situations is more relevant than the collection of additional data, even though this can also be useful.
- Uncertain Exposures
  - Exposures are classified as uncertain when the exposure of the evaluated group cannot be classified as acceptable or unacceptable;
  - When the exposure profile is not appropriately characterized due to lack of data on medical effects, it implies difficulties in determining EL;
  - When uncertainty is due to exposure variability and inadequate statistical techniques chosen for dealing with the data.

The summary shown below shows the continuous improvement cycle in managing changes throughout the field evaluation period.

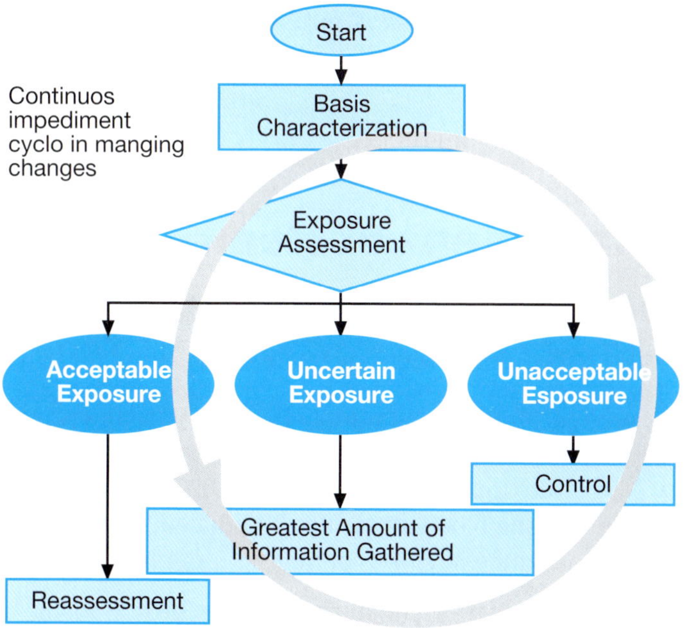

Figure 10.1. Source: Strategy for Assessing and Managing Occupational Exposures; Mulhausen, J. R Ph D/CIH and Damiano, J.-MS/CIH. Second Edition, AIHA Press.

In field work, data analysis can only be duly carried out when it is based on correct statistical tool choices. The premise that statistics should be effective in relation to the execution as well as in the judgement of data becomes more apparent for the hygienist.

Independent of the statistical method used in field data interpretation, attention should be given to the information leading to a conclusion on exposure limit in an evaluated area. There are many published studies carried out over long periods of time; notably that by the American Industrial Hygiene Association – AIHA experts. These statistical methods work with the extreme high part of the Gauss curve, i.e., in the area where the concentration of data should be in at least 95%, as well as the Upper Tolerance Limits. Maintaining the focus on the mathematical average and the reliability interval also positively adds to the Exposure Profile.

The occupational hygienist's objective is to estimate the exposure limit and its variability within the limits of reliability thus allowing good judgement on acceptable conditions of exposure.

In summary, this level of results is obtained when the specialist includes a large range of information that helps at all stages of the field work, whether the information is from personnel involved in the process being evaluated; from existing material and product characteristics in the work area; from the reliability of adopted field sample methodologies for collecting samples, as in the level of sample product toxicity knowledge; from the characteristics of the involved workforce; from applied statistical tools; or from determined exposure limit reliability.

Everything mentioned here is called the Critical and Systemic Vision of the professional: a condition for responsible action from planning to choosing the appropriate tools for samples and statistical analysis of the data, culminating in the practical applicable meaning like personnel experience and continuous formation and information.

10.4. Management of Health Aspects

10.4.1. Solvents in the Context of Human Health – a Hygienic View

When addressing solvents in the occupational health environment, the focus of chronic effects is seen as potential due to their physicochemical characteristics resulting in the possibility of volatile organic compound (VOC) exposure. VOCs can be absorbed through the skin into the blood stream or be inhaled. These entries will become the means of greater relevance in close contact with the human body.

In the physiological sense, a substance is considered absorbed when it has penetrated into the blood stream and as a consequence has been carried to parts throughout the body. Something swallowed and subsequently excreted in the feces does not mean it was necessarily absorbed, even though it remained in the body for hours or even days. The big motivator and challenge for the health professional working in an industrial setting is to understand the relationship mechanism between the product and the bio and physiological reactions of the human organism. Before going further, some concepts are necessary to understand the mechanism called absorption.

- **Absorption through the skin**: This is directly related to contact with organic solvents. Clinical accompaniment (seeing patients and looking at laboratory analysis) allows companies to detect how much of a compound can get into the bloodstream through the skin after direct accidental contamination or from drops falling on to clothing. The use of industrial solvents to remove grease and dirt from hands and arms is frequent, and it is difficult to control this behavioral problem, which can lead to potential sources of dermatitis.
- **Inhalation Absorption**: Publications for medical services at work show more than 90% of all health effects related to the industrial sector, excluding dermatitis, can be attributed to inhalation absorption. There are several hazardous substance forms in the air (cloud and vapor), attributed to solvents that mix with air being and behave like true gases. It is not difficult to calculate how much of a product causes a threat when we establish a scenario of how much an individual working normally is going to inhale in 10 to 12 cubic meters of air during an 8-hour workday.

It is important to understand that not all that is inhaled is subsequently absorbed because part is immediately expelled and another part becomes trapped in the trachea mucus. A part of what is in the mucus is eliminated from the body and part is swallowed, creating easy absorption through the stomach. Volatile gases emitted from several substances and products with the intensity in the case of solvents, travel through the respiratory track freely to the lungs and from there enter the bloodstream as previously described. Due to this, a large majority of known industrial solvents can contaminate the air at a given moment; and there is a number of occupational accident prevention programs directed toward ventilation technologies as hazard reducers.

Another important term is *individual susceptibility* used by authors in different lines of industrial health and hygiene publications. This is not just a definition but also a concept for a reaction to an adverse cause-and-effect since under similar exposure conditions with potentially hazardous effects, individuals present different characteristics. Some may not show evidence of intoxication; others show mild signs, while other have severe or even fatal reactions. The latter may be a characteristic genetic response or a particular anatomic one. For example, the nasal cavity, lung permeability, and other parts are what are called susceptibility in treating causes of occupational health problems. Physiological information added to abnormal functioning of the liver, the organ that detoxifies and secretes dangerous substances, and knowledge of workplace medicine define the level of an exposed individual's susceptibility.

The health professional in the chemical industry environment frequently deals with exposure problems usually resulting from accidents from a high concentration of toxic products. Acute effects generally found with solvents are unconsciousness, shock or collapse, severe lung inflammation or even sudden death, where understanding the aggressive agent is a fundamental importance in the doctor's action in attending these affects. Chronic effects require the professional to have proof of what caused the effect, i.e., evidence of the hazardous agent being found in significant concentrations that the patient had absorbed and the individual's physiological reactions revealed certain experimentally proven disturbances. These toxic effects caused by absorption can be proven using blood and urine analyses. The metabolites, for example, excreted in these fluids is evidence; however, it is important to note that effective proof of an individual being under the effects of toxic product absorption in higher concentrations than those of the personnel who were not exposed is needed. Additional proof that the disturbances were also diagnosed using other occupational medicine procedures, like medical history, physical exam, x-ray studies, or other proven instrumental techniques, in addition to blood and urine analyses which form link-causal diagnosis.

In the solvents industry, benzene stands out due to its scientifically proven chronic toxic effects on vital organs. Small quantities can be detected in the blood immediately in the first stages of human contact. However, many other products, not only solvents, can produce visible effects after long periods of exposure, thus putting individuals on alert, and this does not include those substances which are accumulative as in the case with some metals like lead and mercury.

A controversial question, which is important and no less common for the occupational health professional, is the proof in legal medical problems where the specialized opinion of the doctor weighs in labor decisions in the importance of real understanding by the professional of the differences between *related cause* and *possible cause*. Possible cause is that which probably could produce a hazardous effect. It involves the *possibility* and this subjective characteristic requires the doctor to have complete understanding of toxic effects of products manipulated in the industry. Related cause only exists if a probable cause effectively produced hazard effects. That is, the probability is a tool but decisive information comes from the accredited databases for Health and Hygiene. What remains as a contribution in these cases is the importance of the relationship between the doctor and the hygienist since medicine is not an exact science and it needs to be based on facts and observations, where sustainable practice of hygiene adds value to medical opinion.

10.4.2. The Analysis Laboratory as Medical Diagnostic Support

From the hygienist's point of view, a medical diagnosis is not complete based solely on one observation or test. The full scenario for diagnosing the problem is in the history the doctor completes on the patient including physical exams and lab tests that after interpretation complete the diagnosis.

For reasons specific to the occupational health field cited above basic steps in this description are emphasized in order to develop a laboratory diagnostic medical plan that does not follow the same logic as exact sciences.

1) Lab Test Specification

   Industrial toxicology lab methodologies and procedures can be divided into two main classes: those applied to humans and those applied to the environment. In this section, we will only discuss those applied to humans. The human organism produces adverse responses to reactions against external toxic substance actions as much as it does to the processes of harmful effects that occur on the inside of the body. A blood test can reveal the presence of contaminated cells from an external cause, such as cellular disorder, which is typical in leukemia and anemia. This is why when choosing a lab test it is important to know that very few of them are selective and specific enough to conclude a diagnosis with only one positive result.

2) Normal Values

   Values in an exact science have only one absolute meaning in what is being proposed, as for example, physicochemical constants. In biological science, it is not that simple of a question in relation to analytical values because blood and urine exams are not given as unique results nor are they mentioned as normal fixed values. There are only normal ranges in which understanding is fundamental when eliminating erroneous diagnoses in considering whether a premise is true: in not knowing what is normal, it is impossible to know what is abnormal.

3) Data Tendencies and Quantities

   In hygiene, the number of acceptable and sustainable data determines the success of a field evaluation. In occupational health, this is no different since an isolated lab test is almost always inconclusive when deciding the range of normal and abnormal values. Obtaining data throughout a certain period establishes a perceptible meaningful tendency, which is better than any single test result where values are not normal. The diagnostic concept on data quantity, also known as data mass, leads the professional to a view that averages from values of a group of employees in a common exposed environment is different from a similar group that is not exposed to a determined environment. This reference is not always decisive in revealing whether an individual or group was affected, but it does show that there is an excessive exposure of harmful agents in a location and this means corrective measures need to be taken. The same reasoning for data tendencies and quantity is applied in accompanying individual cases.

4) Critical Analysis of Observed Data From Field Work

There are large amounts of data published in specialized books obtained from laboratories with recognized performance. What is called to our attention is that the health professional cannot conclude an exposure diagnosis based only on information. Knowledge and control of observed data is a recognized practice in the scientific world and is applied in biological experiments, minimizing erroneous conclusions. Within the industrial scenario, when the occupational health area is monitoring a group indicated by the hygienist, where risk exposure exists, a group that is not subject to contamination should also be observed using the same parameters. This practice eliminates doubtful responses on the final diagnosis of the risk group being evaluated.

5) Simple and Complex Exams

There is a tendency for personnel submitted to exposure monitoring exams to have more confidence in the analysis results from sophisticated instrumental exams than those using simpler ones. Industrial toxicology does not always manage to diagnose causes and effects based on sophisticated analyses, and ignoring simpler tests leads to committing errors that are much more harmful to the exposed individual or group's health

6) Most Common Errors

In diagnostic processes of hazardous product exposure effects in humans using lab exams and analyses, the demand for good practices is an important reliability factor. Following is a list of some of the many error possibilities that lead to problems in medical evaluations.

- Errors in Collecting Material

  The most subtle change in collecting material can change the test result. This fact is extremely relevant, mainly in situations where quantitatively contamination species are evaluated. An example includes heavy metals, residual from solvents like benzene, toluene, and others in the blood.

- Cross Contamination

  Sample and analysis reagent handling, storage, and moving procedures in the lab can be a cause of contamination by other stored compounds that are not being analyzed, or there can even be an accentuated result of a species being evaluated just because of its presence in the lab environment.

- Substance Expiration

  It is always desirable to analyze the collected material as quickly as possible. Physical, chemical, and morphological changes can occur in relation to waiting time of an analysis since samples are usually taken from live systems. Preservation methods are acceptable, but there must be great care in these procedures such that they do not need to have chemical conservatives added that risk giving a false result.

- Procedural Errors

  Independent of the level of knowledge the technical team involved in the analytical evaluation may have, procedural errors are inherent in the routine; the team's self-confidence in memorizing standards, and the calibration of instruments used in the lab.

- Data Transfer Errors

  In addition to an error considered high-risk, for example, using the wrong sample, other errors can occur due to distractions while writing or typing like transposing numbers, or transposing data while transferring it from the spreadsheet analysis to the results sheet. Therefore, it is importance to include in the internal organization, a reviewer who has a decisive function in the laboratory's quality and credibility.

- Behavioral Conduct Errors

  The lack of organization and cleanliness and dishonesty in reporting anomalies during analysis procedures are contributing factors to false results. Laboratory accreditation programs in accordance with ISO standard guidelines promote the organization and the rigor of following good practices in the lab's daily routine. This is reflected in the quality of services demanded by the users.

The laboratory is a valuable resource in the context of a prevention and monitoring program for human health risks; however, it should never be a substitute for a medical diagnosis. Whenever there is a conflict of opinions, a step-by-step review of procedures and possible deviations in laboratory results will always be necessary.

10.4.3. Clinical Tests Used in the Occupational Health Practice and Their Meanings

Under this topic, we present some of the more common lab tests in diagnosing health problems caused by solvent exposure using a less exhaustive and easy visualization table format, which can also be used in assessing other compounds

Table 10.2. Overview of Laboratory Urine Exams

Test	Normal Value Range	Meaning of Test Result
Color	Pale to dark yellow	Clear: low specific gravity Dark: high specific gravity
Cloudiness	Clear for recent sample	Cloudiness should not be considered an indicator for abnormality
Acidity	pH 4.8 to 7.5	Amount in recent sample is acidic (pH change to base is response to decomposition)
Density	From 1.001 to 1.030 g/ml	Density is related to liquids entering. Certain kidney illnesses maintain density at 1.010 g/ml.
Sugar (glycose)	Not detected	Foods rich in carbohydrates can lead to sugar in the urine, but its presence does not mean diabetes or dysfunction due to contaminant exposure
Albumin	Detected using specific methods 2 to 8 mg/100 ml	Presence of albumin denotes renal illness. Appears after long periods of post-surgical stagnation or traumas
Red globule particles and leukocytes	0 to 9,000/12 h 0 to 1,500,000/12 h 32,000 to 4,000,000/24 h	Only some particles and red blood cells are expected in normal samples of urine tests
Crystals	Traces (qualitative) 0.001 to 0.010 mg/100 ml (quantitative)	Can increase its concentration in lead exposure
Dissolved solids	Common in the range of 1.002 to 1.020 g/L	It is a very quick, simple method that is useful in measuring renal function
Capable in coating excretion	15 min. 30 to 50% 30 min. 15 to 25% 60 min. 10 to 15% 120 min. 3 to 10% max. 70 to 80% after 2 hours	Values are based on intravenous injection of phenolic sulpho phenalate. Less than % excretion could mean low renal function

Source: Data Banks – http://www.portaldapesquisa.com.br/databases/sites.

leading to toxic effects. Table 10.2 shows a panorama of urine lab exams of the most important diagnostic tests that give a quick answer to contamination by reflecting the vital functioning of kidneys and/or other parts of the urinary tract: urethra, bladder, and glandular structures. These reveal problems that allow quick and effective medical intervention related to effects from some solvents mentioned in published studies.

Solvent Examples	Principales Características de Anomalía	Diagnóstico
Benzene	Presence of red blood cells	Severe poisoning with internal bleeding in the urinary tract
Methyl Bromide	Albumin	Renal Damage
Methyl Chloride	Albumin	Renal Damage
Chlorobenzene	Albumin, red blood cells, and dark coloration	Renal Damage
Chloroform	Difficult to detect this species using simple analysis. Requires more sophisticated techniques like chromatography	Renal dysfunction is shown but it is questionable (medical observation is an important
Dichloroethyl Ether	Undetectable using common techniques	Renal damage recorded only in lab animals. Questionable (Medical Observation is a differential)
Carbon Dissulfide	Albumin	Renal Damage
Glycols	Red blood cells and decrease in renal function	Answers occur after ingestion
Nitrobenzene	Albumin and red blood cells	Renal irritation and color becomes darker
Carbon Tetrachloride	Presence of albumin, crystal particles, blood cells, and reduced function	Typical reactions of renal damage

Table 10.3 shows some blood components and a summary of hematologic abnormalities that may be caused by occupational exposure to some solvents. In reality, there is nothing specific on blood exams that allows a conclusion to be made about anomalies that may possibly appear in an employee exposed to a toxic agent and that show the exposure contributed to that anomaly. Blood exams are generally costly, demand

Table 10.3. Overview of Laboratory Blood Exams

Test	Normal Value Range	Meaning of Test Result	Solvent Examples	Principal Characteristics of Anomaly Cause	Diagnosis
Red Globules	Men 4.5 to 6.0 million/cmm Women 4.5 to 5.0 million/cmm	U.S. applied values	Benzene	Decreases all blood forming elements	May cause death by Bone Marrow Depression
Hemoglobin	Men 14 to 18 g/100 cc Women 12 to 15 g/100 cc	values expressed in % are insignificant	Methyl Chloride	Decrease in all blood forming elements	What is known is only through animal experiments
White Globules	Total 5,000 to 10,000/cmm	Total count varies in hourly rate. Needs statistical basis to guarantee differential	Carbon Disulfide	Shows anemia and leukocytes with drop in white globules	Information in conflict with medical literature. (medical observation is an important diagnosis)
Mean Corpuscular Hemoglobin I	27 to 32 micro-micrograms	Hemoglobin mean per cell	Ethylene Glycol Monomethyl Ether	Decrease in all forming elements and an increase in immature white globule percentage	Diagnosis based on observations of humans
Blood urea	20 to 35 mg/100 cc blood	Increase in kidney illnesses	Nitrobenzene	Causes red globule reduction	Blood degenerative process
Cholesterol Total	150 to 250 mg/100 cc blood	Increase in biliary obstruction and a decrease in liver illnesses	Carbon Tetrachloride	Anemia and leukopenia	Diagnosis questioned (important medical observation)

Source: Data Banks – http://www.portaldapesquisa.com.br/databases/sites.

a lot of time, and may interfere with allowing other exams to be carried out. This is a problem when dealing with a large number of employees. However, regular exams for hemoglobin count and blood smears are acceptable for the majority of abnormalities.

Contrary to urine and blood analysis, hepatic problems may give false negatives when using lab methodologies, even when there is considerable damage. In other words, we need to understand that analytical tests for revealing liver problems have a small range for detecting initial effects caused by chemical exposure. A large number of causes can lead to abnormalities in the liver, but it is always complicated to determine the factor of this cause. Table 4 lists the commonly applied tests for hepatic functions and is followed by a small table of the solvents that may produce these liver dysfunctions, depending on the level of the exposure, resulting in irreversible damage to this vital organ.

Table 10.4. Liver Function Tests and Potential Solvents List

Test	Expected Normal Values	Comment
Bromine Sulpho Naphtalate	Levels below 5% retention after 45 minutes	Illness in liver tissue, biliary or circulatory obstruction, responsible for high retention
Cephalin Flocculation	0 to 1 + flocculation in 48 hours	Increase in flocculation reveals liver or tissue illness associated with protein abnormalities in biliary serum
Biliary Serum	Total 0.2 to 1.0 mg% Direct – 0.1 to 0.7 mg% Indirect – 0.1 to 0.3 mg%	Changes in these concentrations indicate hepatic abnormalities
Thymol Turbidity Test	0 to 4 Maclagan units	Changes in the range indicate liver or tissue illness
Urobilinogen Urine Test	0.5 to 2.0 mg/24 hours or dilution of 1:4 to 1:30	Changes may indicate bile obstruction abnormalities or liver tissue illness
Solvents that can Produce Liver Function Abnormalities		
Methylene Chloride* Carbon Disulfide* Chloroform	Diethylene Dioxide* Nitrobenzene* Carbon Tetrachloride	Ethylene Di-chloride Trichloroethylene Methyl Bromide Trinitrotoluene*

* The scientific community has published positive cause/effect results for liver illnesses.

Source: Data Banks – http://www.portaldapesquisa.com.br/databases/sites.

Special lab tests should be considered by the occupational medical attendant as an arbitrary evaluation tool for diagnosing an occupational illness because it may turn out to be non-occupational. To conclude an illness has a cause/effect outside the work environment, there should be solid, reliable technical arguments. Examples of illnesses with non-occupational origins include inhalation of toxic gases produced by non-regulated heaters or refrigerators; ingestion of poisonous substances from food, or poisonous material from the use of products like turpentine; thinner; etc, in home use for coating dilutions used with organic solvents.

The data presented in the above tables are not exhaustive; there are many tests that help in the medical diagnosis found in specialized literature. The objective here is to show that the occupational health professional currently counts on the analytical support that collaborates the diagnosis and creates a quantitative history that is important in the industrial employee's professional profile.

10.4.4. Carcinogens/Mutagens/Toxins in Human Reproduction - CMR

There is a large-scale world preoccupation with substances that have potential effects like carcinogenic tumors. genetic mutations, or reproductive changes, which are being gradually replaced with other less hazardous types.

It is worth mentioning the classification of a product or substance as CMR – Carcinogenic, Mutagenic, and Toxic to human reproduction, with all its complexity and diverse scientific information in the electronic network world. We usually look at controversial cases when dealing with this subject, where the hygienist, toxicologist, and occupational medical practitioner make decisions and look for highly reliable scientific field data and databanks.

This section starts by focusing on carcinogenic agents defined as those chemical compounds classified by the IARC (International Agency for Research on Cancer) as:

- Group 1 – carcinogenic for humans;
- Group 2A – probably carcinogenic for humans;
- Group 2B – possibly carcinogenic for humans.

Mutagenic agents are those that cause permanent changes to the genetic structure of a living organism and that are also listed in the same manner as the carcinogens. The same is true for toxic agents for human reproduction or teratogenic agents, which, as the name implies, are agents that cause changes in the functions or capacity of living organism reproduction, mainly in men and women.

If we use the line of classification for CMR according to intensity of the effect, we end up using a mathematical mechanism that is usually difficult to understand. Therefore, the preferred classification is descriptive, based on cross information extracted from databases that meet the following definitions.

1. No evidence of carcinogenicity, teratogenicity, or mutagenicity;
2. Carcinogenic, teratogenic, and mutagenic effects confirmed only in animals;
3. Suspect of being carcinogenic, teratogenic, or mutagenic for human beings;
4. Carcinogenic, teratogenic, and mutagenic effects confirmed only in human beings.

In dealing with CMR agents, what is relevant is the efficacy in using managing tools, i.e., in controlling all products, raw materials, and reagents that are potential CMRs. This knowledge is directed at actions to control the employee hygiene and health environment as well as to protect the environment and R&D guidelines focused on replacement; technology implementation of protection points for employees; and the improvement of Safety Data Sheet (MSDS) content.

10.4.5. Neurotoxic Effects

Appropriate data banks previously mentioned give in vitro and in situ test information on experiments and occupational follow-up throughout the years, which announce Central Nervous System (CNS) potential effects. Considering GHS criteria, available information, and proposed and published guidelines in the NIOSH, it is possible to classify solvents as neurotoxic, depending on their CNS potential. It is a support tool for manufacturers and users in protecting human health and product development with less risk.

Table 10.5 presents the criteria for classifying solvents in relation to CNS potential effects in occupational exposure, according to the NIOSH Classification Guidelines under official publication number 87-104.

In relation to the effects from employee exposure, solvents that have a neurotoxic reaction on the CNS generally are classified as Type 1 and Type 2.

Table 10.5. Inductor Solvent Categories in Central Nervous System Disturbances		
Severity	Identification According to World Health Organization (WHO) Work Group – Copenhagen, July 1985	Classification According to Workshop International of Solvents – Raleigh, NC, USA. – October 1985
Minimum	Emotional Syndromes	Type 1
Moderate	Chronic Cerebral Illnesses with medium toxic effect	Type 2A or 2B
Pronounced	Chronic Cerebral Illnesses with severe toxic effect	Type 3

Type 1 – disturbance characterized by fatigue, weak memory, irritability, difficulty in concentrating, and slight stability disturbance.

Type 2A – stability or sustained mood change as well as emotional instability; and impulse control and motivation decrease.

Type 2B – intellectual response function manifests as decreased concentration and memory capacity.

Type 3 – characterized by deterioration of global intellect and memory functions (dementia) that may be irreversible or only somewhat reversible.

Source: WHO 1985/Baker and Seppalainen 1986. (*Current Intelligence Bulletin 48* – March 31, 1987).

10.5. Management of Environmental Aspects

10.5.1. Solvents and Their Eco-system Behavior

The word "solvent" comes from the Latin, solventis, and is commonly used to designate any substance that has the power to dissolve other substances. Solvents may be organic or inorganic and within this concept, water is known as the universal solvent.

Solvents are largely used in the most varied branches of anthropogenic activities and can reach a determined environmental compartment (air, water, soil, and sediment), in any stage of its life cycle, whether it is production or during its use, as well as during its stage of destruction. The ease or difficulty for a solvent to reach an environmental compartment and cause an adverse effect is associated with several factors. Among these factors, we include the amount involved, climatic and meteorological conditions of the medium, and the type of compartment affected. However, the dominant factor for partition in the ecosystem is associated with the physicochemical properties of the solvent in question. (Figure 10.2.).

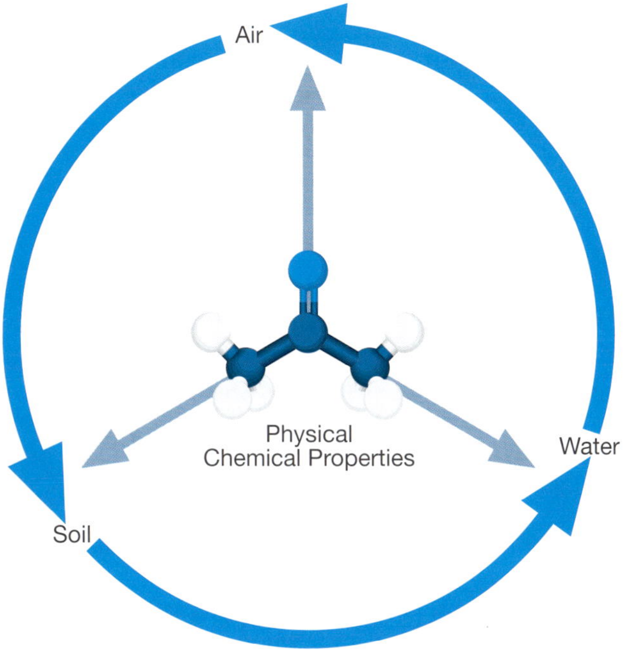

Figure 10.2. Ecosystem Partitions.

Intrinsic characteristics of an environmental compartment may favor a greater or lesser availability for partitioning in the living organism's tissues (biota), in relation to the absorption or adsorption power, which is a necessary condition in expressing an adverse effect.

Behavioral prediction of a solvent, in a determined environmental compartment can be inferred using ecological information obtained using physicochemical properties and/or experimental data.

This part of the chapter will deal with physicochemical properties and the main ecological information used in predicting solvent behavior in the ecosystem.

10.5.2. Physicochemical Properties

A solvent's physicochemical properties are a function of its molecular structure, which is a fundamental piece of information used to understand and predict behavior in ecosystems.

10.5.2.1. Physical State

Solvents in a liquid state, in relation to vapor pressure, can migrate to all other environmental compartments while in a gaseous state its target compartment is the air.

10.5.2.2. pH

The measurement for acidity or basicity is called pH and this allows the evaluation of possible corrosive effects or irritation to the skin and eyes, as well as certain effects on the environment, like corrosion of metal. This is a 14-point scale with 0 being strongly acidic and 14 strongly alkaline.

10.5.2.3. Molecular Weight

Solvents with low molecular weight present greater tendencies of volatility and photodegradation in the atmosphere, while those with greater molecular weight tend to be adsorbed by organic material.

10.5.2.4. Vapour Pressure

Vapor pressure is a volatility measurement of a chemical agent in a pure state and is an important determining factor for volatilization speed starting from contaminated surface soils or bodies of water to the atmosphere. Temperature, wind speed, and soil conditions in a particular location, as well as adsorption characteristics and the solvent's water solubility will affect the volatility rate. In general, a solvent with relatively low vapor pressure and a high affinity for soils or water has less probability of evaporating and reaching the air than a solvent with high vapor pressure and a lower affinity for soil or water.

10.5.2.5. Solubility in Water

This refers to the maximum concentration of a chemical solvent that dissolves in a defined quantity of pure water and in general is in a range of 1 to 100,000mg/L. The solubility of a solvent in water is a function of temperature and the solvent's specific properties. Each individual solvent has a degree of specific solubility. Therefore, different solvents can solubilize in different concentrations in water.

Environmental conditions, like temperature and pH, can influence solubility. Generally, solvents that are very soluble in water have a low adsorption affinity in soils and are easily transported from contaminated soil to bodies of surface and underground water sources.

Volatility is directly related to water solubility; very soluble solvents in water tend to be less volatile. Solubility may be determined using a standard procedure developed by the Organization for Economic Cooperation and Development called OECD 105.

According to Petrus (1995), solubility may be evaluated using the following criteria:

Solubility Range (g/L)	Evaluation
> 1000	Very Soluble
100 a 1000	Easily Soluble
33,3 a 100	Soluble
10 a 33,3	Sufficiently Soluble
1 a 10	Little Solubility
0,1 a 1	Very Little Solubility
< 0,1	Practically Insoluble

10.5.2.6. Density

Density is associated with other properties and can predict the behavior in aquifers. Solvents with low solubility and a density lower than water tend to form a free phase called *Light Non-Aqueous phase Liquids* (LNAPL) that floats on the aqueous surface. Solvents with low solubility and a density higher than water tend to migrate through the different soil layers until they get to an impermeable layer and create *Dense Non-Aqueous Phase Liquids* (DNAPL).

10.5.2.7. Liposolubility

Liposolubility allows the evaluation of a solvent's tendency to accumulate in fatty tissue of living organisms (fish, birds, mammals). It represents the quantity of solvent dissolved in 100 g of standard oil or fat at 37 °C and is expressed in mg/100 g at 37 °C.

10.5.2.8. Partition Coefficient (Kow)

This is specific data for each solvent and is important in estimating ecochemical and ecotoxicological behavior (soil absorption, biological absorption, bioconcentration, and fatty tissue accumulation).

The partition coefficient is a measurement of a solvent between the aqueous phase and an immiscible organic phase used to evaluate the transfer potential starting from an aquatic environment and its possible bioaccumulation.

Live organisms tend to accumulate solvents with high values of Kow in lipidic parts of their tissues. Thus, one way of estimating bioconcentration potential of a solvent is to measure how lipophilic it is. Since it is difficult to measure this directly, the Kow value is used to predict the tendency of a solvent to be distributed in the octanol (fatty representation) and water. The Kow value is directly related to the tendency to concentrate in the biota and is inversely related to water solubility. The Kow can be determined using OECD 107 and OECD 117 standard procedures.

10.5.2.9. Henry's Law Constant

This constant, representing the air/water partition coefficient, takes molecular weight, solubility, and vapor pressure into consideration, and indicates the degree of the solvent volatility in a solution. When the solvent presents high solubility in water and relatively low vapor pressure, it will stay in the water. When vapor pressure is high in relation to its water solubility, the Henry's Law constant is also high and the solvent is volatile. Table 10.15 shows the volatility ranges in relation to the Henry's Law constant values.

Volatility according to Henry's Law constant

Volatility	Value Range (atm m^3/mol)
Non-volatile	Less than 3×10^{-7}
Low Volatility	3×10^{-7} a 3×10^{-5}
Moderate Volatility	1×10^{-5} a 1×10^{-3}
High Volatility	Greater than 1×10^{-3}

Source: ATSDR (1992)

10.5.3. Ecological Information

Ecological information includes information related to probable solvent behavior in an ecosystem, such as mobility, persistence, degradability, bioaccumulation, as well as possible effects on aquatic life and the environment in general (ecotoxicity).

It is important to note that with preparations (solvent mixtures), environmental behavior can be very diverse in relation to each isolated component. The result may be additive, synergistic, or even antagonistic interactions.

Principal information about a solvent in an ecological setting is available in literature or can be inferred using the physicochemical properties, experimentally obtained, and are referred to in items 10.5.3.1. – 10.5.3.14.

10.5.3.1. Mobility

Mobility characterizes the possible exchanges between different environmental compartments and is associated with the following properties:

- Volatility;
- Absorption/Desorption;
- Precipitation;
- Surface Tension;
- Solvent Target Compartment.

10.5.3.2. Volatility

Solvent volatility in an aqueous medium is evaluated starting with the Henry constant (H) representing the partition coefficient between air and water while the soil is associated with vapor pressure (P).

According to Petrus (1995), criteria used to predict behavior may be the following:

- If H ≥ 100 Pa m3/mol, the solvent is considered volatile starting from an aqueous medium;
- If P ≥ 100 Pa, the solvent is considered volatile starting from the soil.

10.5.3.3. Adsorption/Desorption

Adsorption and desorption of a solvent may be estimated evaluating physicochemical properties like the Kow, the organic carbon adsorption coefficient (Koc), or water solubility.

According to Petrus (1995), the adsorption/desorption capacity can be estimated using the following criteria:

- If log Kow ≥ 3 , the solvent will be adsorbed;
- If Kow ≥ 10 m^3/kg, the solvent will be strongly adsorbed;
- If S (solubility) ≥ 1 g/L, the solvent will not be adsorbed, but will easily filter into the soil.

10.5.3.4. Precipitation

Precipitation is evaluated starting with solubility (S) and density relative to water (d 20/4). A solvent with low solubility in water can form sediment, via adsorption in suspended particles or float to the surface of water.

Petrus (1995) establishes the following criteria for predicting precipitation:

- If S < 1 mg/L and if $d_{20/4} > 1$, the solvent precipitates;
- If S < 1 mg/L and if $d_{20/4} < 1$, the solvent floats to the surface.

10.5.3.5. Surface Tension

When a solvent presents decreased surface tension (T), it is considered tensoactive. According to Petrus (1995), if T < 50 mN/m (at a concentration of 1 g/L), the solvent is considered tensoactive.

10.5.3.6. Solvent Target Compartment

A solvent's target compartment is the final location the solvent tends to migrate. This is not necessarily the place the solvent was initially applied and/or accidently discarded. For example, certain solvents disposed of in water can migrate to the atmosphere.

Different components in a mixture or formulation may migrate to distinct target compartments. Solvent mobility is characterized as the possible transfers of the same between different environmental divisions: between water, air, and soil/sediment.

According to Petrus (1995), a solvent's target compartment can be predicted based on water solubility, volatility, and adsorption ability, according to the following criteria:

- If solubility > 1g/L and the solvent is not volatile, the target compartment is water;
- If solubility < 1mg/L and the solvent is adsorbed, the target compartments are soil and sediment;
- If solubility < 1mg/L and the solvent is volatile, the target compartment is air.

Mathematical models may be used to determine solvent distribution among different compartments and its target compartments, as in the example, the Mackay model (Mackay, D. Paterson, S & shin, W.Y. – 1992 – *Generic models for evaluating the regional fate of chemicals – Chemosphere*, 24, 695-717).

- Degradability

  Degradability of a solvent is characterized by the processes associated with abiotic degradation, biodegradability, and persistence.

- Abiotic Degradation

  Abiotic degradation includes the chemical transformations like hydrolysis, photolysis, oxidation/reduction, and ion exchange reactions.

- Hydrolysis

  Hydrolysis is the property of a solvent to react with water. In relation to pH in which it occurs, hydrolysis may be acidic, base, or neutral.

- Photolysis

  Photolysis or photodegradation is the decomposition of a solvent due to light. The idea of photodegradability is generally expressed in half-life, evaluated in days. One half-life represents the necessary time for half of the solvent to degrade.

10.5.3.7. Biotic Degradation or Biodegradability

Biodegradability is the ability of a solvent to biodegrade, i.e., a metabolization (partial or complete decomposition) by microorganisms that use it as a carbon and/or energy source.

Biodegradation may involve relatively small changes in the original molecule, like the substitution or modification of a functional group (primary biodegradation), or the complete destruction of the solvent, with the final result being its conversion into CO_2, H_2O and inorganic salts; a process known as mineralization.

As an objective to predict the behavior and environmental impact, methodologies were established to evaluate biodegradability. In this sense, national and international agencies, like ISO and OECD act in the development and/or revision of these methodologies.

- Immediate Biodegradability

  Immediate biodegradability indicates an ease with which a solvent is degraded by microbial action under similar conditions as those normally found in the natural aquatic medium. Thus, a solvent with a positive immediate biodegradability test result, which normally lasts 28 days, will degrade rapidly in a natural aquatic medium and in biologic wastewater treatment stations, being classified as easily biodegradable. As an example, we can cite OECD tests 301 A-F, OECD test 310, and ISO test 14593.

The European Economic Community (CEE) using immediate biodegradation results as a basis, adopted the following criteria for classifying the degree as well as the ease to biodegrade:

Classification CEE	Immediate Biodegradation Test Results
Easily Biodegradable	$\geq$ 60% releasing of CO_2, or $\geq$ 60% consumption of O_2, or $\geq$ 70% removal of Dissolved Organic Carbon (COD) and $t \leq 10$ days(1)
Not Easily Biodegradable	$<$ 60% releasing of CO_2, or $<$ 60% consumption of O_2, or $<$ 70% removal of COD
Intrinsically Biodegradable	$\geq$ 60% release of CO_2, or $\geq$ 60% consumption of O_2, or $\geq$ 70% removal of DOC and $t > 10$ days

(1) This % should be reached in a period of (t) ≤ a 10-day window, starting from the moment in which 10% biodegradation occurs.

Biodegradability may be inferred from a relation between Biological Oxygen Demand ($BOD_{5\ days}$) and Chemical Oxygen Demand (COD), that when it is more than 0.5, it is considered an indication of quick biodegradation.

- Intrinsic Biodegradability

  Intrinsic biodegradability is applied to solvents that have no evidence of immediate biodegradability. Generally, it is determined by using tests in more favorable experimental conditions for biodegradation, like higher microbial density and longer period of adaptation or acclimation of the inoculum and possibly the provision of other nutrients and substrates containing carbon, for the purpose of inducing the increase of bacterial biomass and cometabolism. The intrinsic biodegradation test is an indicator of the maximum degree in which the compound maybe degraded in favorable environmental conditions, like in biological treatment stations.

As an example, we can cite the Zahn –Wellens test (OECD tests 302 B; ISO 9888, and EPA 835.3200) and the modified Semi-Continuous Activated Sludge (SCAS) test (OECD 302 A; ISO 9887; EPA 835.3210). A negative result on this test indicates probable persistence of the compound in an aquatic medium, and being classified as non-biodegradable.

According to Petrus (1995), classification criteria adopted by the EEC are as follows:

Classification CEE	Intrinsic Biodegradation Test Results
Intrinsically Biodegradable	≥ 70% DOC removal or COD and t ≤ 28 days (total duration)
Partially Biodegradable	< 70% and ≥ 20% DOC removal or COD and t ≤ 28 days (total duration)
Non-biodegradable	< 20% removal of DOC or COD and t ≤ 28 days

10.5.3.8. Persistence

The persistent solvent is one in which its molecule is stable, and not easily destroyed by biological or chemical means, remaining in the environment after use. The stability of a solvent is the function of its chemical structure. Thus, cyclic solvents are generally more stable than aliphatic solvents; and aromatic solvents are more stable than cyclic solvents. Branched structures are more stable than linear structures. The chlorine-carbon bond is more resistant to hydrolysis, and therefore, contributes to the molecule's resistance to biological and photolytic degradation.

Persistence, i.e., high resistance to biotic and/or abiotic degradation, is a characteristic of all non-degrading solvents, or all degraded solvents in non-degradable subproducts, whatever the target compartment where they can migrate.

Solvents that rapidly degrade can be quickly removed from the environment. In case of quick degrading solvent leakage or accidents, their effects will be restricted and will have a short duration period. The absence of quick degradation in the environment implies persistence in a determined compartment for long periods, representing chronic environmental exposure that cannot be solved by simply interrupting its usage. To avoid the accumulation of a solvent, its half-life must be equal to or less than its time of staying in the considered environmental compartment.

A solvent considered to have a "quick degradation" in an aquatic medium is that which can be degraded (biotic and/or abiotically) at a level higher than 70%, within a 28-day period.

10.5.3.9. Bioaccumulation

Bioaccumulation is the ability presented by certain solvents to accumulate in live organisms. Assimilation and retention of a solvent by an organism can lead to an elevated concentration of it with the probability of causing unhealthy effects.

The Kow may be used to predict bioaccumulation.

Petrus (1995) establishes the following criteria for predicting bioaccumulation:

- If log Kow ≥ 3, the solvent is considered as potentially bioaccumulative.
- If log Kow < 3, the solvent is not considered as potentially bioaccumulative.

Solvents with high Kow tend to accumulate in the biota, and are readily adsorbed into the soil, sediment, and organic material; afterwards they are transferred into human beings by means of the food chain. On the other hand, solvents with low Kow tend to distribute themselves in the water and air. Examples of these are volatile organic solvents, like trichloroethylene and tetrachloroethylene that distribute widely in the air and exposure through the food chain has little importance than other means, such as inhalation (ATSDR, 1992).

However, some solvents, like aromatic hydrocarbons, do not significantly accumulate in fish and vertebrates, despite their high Kow. This is due to fish having an ability to metabolize these solvents quickly (ATSDR, 1992).

10.5.3.10. Bioconcentration Factor (BCF)

The Bioconcentration Factor (BCF) is a measurement of chemical distribution magnitude in relation to the equilibrium between a biological medium (like a marine organism tissue) and an external medium (like water). The BCF is determined by dividing the equilibrium concentration (mg/kg) of a solvent in an organism or tissue by its concentration in the external environment.

Normally, fish are the target of bioconcentration studies because of their importance as a food source for humans and the availability of the standard test procedures for these organisms. Estimated BCFs using models (equilibrium or kinetic) or experimentally determined are basic components of studies on environmental risk evaluations or for human health. The standard procedure usually uses OECD 305.

According to Petrus (1995), if the BCF is equal to or greater than 100, the solvent is considered bioaccumulative.

10.5.3.11. Ecotoxicity

Ecotoxicity corresponds to the potential toxic effects of solvents on live organisms, whether they are in an aquatic medium, the soil or in a flora and fauna setting. This is the main reason for evaluating the effects resulting from the presence of these solvents.

The most important ecotoxicological tests, applied worldwide, became normalized by national and international agencies such as the ISO, OECD, and CETESB (the environmental agency in the state of São Paulo).

10.5.3.12. Effects on Aquatic Organisms

Aquatic toxicity studies the toxic effects of solvents on representative organisms in an aquatic environment. The aquatic medium is considered the most important compartment receiver because solvents released into the air or in the soil will reach water through rain, the washing away of soil, and infiltrations.

Studies of solvent effects on aquatic life can be carried out using "in loco" biological tests or in laboratory conditions, with the latter being used the most because it allows a more effective control of possible factors (example: temperature, pH, exposure time, medium, concentration). In relation to the organism test, for technical and economic reasons, it is impossible to test all species that make up a part of the aquatic ecosystem. The most accepted criterion is to choose a species representative of different trophic levels (food chain position). Tests with bacteria, algae, benthic macroinvertebrates, crustacean, (Daphnia for fresh water, shrimp for salt water) freshwater fish (trout, zebrafish/zebra danio) and salt-water fish.

The acute toxicity test generally evaluates severe, quick effects on organisms exposed to the solvent, in a short time, usually from one to four days.

As test criteria that allow the reading of an acute effect of a sample, fundamental organism test reactions were defined. With fish, this reaction is death; in microcrustaceans, mobility; in photobacteria, inhibition of light emission; in pseudomonas, respiratory growth inhibition; and in algae, inhibition of growth or an increase of fluorescence.

Results are expressed in initial median Effective or Lethal Concentration (EC50 or LC50), corresponding to the nominal concentration of a solvent at the beginning of a test that causes an acute effect (lethalness or immobility) in 50% of the test organisms, in a determined time period of exposure.

Acute toxicity is normally determined using the following species

- Fish, such as *Pimephales promelas, Danio rerio, Lepomis macrochirus*. LC50 value in 96 hours is determined by the OECD 203 method or the equivalent;
- Crustaceans, such as *Daphnia Magna, Daphnia Similis, Daphnia pulex*. EC50 value in 24 or 48 hours is determined by the OECD 202 method or the equivalent;
- Algae, such as *Scenedesmus quadricauda, Scenedesmus subspicatus, Chlorella vulgaris*. EC50 value in 72 or 96 hours (growth rate reduction by 50%) is determined by the OECD 201 method or the equivalent;

According to Petrus (1995), classification criteria adopted by the EEC are as follows:

Classification CEE	Acute Aquatic Toxicity Test Results
No Known Harmful Effect	CL or CE 50% > 100 mg/L
Harmful	CL or CE 50% > 10 mg/L and ≤ 100 mg/L
Toxic	CL or CE 50% > 1 mg/L and ≤ 10 mg/L
Very Toxic	CL or CE 50% ≤ 1 mg/L

The chronic toxicity test allows the evaluation of the most subtle adverse effects on exposed organisms, such as growth, survival, reproduction, and development. Test duration may vary from seven days to months, depending on the species.

Data on chronic aquatic toxicity is less available and experimental procedures are less standardized, comparatively to acute toxicity. The most used tests for determining No Observed Effect Concentration (NOEC) corresponding to the greatest nominal concentration of a solvent that does not cause a statistically significant harmful effect on organisms during a set exposure time, under test conditions are OECD 210 (first stage of life for fish) or OECD 211 (*Daphnia reproduction*).

10.5.3.13. Effects on Terrestrial Organisms

For terrestrial ecotoxicity, tests with worms (earthworms), higher plants, birds, and bees are used. Standard test examples for determining terrestrial toxicity are: OECD 207 (acute toxicity in earthworms: Eisenia foetida), OECD 208 (acute toxicity in plants), OECD 216/217 (effects on soil microorganisms), OECD 206 (chronic toxicity in birds).

10.5.3.14. Other Harmful Effects

Diverse chemical processes, important from the point of view of harmful environmental effects, occur in the air, whether it is pure or polluted. These reactions can occur in the troposphere, which includes the region from earth's surface to nearly 15 kilometers altitude or in the stratosphere in the region of 15 to 50 kilometers above the earth.

10.5.4. Ozone Photochemical Formation Potential

High levels of tropospheric ozone are produced due to reactions from pollutants and induced by light. According to Baird, the most important reagents involved in these reactions are volatile organic compounds (VOCs) and nitrogen oxides, with their final compounds being ozone, nitric acid, and partially oxidized organic compounds.

$$\text{VOCs} + \text{NO}^0 + \text{O}_2 + \text{Light} \quad \rightarrow \text{O}_3 + \text{HNO}_3 + \text{Organics compounds}$$

Different photochemical reactivity ranges were developed for classifying organic compounds in relation to their potential to produce ozone. Among these is the Maximum Incremental Reactivity (MIR) range. The MIR concept was developed by Carter (1998) for the *California Air Resources Board*. It is based on the Additional Reactivity corresponding to the number of ozone molecules formed by a carbon atom added to the atmospheric mixture of original VOC, for a given relation of VOC/NOx. The MIR measurement corresponds to the maximum Additional Reactivity of a given VOC, expressed in terms of gO_3/g VOC and Table 10.5 presents these values for some products.

The complete relation of the MIR values can be found in Carter's work at http://pah.cert.ucr.edu/~carter/r98tab.htm

Table 10.5. Additional Reactivity Values – MIR (gram of ozone formed per gram of product)

Substance	MIR (g/g)	Substance	MIR (g/g)
Amyl Acetate	1.16	Cyclobutanone	0.73
Ethyl Acetate	0.80	Dimethyl Ether	1.02
Isobutyl Acetate	1.08	Ethanol	1.92
Isopropyl Acetate	1.21	Ethyl Benzene	2.97
n-Butyl Acetate	1.14	Ethylene Glycol	5.66
Propyl Acetate	0.98	Glycerol	3.76
s-Butyl Acetate	1.72	Metanol	0.99
t-Butyl Acetate	0.21	Methyl Butyl Ether	1.34
Acetylene	1.23	Methyl Isobutyrate	0.42
Acetone	0.48	m-Xylene	11.06
Acetic Acid	0.67	o-Xylene	7.83
Propionic Acid	1.37	Propylene Glycol	2.65
Isopropyl Alcohol	0.81	p-Xylene	4.44
Benzene	1.00	Toluene	4.19

Source: Carter, 1998

10.5.4.1. Ozone Layer Destruction Potential

The ozone layer (O3) is a concentration of ozone gas in the upper atmosphere, between 15 to 50 kilometers up, i.e., in the stratosphere. This layer filters the largest part of biologically harmful ultraviolet sun radiation (UV-B), and any change in this radiation can probably lead to different effects to health and the environment.

Below are the chemical substances originating naturally and anthropogenically that, according to Vienna Convention for the Protection of the Ozone Layer, could change the physical and chemical properties of the Ozone Layer.

a) Carbon Group Substances: carbon monoxide (CO), carbon dioxide (CO_2), methane (CH_4), and the hydrocarbon species without methane.

Hydrocarbon species without methane, that are constituted of a large number of chemical substances, have as many natural sources as anthropogenic sources and play a direct role in the tropospheric photochemistry, in addition to an indirect role in the stratospheric photochemistry.

b) Nitrogen Group Substances: nitrous oxide (N_20), nitrogen oxide (NOx)

NOx sources at ground level represent a decisive direct role in the tropospheric photochemical processes, as well as an indirect role in the photochemistry of the stratosphere, while if there is an NOx opening near the intermediate layer between the troposphere and stratosphere, it can cause changes in ozone concentration in the upper layers of the troposphere and stratosphere.

c) Chloro Group Substances: completely and partially halogenated alkanes, for example: CFC-11, CFC-12, CFC-113, CFC-114, CFC-22, CFC-21.

d) Bromide Group Substances: completely halogenated alkanes, for example: CF_3Br.

e) Hydrogen Group Substances: hydrogen (H_2)

The hydrogen, in which its origin is natural and anthropogenic, plays a less important role in stratospheric photochemistry.

f) water (H_2O)

Water, which comes from a natural source, plays a vital role in photochemistry in the troposphere as much as the stratosphere. Local sources of water vapor in the stratosphere include oxidation of methane, and to a lesser degree, hydrogen.

The potential for risk or destruction of the ozone layer from a chemical substance can be estimated by the Ozone Depletion Potential (ODP). ODP is defined as the depletion of the ozone in the stratosphere caused by emissions of a mass unit of a chemical substance relative to the depletion of the ozone caused by the emission of a mass unit of CFC-11. Any substance with ODP greater than zero can destroy the stratospheric ozone layer. In general, near-zero ODP values are for substances with a half-life of less than one year in the atmosphere. (Verschueren, 2001). Many substances with ozone layer depletion potential also have the potential for global warming.

The Montreal Protocol, an international treaty in which countries that signed it committed themselves to replacing ODP substances, does not include non-halogenated solvents in its relation to ODP.

10.5.5. Potential for Global Warming

Greenhouse gas emissions are considered the main cause of climatic changes, being regulated by the United Nations Framework Convention on Climate Change (UNFCC) and by the subsequent Kyoto Protocol, which does not include any industrial solvent.

10.5.6. Residual Water Treatment Station Effects

Solvent rejects in aqueous mediums can reach urban or industrial residuary water treatment stations, and in relation to characteristics, they are eliminated by chemical and/or biological processes, such as volatilization, adsorption, or biodegradation.

In biological processes from treatment stations, like activated sludge, a solvent can possibly provoke adverse effects, in function of the rejected quantity, of its toxicity to organisms that constitute activated sludge and its biodegradability. Examples of adverse effects are shock loading and partial or total inhibition of the biological system, with a loss in treatment efficiency.

Respirometry tests for inhibition of oxygen consumption by organisms in activated sludge (OECD 209 or ISO 8192) and treatment simulation tests of activated sludge in laboratory pilot units (OECD 303A or ISO 11733) can be used to evaluate the effects and treatability of solvents at Biological treatment stations.

Interesting Links

The web offers several data and information links related to environmental, hygiene, and occupational health behavior of solvents. Here is a small example:

http://www.syrres.com/esc/chemfate.htm – Physicochemical Properties.

http://www.syrres.com/esc/physdemo.htm – Physicochemical Properties.

http://toxnet.nlm.nih.gov/cgi-bin/sis/htmlgen?HSDB.htm – toxicological data, physicochemical properties, and ecological information.

http://www.syrres.com/esc/biodeg.htm – biodegradability data.

http://cfpub.epa.gov/ecotox/quick_query.htm – ecotoxicity data.

http://pah.cert.ucr.edu/~carter/r98tab.htm – photochemical reactivity data.

http://www.syrres.com/esc/ozone.asp – data on ODP and GWP (Global Warming Potential).

http://hq.unep.org/ozone/Montreal-Protocol/Montreal – Montreal Protocol.

http://www2.mst.dk/common/Udgivramme/Frame.asp?pg=http://www2.mst.dk/udgiv/Publications/2001/87-7944-596-9/html/kap11_eng.htm – ecological information for more commonly used solvent groups.

http://chemfinder.cambridgesoft.com/ – supplies links to several websites with data on physicochemical toxicological and ecotoxicological properties.

http://www.portaldapesquisa.com.br/databases/sites

Bibliographical References

Associação Brasileira de Higienistas Ocupacionais (ABHO) – I Congresso Panamericano de Higiene Ocupacional. Anais/Curso Estratégia de Amostragem de Agentes Químicos.

American Conference of Governmental and Industrial Hygienists (ACGIH) – Limites de Exposição Ocupacional (TLVs®) para Substâncias Químicas e Agentes Físicos & Índices Biológicos de Exposição (BEIs®). Tradução ABHO, 2005.

Agency for toxic substances and disease registry. (ATSDR). *Public health assessment guidance manual*. Boca Raton: Lewis Publishers, 1992.

Cartr, W. *Updated maximum incremetal reactivity scale for regulatory application*. 1998.

L. Cikui; G. David. *Prediction of physical and chemical properties by quantitative structur-property relationships*. American Laboratory 1997.

Comstock, B. S. A review of psychological measures relevant to central nervous system toxicity, with specific reference to solvent inhalation. *Clin. Toxicol.*, 11: 317-24, 1977.

J. M. Haguenoer; D. Furon. *Toxicologie et hygiene industrielles*. Paris, Technique et Documentation. v. 10, p. 91-198. 1983.

J. R. Mulhausen; J. Damiano. *Strategy for Assessing and Managing Occupational Exposures*. Second Edition, AIHA Press.

M. MacLeod; D. Mackay. *Modeling transport and deposition of contaminants to ecosystems of concern*: a case study for the Laurentian Great Lakes. 2003.

Niosh Publication nb. 87-104. *Current Intelligence Bulletin 48/Organic Solvent Neurotoxicity*, march 31, 1987.

Organizações das Nações Unidas (ONU). Convenção de Viena para a Proteção da Camada de Ozônio. 1985.

Organizações das Nações Unidas (ONU). Protocolo de Montreal sobre substâncias que destroem a camada de ozônio. 1989.

P. A. Zagatto; E. Bertoletti. *Ecotoxicologia aquática* – princípios e aplicações. 2006.

R. Petrus. *Fiches de données de sécurité pour les produits chimiques dangereux*. Manuscrit définit – Rhone Poulenc Chimie. 1995.

S. Rivaldo. *Toxicologia Industrial, Saúde e Trabalho on line*.

K. Verschueren. *Handbook of Environmental Data on Organic Chemicals*, 4. ed. 2001.

Z. Rebouças. *Avaliação da Exposição de Agentes Químicos e Físicos*. ABIQUIM, agosto, 2004.

Glossary

H	Relative Enthalpy
ΔH°	Reaction Enthalpy at 25°C and 1 atm
$\delta\text{-}^{-}$	Negative Polarity
$\delta\text{-}^{+}$	Positive Polarity
pe	Dipole Moment
Q	negative charge
l	Distance between charge centers
atm	atmosphere
°C	Degrees Celsius
p	Vapor Pressure
T_c	Critical Temperature
p_c	Critical Pressure
C_nH_{2n}	Alkane Group
OH	Hydroxyl Group
O	Oxygen
N	Nitrogen
F	Fluorine
bp	Boiling Point
mp	Melting Point
Na^+	Sodium Ions
Cl^-	Chloride Ions
K	Kelvin
Pa	Pascal
q	Charges
r	Distance between charges in a vacuum
V	Potential Energy
ε_0	Permissivity in a vacuum
ε	Medium Permissivity
ε_r	Relative Permissivity of the Dielectric Constant
k	Boltzman Constant
ρ	Density
η	Viscosity
M	Molar Mass
P_m	Polarization
T	Temperature
E	Energy
R	Universal Constant of Gases
n_{21}	Relative Refractive Index
v_1	Speed of Light in Medium 1
v_2	Speed of Light in Medium 2
c	Speed of Light in a vacuum
n	Absolute Refractive Index
θ_1	Angle of Incidence
θ_2	Angle of Refraction
γ	Surface Tension
w	Work
A_{sup}	Surface Area
s	Seconds
m	Mass
V	Volume
d	Relative Density
ρ_0	Standard Absolute Density
VOC	Volatile Organic Compounds

Index